D0153668

Ming Yang

PRINCIPLES OF

NUMERICAL ANALYSIS

PRINCIPLES OF NUMERICAL ANALYSIS

ALSTON S. HOUSEHOLDER

DOVER PUBLICATIONS, INC. • MINEOLA, NEW YORK

Copyright

Bibliographical Note

This Dover edition, first published in 1974 and reissued in 2006, is an unabridged, slightly corrected republication of the work originally published by the McGraw-Hill Book Company, Inc., New York, in 1953.

International Standard Book Number: 0-486-45312-X

Manufactured in the United States of America
Dover Publications, Inc., 31 East 2nd Street, Mineola, N.Y. 11501

TO

B, J, AND J

PREFACE

This is a mathematical textbook rather than a compendium of computational rules. It is hoped that the material included will provide a useful background for those seeking to devise and evaluate routines for numerical computation.

The general topics considered are the solution of finite systems of equations, linear and nonlinear, and the approximate representation of functions. Conspicuously omitted are functional equations of all types. The justification for this omission lies first in the background presupposed on the part of the reader. Second, there are good books, in print and in preparation, on differential and on integral equations. But ultimately, the numerical "solution" of a functional equation consists of a finite table of numbers, whether these be a set of functional values, or the first n coefficients of an expansion in terms of known functions. Hence, eventually the problem must be reduced to that of determining a finite set of numbers and of representing functions thereby, and at this stage the topics in this book become relevant.

The endeavor has been to keep the discussion within the reach of one who has had a course in calculus, though some elementary notions of the probability theory are utilized in the allusions to statistical assessments of errors, and in the brief outline of the Monte Carlo method. The book is an expansion of lecture notes for a course given in Oak Ridge for the University of Tennessee during the spring and summer quarters of 1950.

The material was assembled with high-speed digital computation always in mind, though many techniques appropriate only to "hand" computation are discussed. By a curious and amusing paradox, the advent of high-speed machinery has lent popularity to the two innovations from the field of statistics referred to above. How otherwise the continued use of these machines will transform the computer's art remains to be seen. But this much can surely be said, that their effective use demands a more profound understanding of the mathematics of the problem, and a more detailed acquaintance with the potential sources of error, than is ever required by a computation whose development can be watched, step by step, as it proceeds. It is for this reason that a textbook on the mathematics of computation seems in order.

Help and encouragement have come from too many to permit listing all by name. But it is a pleasure to thank, in particular, J. A. Cooley, C. C. Hurd, D. A. Flanders, J. W. Givens, A. de la Garza, and members of the Mathematics Panel of Oak Ridge National Laboratory. And for the painstaking preparation of the copy, thanks go to Iris Tropp, Gwen Wicker, and above all, to Mae Gill.

A. S. Householder

CONTENTS

CHAPTER 1

THE ART OF COMPUTATION

1. The Art of Computation

We are concerned here with mathematical principles that are sometimes of assistance in the design of computational routines. It is hardly necessary to remark that the advent of high-speed sequenced computing machinery is revolutionizing the art and that it is much more difficult to explain to a machine how a problem is to be done than to explain to most human beings. Or that the process that is easiest for the human being to carry out is not necessarily the one that is easiest or quickest for the machine. Not only that, but a process may be admirably well adapted to one machine and very poorly adapted to another. Consequently, the robot master has very few tried and true rules at his disposal, and is forced to go back to first principles to construct such rules as seem to conform best to the idiosyncrasy of his particular robot.

If a computation requires more than a very few operations, there are usually many different possible routines for achieving the same end result. Even so simple a computation as ab/c can be done $(ab)/c$, $(a/c)b$, or $a(b/c)$, not to mention the possibility of reversing the order of the factors in the multiplication. Mathematically these are all equivalent; computationally they are not (cf. §1.2 and §1.4). Various, and sometimes conflicting, criteria must be applied in the final selection of a particular routine. If the routine must be given to someone else, or to a computing machine, it is desirable to have a routine in which the steps are easily laid out, and this is a serious and important consideration in the use of sequenced computing machines. Naturally one would like the routine to be as short as possible, to be self-checking as far as possible, to give results that are at least as accurate as may be required. And with reference to the last point, one would like the routine to be such that it is possible to assert with confidence (better yet, with certainty) and in advance that the results will be as accurate as may be desired, or if an advance assessment is out of the question, as it often is, one would hope that it can be made at least upon completion of the computation.

1.1. Errors and Blunders. The number 0.33, when expressing the result of the division $1 \div 3$, is correctly obtained even though it deviates by 1 per cent from the true quotient. The number 0.334, when expressing the result of the same division, deviates by only 0.2 per cent from the true quotient, and yet is incorrectly obtained. The deviation of 0.33 from the true quotient will be called an error. If the division is to be carried out to three places but not more, then 0.333 is the best representation possible and the replacement of the final "3" by a final "4" will be called a blunder.

Blunders result from fallibility, errors from finitude. Blunders will not be considered here to any extent. There are fairly obvious ways to guard against them, and their effect, when they occur, can be gross, insignificant, or anywhere in between. Generally the sources of error other than blunders will leave a limited range of uncertainty, and generally this can be reduced, if necessary, by additional labor. It is important to be able to estimate the extent of the range of uncertainty.

Four sources of error are distinguished by von Neumann and Goldstine, and while occasionally the errors of one type or another may be negligible or absent, generally they are present. These sources are the following:

1. Mathematical formulations are seldom exactly descriptive of any real situation, but only of more or less idealized models. Perfect gases and material points do not exist.

2. Most mathematical formulations contain parameters, such as lengths, times, masses, temperatures, etc., whose values can be had only from measurement. Such measurements may be accurate to within 1, 0.1, or 0.01 per cent, or better, but however small the limit of error, it is not zero.

3. Many mathematical equations have solutions that can be constructed only in the sense that an infinite process can be described whose limit is the solution in question. By definition the infinite process cannot be completed, so one must stop with some term in the sequence, accepting this as the adequate approximation to the required solution. This results in a type of error called the truncation error.

4. The decimal representation of a number is made by writing a sequence of digits to the left, and one to the right, of an origin which is marked by the decimal point. The digits to the left of the decimal are finite in number and are understood to represent coefficients of increasing powers of 10 beginning with the zeroth; those to the right are possibly infinite in number, and represent coefficients of decreasing powers of 10. In digital computation only a finite number of these digits can be taken account of. The error due to dropping the others is called the round-off error.

In decimal representation 10 is called the base of the representation. Many modern computing machines operate in the binary system, using the base 2 instead of the base 10. Every digit in the two sequences is either 0 or 1, and the point which marks the origin is called the binary point, rather than the decimal point. Desk computing machines which use the base 8 are on the market, since conversion between the bases 2 and 8 is very simple. Colloquial languages carry the vestiges of the use of other bases, *e.g.*, 12, 20, 60, and in principle, any base could be used.

Clearly one does not evaluate the error arising from any one of these sources, for if he did, it would no longer be a source of error. Generally it cannot be evaluated. In particular cases it can be evaluated but not represented (*e.g.*, in the division $1 \div 3$ carried out to a preassigned number of places). But one does hope to set bounds for the errors and to ascertain that the errors will not exceed these bounds.

The computor is not responsible for sources 1 and 2. He is not concerned with formulating or assessing a physical law nor with making physical measurements. Nevertheless, the range of uncertainty to which they give rise will, on the one hand, create a limit below which the range of uncertainty of the results of a computation cannot come, and on the other hand, provide a range of tolerance below which it does not need to come.

With the above classification of sources, we present a classification of errors as such. This is to some extent artificial, since errors arising from the various sources interact in a complex fashion and result in a single error which is no simple sum of elementary errors. Nevertheless, thanks to a most fortunate circumstance, it is generally possible to estimate an over-all range of uncertainty as though it were such a simple sum (§1.2). Hence we will distinguish propagated error, generated error, and residual error.

At the outset of any computation the data may contain errors of measurement, round-off errors due to the finite representation in some base of numbers like $\frac{1}{3}$, or even numbers requiring a finite but large number of places for exact representation. These initial errors carry through the computation and lead to an uncertainty at every step. It is important to know how these initial errors are propagated through the computation and to what extent they render the results uncertain.

In addition to this, at every step, or nearly every step, new errors may arise as a result of round-off, these combine with the errors already propagated, and the total is propagated through what computations remain. Finally, when the computation is terminated, a truncation error may remain and further enlarge the region of uncertainty. Roughly the extent to which errors are propagated and the uncertainty due to residual error depend upon the mathematical formulation of the computational procedure, while the generation is more dependent upon the detailed ordering of the computational steps.

Any computation, however elaborate, consists of a finite number of elementary operations carried out in some sequence. The elementary operations are usually additions and subtractions, multiplications and divisions, comparisons, possibly table look-ups, and the like. An unambiguous description of the sequence in which the operations are performed or to be performed, with a specification of the data upon which each is to operate, constitutes a routine. If multiplication and division are elementary operations, there are six possible routines for computing ab/c. Hence a routine is by no means defined when a mathematical formula, or sequence of them, is written down.

A routine of any complexity breaks up naturally into parts or subroutines. A subroutine may have for its purpose the computation of an intermediate quantity, of no interest in itself but serving as a datum or operand for one or more subsequent subroutines. Thus, in order to calculate ab/c, one must first calculate ab, or a/c, or b/c. Or a subroutine may operate upon intermediate results to produce a final result.

Suppose a subroutine is intended to compute a function $f(x, y, \ldots)$ for given values of its arguments. If f is a rational function or a polynomial, there need be no residual error in the computation, but only propagated and generated errors. If f is not a rational function, some rational approximation must be devised. One type of rational approximation is a Taylor series. If a Taylor series is used, only a finite number of terms will be computed. The residual error in the computation is the sum of all neglected terms. Hence the residual error is fixed by the mathematical formulation of the problem, together with the specification of the number of terms to be used in the computation. But an error may be generated and propagated in each computed term.

Another type of approximation, *e.g.*, in solving an equation by Newton's method, is the following: From an initial approximation f_0, one defines a sequence of approximations by a relation $f_{i+1} = \phi(f_i)$, where ϕ is some function which can be evaluated. If the mathematical sequence converges, one may take as a criterion for terminating the sequence the condition that successive terms shall differ by less than some assigned quantity. Clearly the assigned quantity must not be smaller than the smallest quantity representable by the machine, and normally it will be some integral multiple of this. It will have to depend upon the error generated by the routine for evaluating ϕ. The residual error is the difference between the f_i which one accepts and the true value of f, and it is limited by the quantity used in the criterion which is, in turn, limited by the precision of the machine operations. For illustration, a routine for computing square roots will be considered in §1.5.

1.2. Composition of Error. Let x^*, y^*, . . . designate numbers which might occur as data or as results of a particular computation. That is to say, x^* has a form

(1.2.1) $$x^* = \pm(x_1\beta^{-1} + x_2\beta^{-2} + \cdots + x_\lambda\beta^{-\lambda})\beta^\sigma,$$

where β is the base, usually 2 or 10, λ is a positive integer, and σ any integer, possibly zero. It may be that λ is fixed throughout the course of the computation, or it may vary, but in any case it is limited by practical considerations. Such a number will be called a representation. It may be that x^* is obtained by "rounding off" a number whose true value is x (for example, $x = 1/3$, $x^* = 0.33$), or that x^* is the result of measuring physically a quantity whose true value is x, or that x^* is the result of a computation intended to give an approximation to the quantity x.

Suppose one is interested in performing an operation ω upon a pair of numbers x and y. That is to say, $x\omega y$ may represent a product of x and y, a quotient of x by y, the yth power of x, ... In the numerical computation, however, one has only x^* and y^* upon which to operate, not x and y (or at least these are the quantities upon which one does, in fact, operate). Not only this, but often one does not even perform the strict operation ω, but rather a pseudo operation ω^*, which yields a rounded-off product, quotient, power, etc. Hence, instead of obtaining the desired result $x\omega y$, one obtains a result $x^*\omega^*y^*$.

The error in the result is therefore

(1.2.2) $$x\omega y - x^*\omega^*y^* = (x\omega y - x^*\omega y^*) + (x^*\omega y^* - x^*\omega^*y^*).$$

Since x^* and y^* are numbers, the operation ω can be applied to them, and $x^*\omega y^*$ is well defined, except for special cases as when ω represents division and $y^* = 0$. But the expression in the first parentheses on the right represents propagated error, and that in the second parentheses represents generated error, or round-off. Hence the total error in the result is the sum of the error propagated by the operation and that generated by the operation.

It may happen that the two errors are opposite in sign, though often the sign is not known, but only the magnitude. In any case,

(1.2.3) $$|x\omega y - x^*\omega^*y^*| \leq |x\omega y - x^*\omega y^*| + |x^*\omega y^* - x^*\omega^*y^*|,$$

and one can say at least that the total error does not exceed the sum of the two errors.

That propagated and generated errors depend upon the details of the routine, such as the value of λ for any representation, and even the order in which certain operations are carried out, is easily seen. Thus, to consider only round-off, if ϕ is some operation, it may be that mathematically

$$(x^*\omega y^*)\phi z^* = x^*\omega(y^*\phi z^*).$$

That is, the two operations may be associative mathematically. Nevertheless,

$$(x^*\omega y^*)\phi z^* - (x^*\omega^* y^*)\phi^* z^* = [(x^*\omega y^*)\phi z^* - (x^*\omega^* y^*)\phi z^*] \\ + [(x^*\omega^* y^*)\phi z^* - (x^*\omega^* y^*)\phi^* z^*],$$

whereas

$$x^*\omega(y^*\phi z^*) - x^*\omega^*(y^*\phi^* z^*) = [x^*\omega(y^*\phi z^*) - x^*\omega(y^*\phi^* z^*)] \\ + [x^*\omega(y^*\phi^* z^*) - x^*\omega^*(y^*\phi^* z^*)].$$

Thus the errors generated by performing the operations in the two possible ways have different expressions, and cannot be assumed equal without proof.

1.3. Propagated Error and Significant Figures. Let

$$x = x^* + \xi, \qquad y = y^* + \eta, \ . \ . \ . \ , \tag{1.3.1}$$

and consider the problem of evaluating the function $f(x, y, \ldots)$. If the function can be expanded in Taylor's series, then

$$f(x, y, \ldots) - f(x^*, y^*, \ldots) = \xi f_x + \eta f_y + \cdots \\ + \tfrac{1}{2}(\xi^2 f_{xx} + 2\xi\eta f_{xy} + \eta^2 f_{yy} + \cdots) + \cdots, \tag{1.3.2}$$

where the partial derivatives are to be evaluated at $x^*, y^*, \ldots$. This represents the error in f arising from errors in the arguments, *i.e.*, the propagated error. Generally one expects the errors $\xi, \eta, \ldots$ to be "small" so that the terms of second and higher power can be neglected. If so, then the propagated error Δf satisfies, approximately,

$$|\Delta f| \leq |\xi f_x| + |\eta f_y| + \cdots. \tag{1.3.3}$$

This is strictly true when f is a simple sum:

$$f = \pm x \pm y \pm \cdots.$$

Hence the error in a sum does not exceed the sum of the errors in the terms.

One can, by direct differentiation, write down any number of special relations (1.3.3) based upon the assumption that the errors in the arguments are small. Nevertheless, for the detailed analysis one must go to the individual elementary operations.

Consider the case of the product and quotient. For the first we have

$$xy - x^*y^* = x^*\eta + y^*\xi + \xi\eta.$$

It is sometimes convenient to consider the relative error, which is the ratio of the error to the magnitude. Hence

$$\frac{xy - x^*y^*}{x^*y^*} = \frac{\xi}{x^*} + \frac{\eta}{y^*} + \frac{\xi}{x^*}\frac{\eta}{y^*}. \tag{1.3.4}$$

Usually one says that the relative error in a product is the sum of the relative errors in the factors, and this is approximately true if these relative errors are both small, but only then.

For the quotient

$$\frac{x}{y} - \frac{x^*}{y^*} = \frac{y^*\xi - x^*\eta}{y^*(y^* + \eta)},$$

and the relative error is

$$\frac{x/y - x^*/y^*}{x^*/y^*} = \frac{\xi/x^* - \eta/y^*}{1 + \eta/y^*}. \tag{1.3.5}$$

If the relative error η/y^* is negligible, then the relative error in the quotient does not exceed in magnitude the sum of the magnitudes of the relative errors in the terms. Nevertheless, if $\eta/y^* < 0$ and not numerically small, the conclusion does not follow.

If x^*, given by (1.2.1), represents a number whose true value is x, and if

$$|x^* - x| \leq \beta^{\sigma-\lambda}/2, \tag{1.3.6}$$

the digits $x_1, \ldots, x_\lambda$ may be said to be reliable or significant. If $x_1 \neq 0$, then x^* is said to contain λ significant figures. However, if the inequality (1.3.6) is not known to hold, the last digit, x_λ, is "in doubt." When a number such as x^* is written in its usual form as a sequence of digits with a decimal point, it is usually understood that all digits are significant, from the first non-null digit to the last digit written, unless limits of error are specified. Some textbooks give various rules of thumb for determining the number of significant digits in computed quantities, and the implication is sometimes left that nonsignificant digits should be dropped before making subsequent computations with these numbers, or at least that nothing is to be gained by retaining the doubtful digits. But this is clearly not the case, since dropping these digits generally enlarges the region of uncertainty.

1.4. Generated Error. In many of the newer automatic computing machines, the "built-in" arithmetic operations are designed to yield correct results only when the operands as well as the results are numbers of a form (1.2.1), where λ is fixed, and $\sigma = 0$. Such a number von Neumann and Goldstine call a "digital number." Any digital number is therefore less than unity.

Given two digital numbers a^* and b^*, the machine will correctly form the sum only if

$$|a^* + b^*| < 1,$$

and will correctly form the difference if

$$|a^* - b^*| < 1.$$

But if the one condition or the other holds, the machine will correctly form the sum or difference, as the case may be, and no round-off is generated. Hence if $a^* + b^*$ or $a^* - b^*$ is digital, $a^* + b^*$ or $a^* - b^*$ can be formed, and formed correctly, without generating any new error.

If a^* and b^* are digital, then necessarily

$$|a^*b^*| < 1.$$

However, the true product a^*b^* is a number of 2λ places. If the machine holds only λ places, it will not form the true product a^*b^*, but a pseudo product. It can be represented $a^* \times b^*$. It may be that the machine merely drops off the last λ places from a^*b^*. Or it may be that the machine first forms $a^*b^* + \beta^{-\lambda}/2$ and then drops the last λ places. In the latter event the pseudo product satisfies

$$|a^*b^* - a^* \times b^*| \leq \epsilon = \beta^{-\lambda}/2, \tag{1.4.1}$$

where ϵ is introduced to simplify notation. Let us assume that (1.4.1) holds.

For division, the quotient a^*/b^* will usually require infinitely many places for its true representation, even though a^* and b^* are both digital. The machine, however, can retain only the first λ, and it may compute the first λ places of a^*/b^* and drop the rest, or it may compute the first λ places of $a^*/b^* + \beta^{-\lambda}/2$. In the latter event, the retained λ places represent the pseudo quotient $a^* \div b^*$, which satisfies

$$|a^*/b^* - a^* \div b^*| \leq \epsilon. \tag{1.4.2}$$

Given a series of n products $a_i^*b_i^*$ to be added together, we have

$$|\Sigma a_i^* \times b_i^* - \Sigma a_i^*b_i^*| \leq n\epsilon. \tag{1.4.3}$$

However, instead of recording each product as a digital number and adding the results, it may be possible for the machine to retain and accumulate the true products of a^* and b^*, rounding off the sum as a digital number. This pseudo operation may be designated $\Sigma^* a_i^*b_i^*$, and for this we have

$$|\Sigma^* a_i^*b_i^* - \Sigma a_i^*b_i^*| \leq \epsilon. \tag{1.4.4}$$

While

$$a^* \times b^* = b^* \times a^*, \tag{1.4.5}$$

equalities in terms of arithmetic operations do not always hold strictly when these are replaced by pseudo operations, as was already shown in general. In particular,

$$|(a^* + b^*) \times c^* - (a^* \times c^* + b^* \times c^*)| \leq 3\epsilon,$$

since each pseudo multiplication could give rise to an error ϵ. However, the two quantities being compared can differ only by a digit in the last place, which is to say by an integral multiple of 2ϵ. Consequently we can improve this by saying that

$$|(a^* + b^*) \times c^* - (a^* \times c^* + b^* \times c^*)| \leq 2\epsilon. \tag{1.4.6}$$

In order to examine the effect of grouping in a continued product, we note that

$$a^* \times (b^* \times c^*) - a^*b^*c^* = [a^* \times (b^* \times c^*) - a^*(b^* \times c^*)] + [a^*(b^* \times c^*) - a^*b^*c^*]$$

so that

$$|a^* \times (b^* \times c^*) - a^*b^*c^*| \leq (1 + |a^*|)\epsilon \leq 2\epsilon.$$

If we now interchange a and c and add results, we have

$$|a^* \times (b^* \times c^*) - (a^* \times b^*) \times c^*| \leq (2 + |a^*| + |c^*|)\epsilon.$$

But if $a^* = c^* = 1$, the left-hand side is zero, and otherwise the left-hand side is less than 4ϵ. But it must be an integral multiple of 2ϵ, so that

$$|a^* \times (b^* \times c^*) - (a^* \times b^*) \times c^*| \leq 2\epsilon. \tag{1.4.7}$$

The two pseudo products either agree or differ by one in the last place.

Finally consider

$$(a^* \div b^*) \times b^* - a^* = [(a^* \div b^*) \times b^* - (a^* \div b^*)b^*] + (a^* \div b^* - a^*/b^*)b^*.$$

Since this is less than $(1 + |b^*|)\epsilon$, it is actually less than 2ϵ, and since it is an integral multiple of 2ϵ, it vanishes. Hence

$$(a^* \div b^*) \times b^* = a^*. \tag{1.4.8}$$

However

$$(a^* \times b^*) \div b^* - a^* = [(a^* \times b^*) \div b^* - (a^* \times b^*)/b^*] + [(a^* \times b^*) - a^*b^*]/b^*,$$

so that

$$|(a^* \times b^*) \div b^* - a^*| \leq (1 + |b^*|^{-1})\epsilon. \tag{1.4.9}$$

If $|b^*|$ is small, $|b^*|^{-1}$ will be large, and the error can be large. However, bear in mind that this comparison is made for pseudo operations in which the rounded-off product is used. If the machine retains the complete product to use as the dividend, the error will not arise.

More generally, and in the same way, we have

$$|(a^* \div b^*) \times c^* - (a^*/b^*)c^*| \leq (1 + |c^*|)\epsilon, \tag{1.4.10}$$

while

$$|(a^* \times b^*) \div c^* - a^*b^*/c^*| \leq (1 + |c^*|^{-1})\epsilon. \tag{1.4.11}$$

Interchange of a^* and c^* in (1.4.10) gives

$$|(c^* \div b^*) \times a^* - (c^*/b^*)a^*| \leq (1 + |a^*|)\epsilon,$$

so that by addition

$$|(a^* \div b^*) \times c^* - (c^* \div b^*) \times a^*| \leq (2 + |a^*| + |c^*|)\epsilon.$$

But the left member vanishes when $|a^*| = |c^*| = 1$, and otherwise the right member is less than 4ϵ. Hence in any event

$$(1.4.12) \qquad |(a^* \div b^*) \times c^* - (c^* \div b^*) \times a^*| \leq 2\epsilon.$$

1.5. Complete Error Analyses. It was shown above that the magnitude of the error in the result of any subroutine cannot exceed the sum of the magnitudes of the propagated, generated, and residual errors. In particular instances a routine can be devised such that the error due to one of two sources necessarily is of one sign and that due to the other is of the opposite sign. In such a situation the resultant error cannot exceed the larger of the two individual errors. But to devise a routine having this property, or even to discover that a given routine has it, may take a disproportionate amount of time and not be worth the effort. Generally one must assume that the errors can build up, and attempt to devise a routine that will keep all errors as small as possible. And in any event one must somehow balance the time to be spent on an analysis against the time required for the computation and the allowed tolerances.

Some computations are of such frequent occurrence that it may be worth while devoting considerable time and effort to the design of an optimal routine and a precise error analysis for that routine. We give an example or two.

Consider a binary computing machine with λ magnitude digits and one sign digit, representing only numbers of magnitude less than unity. This machine will compute a (2λ)-digit product and accept a (2λ)-digit dividend, but the digital pseudo products and pseudo quotients will be supposed to satisfy, in general,

$$0 \leq ab - a \times b < 2^{-\lambda},$$
$$0 \leq a/b - a \div b < 2^{-\lambda},$$

the latter relation presupposing the division to be possible.

We require an optimal routine and precise error analysis for $\sqrt{a}$, using Newton's method. This means that the number a whose square root is required is a digital number, $0 \leq a < 1$, and the routine is to yield a digital number x for which the maximal $(x - \sqrt{a})$ is to be as small as we can make it. Indeed if the routine is properly constructed, then there should be some half-open interval of length $2^{-\lambda}$ which contains both $\sqrt{a}$ and x.

By Newton's method one takes some positive $x'_0 \leq 1$ and forms

$$x'_{i+1} = x'_i - (x'_i - a/x'_i)/2, \tag{1.5.1}$$

and the sequence can be shown to approach $\sqrt{a}$ as a limit. Moreover, if one takes $x'_0 = 1$, then surely

$$x'_i > \sqrt{a} \tag{1.5.2}$$

when $i = 0$, and one can show inductively that this relation holds for every i. In fact, one verifies directly that

$$x'^2_{i+1} - a = (x'_i - a/x'_i)^2/4.$$

Nevertheless, the sequence one actually forms in the machine is not strictly defined by (1.5.1), but instead by

$$x_{i+1} = x_i - (x_i - a \div x_i) \div 2. \tag{1.5.3}$$

(All numbers will be digital numbers, and the asterisk can be omitted.) There is no a priori assurance that the numbers x_i will satisfy (1.5.2), and this point must be investigated.

We first show that, if y and z are any two digital numbers satisfying

$$(z - a \div z) \div 2 > (y - a \div y) \div 2 > 0, \tag{1.5.4}$$

then

$$z > y > \sqrt{a}. \tag{1.5.5}$$

The second inequality in (1.5.4) is equivalent to

$$\begin{gathered}(y - a \div y) \div 2 \geq 2^{-\lambda}, \\ y - a \div y \geq 2^{-\lambda+1}, \\ y \geq 2^{-\lambda+1} + a \div y > 2^{-\lambda} + a/y > a/y,\end{gathered}$$

whence $y^2 > a$, which proves the second inequality in (1.5.5). By the first inequality in (1.5.4),

$$(z - a \div z) \div 2 \geq (y - a \div y) \div 2 + 2^{-\lambda}.$$

But

$$\begin{gathered}(z - a \div z)/2 \geq (z - a \div z) \div 2, \\ (y - a \div y) \div 2 \geq (y - a \div y)/2 - 2^{-\lambda-1},\end{gathered}$$

since in forming the pseudo quotient by 2 (which is a single shift to the right) the error is either 0 or $2^{-\lambda-1}$. Hence

$$\begin{gathered}(z - a \div z)/2 \geq (y - a \div y)/2 + 2^{-\lambda-1}, \\ z - a \div z \geq y - a \div y + 2^{-\lambda}.\end{gathered}$$

Again,

$$\begin{gathered}a \div z - a/z > -2^{-\lambda}, \\ a \div y - a/y \leq 0,\end{gathered}$$

whence

$$z - a/z > y - a/y.$$

But the function $f(z) = z - a/z$ is properly monotonically increasing. Hence $f(z) > f(y)$ implies the first inequality in (1.5.5).

This implies, in particular, that, if

$$(x_i - a \div x_i) \div 2 = x_i - x_{i+1} > 0,$$

then $x_i > \sqrt{a}$. If it should happen that

$$(x_i - a \div x_i) \div 2 \leq 0, \tag{1.5.6}$$

then clearly we should take x_i and not x_{i+1} as x. On the other hand, if

$$(x_{i-1} - a \div x_{i-1}) \div 2 \geq 2^{-\lambda}, \tag{1.5.7}$$

we shall take at least one more step in the iteration.

Suppose the equality holds in (1.5.7). Then

$$x_i = x_{i-1} - 2^{-\lambda},$$
$$(x_{i-1} - a \div x_{i-1})/2 \geq (x_{i-1} - a \div x_{i-1}) \div 2 = 2^{-\lambda},$$

whence

$$x_{i-1} - a \div x_{i-1} \geq 2^{-\lambda+1}.$$

Hence

$$x_{i-1} \geq a \div x_{i-1} + 2^{-\lambda+1} > a/x_{i-1} + 2^{-\lambda},$$
$$x_i > a/x_{i-1},$$
$$x_{i-1} > a/x_i,$$
$$x_i + 2^{-\lambda} > a/x_i,$$
$$x_i - a/x_i > -2^{-\lambda}.$$

This holds a fortiori if the inequality holds in (1.5.7). Hence in all cases

$$x - a/x > -2^{-\lambda},$$
$$(x + 2^{-\lambda-1})^2 > a + 2^{-2\lambda-2},$$
$$x > (a + 2^{-2\lambda-2})^{1/2} - 2^{-\lambda-1}. \tag{1.5.8}$$

This gives a lower bound for the computed value.

Next suppose

$$(x_i - a \div x_i) \div 2 \leq 0.$$

Then

$$x_i - a \div x_i \leq 2^{-\lambda},$$
$$x_i \leq a \div x_i + 2^{-\lambda} \leq a/x_i + 2^{-\lambda}.$$

Hence in all cases

$$x \leq a/x + 2^{-\lambda},$$
$$(x - 2^{-\lambda-1})^2 \leq a + 2^{-2\lambda-2},$$
$$x \leq (a + 2^{-2\lambda-2})^{1/2} + 2^{-\lambda-1}. \tag{1.5.9}$$

Inequalities (1.5.8) and (1.5.9) define a half-open interval of length $2^{-\lambda}$

and center $(a + 2^{-2\lambda-2})^{1/2}$ upon which x must lie. These inequalities can be written

$$(x^2 - 2^{-\lambda}x)^{1/2} \leq \sqrt{a} < (x^2 + 2^{-\lambda}x)^{1/2}. \tag{1.5.10}$$

In the worst case, when $a = 0$, $x = 2^{-\lambda}$, and the error is $2^{-\lambda}$. In all other cases the error is less. The case $a = 0$, $x = 2^{-\lambda}$ could arise in machine computation when a is an intermediate result, obtained from previous computation.

If a itself can be in error by an amount α, then by §1.3 the propagated error is approximately $\alpha/(2\sqrt{a})$, when $a \neq 0$. Hence the total error cannot exceed $2^{-\lambda} + \alpha/(2\sqrt{a})$, when $a \neq 0$. If α represents the maximum possible difference between $a + 2^{-2\lambda-2}$ and the true value for a, then the maximum error can be written as $2^{-\lambda-1} + \alpha/(2x)$.

This case is especially favorable because of the fact that round-off errors do not build up in the course of the computation. At each step the best available approximation is used as the basis for obtaining a better one, and one continues to get improvement until a stage is reached at which the error inherent in a single step is as great as the truncation error.

By way of comparison, consider a computation based upon Taylor's series, where the round-off can accumulate. We require the evaluation of both $\sin x$ and $\cos x$, for $|x| \leq \pi/4 < 1$, to be followed by a check based upon the identity $\sin^2 x + \cos^2 x = 1$. Clearly the computed values of $\sin x$ and $\cos x$ will not necessarily satisfy the identity strictly. How close can we get to the true values of $\sin x$ and $\cos x$, and how closely can we expect our approximations to satisfy the identity?

We shall, in fact, describe a routine for computing s, an approximation to $\sin x$, and c, an approximation to $1 - \cos x$. Let $w_1 = x$, a digital number, and

$$w_n = (x/n)w_{n-1},$$
$$w_n^* = (x \div n) \times w_{n-1}^*.$$

For these operations it is assumed that

$$|a^*b^* - a^* \times b^*| \leq 2^{-40}.$$

The terms w_n are the terms which appear in the expansions, while w_n^* are the terms we actually obtain in the computation. For this machine $\lambda = 39$, and the machine accepts a (2λ)-digit dividend so that divisions $x \div n$ are performed by dividing $(2^{-\lambda}x)$ by $(2^{-\lambda}n)$. Let

$$\epsilon_n = 2^{-40}|w_n - w_n^*|.$$

Then

$$w_n^* - w_n = [(x \div n) \times w_{n-1}^* - (x \div n)w_{n-1}^*] + [(x \div n) - (x/n)]w_{n-1}^* + (x/n)(w_{n-1}^* - w_{n-1}).$$

The division steps satisfy

$$0 \le x/n - x \div n \le 2^{-39}(n-1)/n.$$

Also

$$|w_n^*| \le 1/n!.$$

Hence

$$\epsilon_n \le 1 + 2(n-1)/n! + \epsilon_{n-1}/n.$$

The residual error is less than the first neglected term, and for $n \ge 15$, $|w_n| < 2^{-40} = 2^{-\lambda-1}$. Hence on solving recursively and adding the generated errors (the ϵ's) and the residual errors, we have

$$|c - (1 - \cos x)| < 1.197 \times 2^{-37},$$
$$|s - \sin x| < 1.140 \times 2^{-37}.$$

For the check let

$$\cos x = 1 - c + \epsilon, \qquad \sin x = s + \epsilon',$$

where ϵ and ϵ' are bounded by the right members of the above inequalities. Then

$$2c - c^2 - s^2 = 2\epsilon \cos x + 2\epsilon' \sin x - \epsilon^2 - \epsilon'^2.$$

Hence

$$\begin{aligned} |2c - c^2 - s^2| &\le 2\epsilon'(\cos x + \sin |x|) + 2(\epsilon - \epsilon') \cos x \\ &\le 2\epsilon' \sqrt{2} + 2(\epsilon - \epsilon') \\ &< 1.669 \times 2^{-36}. \end{aligned}$$

Hence

$$\begin{aligned} |2c - c \times c - s \times s| &\le |2c - c^2 - s^2| + |c \times c - c^2| + |s \times s - s^2| \\ &< 2^{-35}. \end{aligned}$$

Thus in applying the check, we compute a quantity which should vanish if there were no errors due to truncation or round-off, but on the basis of this analysis we can say only that the computed value must be less than 2^{-35}. If a larger value occurs, it must be due to blunders.

This analysis shows only that the quantity computed for the check cannot be so great as 2^{-35} in magnitude. It does not show that quantities as great as $15 \cdot 2^{-40}$ could, in fact, occur, nor even quantities as great as 2^{-36}. Hence a more detailed analysis, which pays attention to the possible signs of the errors at each step, might yield a somewhat smaller bound.

1.6. Statistical Estimates of Error. The discussion in §1.3 shows what is intuitively obvious to begin with: that given n numbers, each of which can be in error by as much as ϵ, their sum can be in error by as much as $n\epsilon$. However, the occurrence of this greatest possible error will be extremely infrequent in practice. Even assuming that the error in each term is maximal, which is improbable in itself, the probability is only 2^{-n} that the maximal error of $n\epsilon$ would occur in the sum, since that would require that the individual errors all be of the same sign. If one can assert of each term x in the sum that its error can have any value between 0 and ϵ, say with uniform probability, then the probability of occurrence of the maximal error $n\epsilon$ becomes much smaller than 2^{-n}.

As with sums, so with any other computation or sequence of computations, the probability of occurrence of a maximal error may be extremely small. It is reasonable to inquire, therefore, as to the probability that the accumulated error in a given computation will exceed some assigned limit. A probabilistic approach is the more clearly indicated if one considers the fact that limits of error in measured quantities can seldom be assigned with certainty. At best one can assign a probability to the assertion that the error of measurement does not exceed some given amount.

In principle a statistical estimate of errors can be made by going through the same steps as in a strict estimate, except that at each step one requires a distribution of errors in the data and seeks a distribution, rather than strict limits, for the errors in the result. Unfortunately, besides the fact that the computation of these distributions is intrinsically difficult, questions of statistical independence are especially troublesome. Consequently we mention this approach only in passing and point out that there is a growing literature on the subject, a few titles of which are listed among the references.

1.7. Bibliographic Notes. The subject of errors is given at least casual discussion in most standard texts, either as a separate topic or in connection with particular computations. Most papers in the periodicals deal with errors in particular computations. The four sources of error are distinguished, digital numbers defined, and the error formulas of §1.4 are given in von Neumann and Goldstine (1947), where the major topic is errors in matrix computations. Turing (1948) also discusses matrix operations in some detail. Rademacher (1947) and Harrison (1951) discuss errors in the numerical solution of differential equations, but these papers are also of more general interest. Dwyer (1951) discusses errors at length.

On statistical assessments Inman (1950) discusses the problem in general, Rademacher (1947) makes applications to differential equations, while Huskey (1949) finds a failure which is explained by Hartree. Goldstine and von Neumann (1951) give statistical estimates in matrix computations.

Papers yet to be published by Goldstine, Murray and von Neumann, and J. W. Givens, each concerned with the problem of finding proper values of matrices, will contain elaborate error analyses. It seems safe to predict that an increasing number of detailed analyses of specific routines, such as the ones given here for the square root and the circular functions, will be issued for limited distribution by groups operating high-speed computing machines.

On the operation of automatic digital computers see Berkeley (1949), Wilks, Wheeler and Gill (1951), Engineering Research Associates (1950), and the (Harvard) Computation Laboratory (1946, 1949). In addition to these references, which are listed in the bibliography, there are numerous reports and memoranda issued by organizations which build or operate particular machines: IBM Corporation, MIT, Harvard Computation Laboratory, University of Illinois, NBS, Institute for Advanced Study, and a number of others. A section of the periodical *MTAC* is devoted to electronic computers. And finally, abstracts or complete papers presented at meetings of the Association for Computing Machinery are obtainable.

CHAPTER 2

MATRICES AND LINEAR EQUATIONS

2. Matrices and Linear Equations

The numerical solution of an integral equation, of a partial differential equation, or of an ordinary differential equation with two-point boundary conditions is generally obtained by solving an approximating linear algebraic system. Moreover in order to solve a nonlinear problem, one may replace it by a sequence of linear systems providing progressively improved approximations. For these reasons, and because of the theoretical simplicity, we start with linear systems of equations. For studying linear systems of equations a geometric terminology, with the compact symbolism of vectors and matrices, is extremely useful. A résumé of the basic principles is therefore included.

2.01. *Vectors and Coordinate Systems.* Any n vectors $\mathbf{e}_1, \ldots, \mathbf{e}_n$ are said to be linearly dependent in case any of them can be expressed as a linear combination of the others. A more symmetric statement of the same property is that there are scalars $\alpha_1, \ldots, \alpha_n$ not all zero satisfying

$$\alpha_1\mathbf{e}_1 + \cdots + \alpha_n\mathbf{e}_n = \mathbf{0}.$$

The equivalence of the two statements is made clear by considering that, if $\alpha_i \neq 0$, then we could solve for $\mathbf{e}_i$ in terms of the other vectors. As an example, two vectors are linearly dependent in case they are parallel, or if one is the null vector.

A vector space is n-dimensional in case there are n linearly independent vectors in the space, but any $n + 1$ vectors are linearly dependent. Let $\mathbf{e}_1, \ldots, \mathbf{e}_n$ be linearly independent, and let $\mathbf{x}$ be any vector in the space. These $n + 1$ vectors are linearly dependent. Hence we can find scalars $\xi', \xi'_1, \ldots, \xi'_n$, not all zero, such that

$$\xi'\mathbf{x} + \xi'_1\mathbf{e}_1 + \cdots + \xi'_n\mathbf{e}_n = \mathbf{0}.$$

But certainly, then, $\xi' \neq 0$, since if we had $\xi' = 0$, the relation would express the linear dependence of the set of $\mathbf{e}$'s, whereas they are taken to be linearly independent. We can therefore solve for $\mathbf{x}$ and write

$$\mathbf{x} = \sum_i \xi_i\mathbf{e}_i. \tag{2.01.1}$$

Hence every vector of the space is expressible as a linear combination of any n linearly independent vectors of the space; these vectors constitute a basis or coordinate system for the space; the numbers ξ_i are the coordinates of $\mathbf{x}$ in that system, and the vectors $\xi_i\mathbf{e}_i$ its components. The set of n coordinates ξ_i constitutes a "numerical vector" x. The numerical vector x specifies $\mathbf{x}$ completely in a given coordinate system. The individual ξ_i will be called the elements of x.

Suppose $\mathbf{f}_1, \ldots, \mathbf{f}_n$ are also linearly independent vectors in the space. Each $\mathbf{f}_j$ is expressible as a linear combination of $\mathbf{e}_1, \ldots, \mathbf{e}_n$:

$$\mathbf{f}_j = \sum_i \mathbf{e}_i\psi_{ij}. \tag{2.01.2}$$

But also each $\mathbf{e}_i$ is expressible as a linear combination of $\mathbf{f}_1, \ldots, \mathbf{f}_n$:

$$\mathbf{e}_i = \sum_j \mathbf{f}_j\epsilon_{ji}, \tag{2.01.3}$$

and either set of relations must be obtainable from the other by treating it as a set of n equations in n unknowns and solving in the usual manner.

If, in Eq. (2.01.1), we replace each $\mathbf{e}_i$ by its expression (2.01.3), we obtain

$$\mathbf{x} = \sum_i \sum_j \mathbf{f}_j\epsilon_{ji}\xi_i. \tag{2.01.4}$$

Hence if

$$\xi'_j = \sum_i \epsilon_{ji}\xi_i, \tag{2.01.5}$$

the ξ'_j are the coordinates of the geometric vector $\mathbf{x}$ in the $\mathbf{f}_j$ coordinate system, and the set of coordinates ξ'_j constitute the numerical vector x' which represents $\mathbf{x}$ in that system of coordinates.

It is convenient to arrange the coefficients ϵ_{ji} in the rectangular array

$$\begin{pmatrix} \epsilon_{11} & \epsilon_{12} & \cdots & \epsilon_{1n} \\ \epsilon_{21} & \epsilon_{22} & \cdots & \epsilon_{2n} \\ \cdots & \cdots & \cdots & \cdots \\ \epsilon_{n1} & \epsilon_{n2} & \cdots & \epsilon_{nn} \end{pmatrix}, \tag{2.01.6}$$

and the coordinates ξ_i in the column

$$\begin{pmatrix} \xi_1 \\ \xi_2 \\ \cdots \\ \xi_n \end{pmatrix}, \tag{2.01.7}$$

with the coordinates ξ'_j similarly arranged. These arrangements permit a simple rule for obtaining each ξ' from the ϵ's and ξ's. The rule is more easily observed, by referring to (2.01.5), than stated. The array (2.01.6) is called a matrix, and the column (2.01.7) a (numerical) vector, and we designate these E and x, respectively. The rule, then, is the rule for multiplying a matrix by a vector,

(2.01.8) $$x' = Ex,$$

where x' is the column of the ξ'_j.

Besides the vectors $\mathbf{e}_i$ and $\mathbf{f}_j$, let $\mathbf{g}_1, \ldots, \mathbf{g}_n$ represent also a set of linearly independent vectors in the same space, and let

$$\mathbf{f}_j = \sum_k \mathbf{g}_k \phi_{kj}.$$

Then

$$\mathbf{e}_i = \sum_j \mathbf{f}_j \epsilon_{ji} = \sum_k \sum_j \mathbf{g}_k \phi_{kj} \epsilon_{ji}.$$

Hence if F represents the matrix of the ϕ_{kj},

$$F = (\phi_{kj}) = \begin{pmatrix} \phi_{11} & \cdots & \phi_{1n} \\ \cdot & \cdots & \cdot \\ \phi_{n1} & \cdots & \phi_{nn} \end{pmatrix},$$

and if P represents the matrix

(2.01.9) $$P = \left(\sum_j \phi_{kj}\epsilon_{ji}\right) = \begin{pmatrix} \sum_j \phi_{1j}\epsilon_{j1} & \cdots & \sum_j \phi_{1j}\epsilon_{jn} \\ \cdot & \cdots & \cdot \\ \sum_j \phi_{nj}\epsilon_{j1} & \cdots & \sum_j \phi_{nj}\epsilon_{jn} \end{pmatrix},$$

we shall say that P is the product of the matrices F and E:

(2.01.10) $$P = FE,$$

with (2.01.9) giving the rule for multiplication. This is consistent with the rule for multiplying a matrix by a vector. Note that the product EF is, in general, not the same as the product FE.

In the particular case when

$$\mathbf{g}_i = \mathbf{e}_i,$$

then

$$\phi_{kj} = \psi_{kj},$$

and it must then be true that

$$\sum_j \psi_{kj}\epsilon_{ji} = \delta_{ki},$$

where δ_{ki} is the Kronecker delta, defined by

(2.01.11) $$\delta_{ki} \begin{array}{l} = 0 \text{ when } k \neq i, \\ = 1 \text{ when } k = i. \end{array}$$

In this event the matrix P has the simple form

$$\begin{pmatrix} 1 & 0 & 0 & \cdots \\ 0 & 1 & 0 & \cdots \\ 0 & 0 & 1 & \cdots \\ \cdot & \cdot & \cdot & \cdots \end{pmatrix}, \tag{2.01.12}$$

and is called the identity matrix I. Since, in that case,

$$I = FE,$$

one says that the matrices F and E are reciprocals:

$$F = E^{-1}.$$

This is one case in which the order of multiplication is immaterial:

$$I = EE^{-1} = E^{-1}E.$$

The matrices introduced so far have been square matrices, but matrices may also be rectangular, and in particular the vectors x and x' discussed above may be regarded as matrices of n rows and one column each. We may also have a matrix of one row and n columns. Such a matrix would be called a row vector, while the vectors x and x' are column vectors.

We can extend the notational scheme by writing

$$\mathbf{e} = (\mathbf{e}_1 \quad \mathbf{e}_2 \quad \cdots \quad \mathbf{e}_n), \tag{2.01.13}$$

and treating $\mathbf{e}$ formally as though it were a (numerical) row vector, which enables us to write (2.01.3) in abbreviated form

$$\mathbf{e} = \mathbf{f}E. \tag{2.01.14}$$

Then, also,

$$\mathbf{f} = \mathbf{g}F,$$

whence, by formal substitution,

$$\mathbf{e} = \mathbf{g}FE.$$

This is consistent with previous results, where we found that

$$\mathbf{e} = \mathbf{g}P$$

with

$$P = FE.$$

2.02. *Linear Transformations.* A transformation of vectors (as distinguished from a change of coordinates) is a natural generalization of the notion of a function. A transformation of vectors is a rule which, to every vector of the space under consideration, associates a unique vector in the space, and this vector is called its transform. The transformation is linear if to the sum of two vectors corresponds the sum of the transforms and to a scalar multiple of a vector corresponds the same scalar multiple of the transform. In many contexts, especially in discussions of quantum mechanics, it is customary to speak of the transformation as being an operation performed by an operator upon the vector and yielding the transformed vector (the result of the transformation).

An equivalent definition, and the one which will be used here, is the following: If $T(\mathbf{x})$ designates the transform of $\mathbf{x}$, then the transformation is linear in case

$$T(\mathbf{x}) = \Sigma \xi_i T(\mathbf{e}_i),$$

when $\mathbf{x}$ is given by (2.01.1). In accordance with the abbreviated notation this can be written

(2.02.1) $$T(\mathbf{x}) = T(\mathbf{e})x.$$

Since each $T(\mathbf{e}_i)$ is in the space of the $\mathbf{e}_i$, it can be expressed

$$T(\mathbf{e}_i) = \Sigma \mathbf{e}_j \tau_{ji},$$

or in abbreviated form,

(2.02.2) $$T(\mathbf{e}) = \mathbf{e}T,$$

where T is the matrix

(2.02.3) $$T = (\tau_{ij}).$$

But from (2.02.1)

(2.02.4) $$T(\mathbf{x}) = \mathbf{e}Tx.$$

Hence Tx is the numerical vector representing $T(\mathbf{x})$ in the coordinate system of the $\mathbf{e}_i$, and the matrix T represents the transformation (more strictly, the operator) in that same coordinate system.

In a different coordinate system, however, $\mathbf{x}$, $T(\mathbf{x})$, and the transformation itself are otherwise represented. In fact, if

$$\mathbf{x} = \mathbf{e}x,$$

and

$$\mathbf{e} = \mathbf{f}E,$$

then

$$\mathbf{x} = \mathbf{f}Ex = \mathbf{f}x',$$

so that, as we have already seen, $x' = Ex$ represents $\mathbf{x}$ in the coordinate system $\mathbf{f}$. Now

$$T(\mathbf{x}) = \mathbf{e}Tx = \mathbf{f}ETx,$$

and

$$x = E^{-1}x'.$$

Hence

$$T(\mathbf{x}) = \mathbf{f}ETE^{-1}x'.$$

Hence

(2.02.5) $$T' = ETE^{-1}$$

is the matrix which represents the transformation in the coordinate system **f**, since this is the matrix which, when applied to the numerical vector x' (which represents **x** in that system), will yield the numerical vector representing $T(\mathbf{x})$ in that system.

2.03. *Determinants and Outer Products.* An outer product of two vectors **a** and **b** is a new type of geometric entity, defined to have the following properties:

$$(1_2) \qquad [\mathbf{a}, \mathbf{b}] = -[\mathbf{b}, \mathbf{a}];$$

$$(2_2) \qquad [\alpha\mathbf{a}, \mathbf{b}] = \alpha[\mathbf{a}, \mathbf{b}];$$

$$(3_2) \qquad [\mathbf{a}, \mathbf{b}] + [\mathbf{a}, \mathbf{c}] = [\mathbf{a}, \mathbf{b} + \mathbf{c}].$$

It can be pictured as a two-dimensional vector, whose magnitude, taken positively or negatively, is the area of the parallelogram determined by the vectors in the product. It follows immediately from (1_2) that

$$[\mathbf{a}, \mathbf{a}] = 0.$$

Hence

$$[\mathbf{a}, \mathbf{b}] = [\mathbf{a}, \mathbf{b}] + \alpha[\mathbf{a}, \mathbf{a}] = [\mathbf{a}, \mathbf{b}] + [\mathbf{a}, \alpha\mathbf{a}] = [\mathbf{a}, \mathbf{b} + \alpha\mathbf{a}].$$

Hence for any scalars α and β,

$$[\mathbf{a}, \mathbf{b}] = [\mathbf{a}, \mathbf{b} + \alpha\mathbf{a}] = [\mathbf{a} + \beta\mathbf{b}, \mathbf{b}].$$

If $\mathbf{e}_1$, $\mathbf{e}_2$ are any linearly independent vectors in the space of **a** and **b**, then

$$(2.03.1) \qquad [\mathbf{a}, \mathbf{b}] = |a \quad b|[\mathbf{e}_1, \mathbf{e}_2],$$

where

$$(2.03.2) \qquad |a \quad b| = \alpha_1\beta_2 - \alpha_2\beta_1$$

is called the determinant of the numerical vectors a and b. The evaluation is immediate:

$$\begin{aligned}[\mathbf{a}, \mathbf{b}] &= [\mathbf{a}, \beta_1\mathbf{e}_1 + \beta_2\mathbf{e}_2] = \beta_1[\mathbf{a}, \mathbf{e}_1] + \beta_2[\mathbf{a}, \mathbf{e}_2] \\ &= \beta_1[\alpha_1\mathbf{e}_1 + \alpha_2\mathbf{e}_2, \mathbf{e}_1] + \beta_2[\alpha_1\mathbf{e}_1 + \alpha_2\mathbf{e}_2, \mathbf{e}_2] \\ &= \beta_1\alpha_2[\mathbf{e}_2, \mathbf{e}_1] + \beta_2\alpha_1[\mathbf{e}_1, \mathbf{e}_2] \\ &= (\alpha_1\beta_2 - \alpha_2\beta_1)[\mathbf{e}_1, \mathbf{e}_2].\end{aligned}$$

The determinant is a number, and its relation to the outer product is similar to that of the coordinates to a vector.

It is a simple geometric exercise to show that (3_2) holds in the parallelogram interpretation when **a**, **b**, and **c** are all in the same 2 space. When they are not, the relation serves to specify the rule of composition.

For outer products of n vectors the defining relations are sufficiently typified by the case $n = 3$:

$$(1_3) \qquad [\mathbf{a}, \mathbf{b}, \mathbf{c}] = -[\mathbf{a}, \mathbf{c}, \mathbf{b}] = -[\mathbf{c}, \mathbf{b}, \mathbf{a}] = \cdots;$$

$$(2_3) \qquad [\alpha\mathbf{a}, \mathbf{b}, \mathbf{c}] = \alpha[\mathbf{a}, \mathbf{b}, \mathbf{c}];$$

$$(3_3) \qquad [\mathbf{a}, \mathbf{b}, \mathbf{c}] + [\mathbf{a}, \mathbf{b}, \mathbf{d}] = [\mathbf{a}, \mathbf{b}, \mathbf{c} + \mathbf{d}].$$

From these we deduce that

$$[\mathbf{a}, \mathbf{a}, \mathbf{c}] = \cdots = 0;$$
$$[\mathbf{a}, \mathbf{b}, \mathbf{c}] = [\mathbf{a} + \beta\mathbf{b}, \mathbf{b}, \mathbf{c}] = [\mathbf{a}, \mathbf{b}, \mathbf{c} + \alpha\mathbf{a}] = \cdots;$$

and if $\mathbf{e}_1$, $\mathbf{e}_2$, $\mathbf{e}_3$ are linearly independent vectors in the space of **a**, **b**, and **c**, then

$$[\mathbf{a}, \mathbf{b}, \mathbf{c}] = |a \quad b \quad c|[\mathbf{e}_1, \mathbf{e}_2, \mathbf{e}_3], \tag{2.03.3}$$

where $|a \quad b \quad c|$ is called the determinant of the numerical vectors a, b, and c, and its value will now be obtained. Note first that, if

$$\mathbf{a}' = \alpha_1\mathbf{e}_1 + \alpha_2\mathbf{e}_2, \qquad \mathbf{b}' = \beta_1\mathbf{e}_1 + \beta_2\mathbf{e}_2,$$

then

$$[\mathbf{a}, \mathbf{b}, \mathbf{e}_3] = [\mathbf{a}', \mathbf{b}', \mathbf{e}_3] = (\alpha_1\beta_2 - \alpha_2\beta_1)[\mathbf{e}_1, \mathbf{e}_2, \mathbf{e}_3].$$

In fact, the identical steps that led to (2.03.1) and (2.03.2) will, if applied to the second member of this last equality, yield the third member. Now by an obvious modification we obtain

$$[\mathbf{a}, \mathbf{b}, \mathbf{e}_2] = (\alpha_3\beta_1 - \alpha_1\beta_3)[\mathbf{e}_1, \mathbf{e}_2, \mathbf{e}_3],$$

and again

$$[\mathbf{a}, \mathbf{b}, \mathbf{e}_1] = (\alpha_2\beta_3 - \alpha_3\beta_2)[\mathbf{e}_1, \mathbf{e}_2, \mathbf{e}_3].$$

By putting these together we obtain

$$|a \quad b \quad c| = \gamma_1(\alpha_2\beta_3 - \alpha_3\beta_2) + \gamma_2(\alpha_3\beta_1 - \alpha_1\beta_3) + \gamma_3(\alpha_1\beta_2 - \alpha_2\beta_1). \tag{2.03.4}$$

When we write the determinant explicitly in the form

$$|a \quad b \quad c| = \begin{vmatrix} \alpha_1 & \beta_1 & \gamma_1 \\ \alpha_2 & \beta_2 & \gamma_2 \\ \alpha_3 & \beta_3 & \gamma_3 \end{vmatrix},$$

we see that in the expansion of the determinant in terms of the γ's, the coefficient of each γ_i is, except for sign, that second-order determinant that remains after deleting the row and column containing γ_i. The sign is that power of -1 whose exponent is obtained by adding together the number of the row and column. Thus γ_2 is in the second row, third column, and the sign is $(-1)^{2+3}$. The coefficient of γ_i with its proper sign is called the cofactor of γ_i. By interchanging rows and columns and going through the same process, we find

$$\begin{aligned} |a \quad b \quad c| &= \alpha_1 A_1 + \alpha_2 A_2 + \alpha_3 A_3 \\ &= \beta_1 B_1 + \beta_2 B_2 + \beta_3 B_3 \\ &= \gamma_1 \Gamma_1 + \gamma_2 \Gamma_2 + \gamma_3 \Gamma_3, \end{aligned}$$

where the capital letters signify cofactors formed by the same rule. It is also true that, for example,

$$0 = |a \quad b \quad a| = \alpha_1\Gamma_1 + \alpha_2\Gamma_2 + \alpha_3\Gamma_3,$$

and, in general, when the elements of any column are multiplied by the cofactors of some other column and the products summed, the sum vanishes.

Finally, we have the expansions

$$\begin{aligned} |a \quad b \quad c| &= \alpha_1 A_1 + \beta_1 B_1 + \gamma_1\Gamma_1 \\ &= \alpha_2 A_2 + \beta_2 B_2 + \gamma_2\Gamma_2 \\ &= \alpha_3 A_3 + \beta_3 B_3 + \gamma_3\Gamma_3. \end{aligned}$$

These are the expansions we should get if we were to rewrite the determinant, writing the rows of the original as columns of the new one, and these equations say in effect that this exchange of rows for columns leaves the value of the determinant unaltered. It is quite clear that such an exchange leaves the value of a second-order determinant unaltered. Hence it is clear that in the expansion of either third-order determinant, the original or the transposed, the coefficient of any element is the same. The theorem follows because, when the determinant is expressed as an explicit function of its elements, each term contains as a factor one and only one element from each row and one and only one element from each column.

The recursive extension to successively higher dimensions can be made by following the same pattern, and the formulas need not be written explicitly. For each extension the expansion is made first along a particular column, and one observes that it is equally possible to expand along any row.

For determinants of order 4 or greater another type of expansion is possible, called the Laplace expansion. To describe this it is convenient to introduce the symbolism

$$|\alpha_i \quad \beta_j| = \begin{vmatrix} \alpha_i & \beta_i \\ \alpha_j & \beta_j \end{vmatrix}.$$

If, now,

$$\begin{aligned} \mathbf{a} &= \alpha_1\mathbf{e}_1 + \alpha_2\mathbf{e}_2 + \alpha_3\mathbf{e}_3 + \alpha_4\mathbf{e}_4, \\ &\cdots\cdots\cdots\cdots\cdots\cdots \\ \mathbf{d} &= \delta_1\mathbf{e}_1 + \delta_2\mathbf{e}_2 + \delta_3\mathbf{e}_3 + \delta_4\mathbf{e}_4, \end{aligned}$$

then we can see that

$$[\mathbf{a}, \mathbf{b}, \mathbf{e}_3, \mathbf{e}_4] = |\alpha_1 \quad \beta_2|[\mathbf{e}_1, \mathbf{e}_2, \mathbf{e}_3, \mathbf{e}_4],$$

and

$$[\mathbf{e}_1, \mathbf{e}_2, \mathbf{c}, \mathbf{d}] = |\gamma_3 \quad \delta_4|[\mathbf{e}_1, \mathbf{e}_2, \mathbf{e}_3, \mathbf{e}_4].$$

Hence we shall find that in the expansion of $|a \quad b \quad c \quad d|$ there will appear terms which make up the product $|\alpha_1 \quad \beta_2||\gamma_3 \quad \delta_4|$. But if we interchange $\mathbf{e}_2$ and $\mathbf{e}_3$, say, we shall find that there are also terms making up the product $-|\alpha_1 \quad \beta_3||\gamma_2 \quad \delta_4|$, and there is no term common to the two products. When all possible interchanges are made that yield distinct products, we find

$$|a \quad b \quad c \quad d| = |\alpha_1 \quad \beta_2||\gamma_3 \quad \delta_4| - |\alpha_1 \quad \beta_3||\gamma_2 \quad \delta_4| + |\alpha_1 \quad \beta_4||\gamma_2 \quad \delta_3| \\ + |\alpha_2 \quad \beta_3||\gamma_1 \quad \delta_4| - |\alpha_2 \quad \beta_4||\gamma_1 \quad \delta_3| + |\alpha_3 \quad \beta_4||\gamma_1 \quad \delta_2|.$$

To determine the sign in each case we note, for example, that in the expansion $|\alpha_3 \quad \beta_4||\gamma_1 \quad \delta_2|$ would appear as the coefficient of $[\mathbf{e}_3, \mathbf{e}_4, \mathbf{e}_1, \mathbf{e}_2]$, but that this is equal to $[\mathbf{e}_1, \mathbf{e}_2, \mathbf{e}_3, \mathbf{e}_4]$, whence the sign is plus.

Note finally that the determinant of the product of two matrices is the product of the determinants. In fact, if

$$\mathbf{e} = \mathbf{f}E, \qquad \mathbf{f} = \mathbf{g}F,$$

then

$$\mathbf{e} = \mathbf{g}FE.$$

But on the one hand,

$$[\mathbf{e}_1, \ . \ . \ . \ , \mathbf{e}_n] = |E|[\mathbf{f}_1, \ . \ . \ . \ , \mathbf{f}_n] \\ = |E||F|[\mathbf{g}_1, \ . \ . \ . \ , \mathbf{g}_n],$$

and on the other hand,

$$[\mathbf{e}_1, \ . \ . \ . \ , \mathbf{e}_n] = |FE|[\mathbf{g}_1, \ . \ . \ . \ , \mathbf{g}_n].$$

Hence

$$|FE| = |F||E|.$$

2.04. *Length and Orthogonality.* Geometrically the scalar product $\mathbf{xy}$ of the vectors $\mathbf{x}$ and $\mathbf{y}$ is equal to the product of their lengths into the cosine of the angle between them, or the product of the length of one vector by the projection of the other upon it. It is clear geometrically that the projection of a broken line upon a given line is the sum of the projections of the separate segments and is equal also to the projection of the single segment which joins the two ends of the broken line. Hence, if $\mathbf{x}$ is given by (2.01.1), its projection upon $\mathbf{y}$ is the sum of the projections of the segments $\xi_i\mathbf{e}_i$ upon $\mathbf{y}$. Hence

$$\mathbf{xy} = (\Sigma\xi_i\mathbf{e}_i)\mathbf{y}.$$

But by the same rule, if

$$\mathbf{y} = \Sigma\eta_j\mathbf{e}_j,$$

then

$$\mathbf{xy} = \mathbf{x}\Sigma\eta_j\mathbf{e}_j,$$

and therefore

$$\mathbf{xy} = \Sigma\Sigma\xi_i\eta_j\mathbf{e}_i\mathbf{e}_j. \tag{2.04.1}$$

Now the scalar products

$$\gamma_{ij} = \mathbf{e}_i\mathbf{e}_j = \gamma_{ji} \tag{2.04.2}$$

are known once the vectors $\mathbf{e}_i$ are themselves known. Hence the scalar product of any two vectors can be calculated from (2.04.1).

If each column of a matrix M is written as a row, the order remaining the same, the resulting matrix is known as the transpose of the original and is designated M^{T}. In particular, if x is the column vector of the ξ_i, x^{T} is the row vector of the ξ_i. With this understanding, if G is the matrix

$$G = (\gamma_{ij}) = \mathbf{e}^{\mathsf{T}}\mathbf{e}, \tag{2.04.3}$$

this is said to define the metric in the space, and Eq. (2.04.1) becomes

$$\mathbf{xy} = x^{\mathsf{T}}Gy = y^{\mathsf{T}}Gx. \tag{2.04.4}$$

The matrix G is equal to its own transpose and is said to be symmetric. When the metric G is known and fixed throughout the discussion, one often uses the notation

$$(x, y) = (y, x) = x^{\mathsf{T}}Gy. \tag{2.04.5}$$

An often-used type of coordinate system is one in which each reference vector $\mathbf{e}_i$ is of unit length and orthogonal to all the others. In this case

$$G = I, \qquad \mathbf{xy} = x^{\mathsf{T}}y = y^{\mathsf{T}}x.$$

With reference to the coordinate system $\mathbf{f}$ given by (2.01.14), the metric is defined by

$$H = \mathbf{f}^{\mathsf{T}}\mathbf{f}.$$

Now

$$G = \mathbf{e}^{\mathsf{T}}\mathbf{e} = E^{\mathsf{T}}\mathbf{f}^{\mathsf{T}}\mathbf{f}E,$$

$$G = E^{\mathsf{T}}HE. \tag{2.04.6}$$

This relates the metrics for the two coordinate systems.

If both $\mathbf{e}$ and $\mathbf{f}$ are unit-orthogonal systems, then

$$G = H = I$$

and hence

$$I = E^{\mathsf{T}}E.$$

In this case

$$E^{\mathsf{T}} = E^{-1}$$

and the matrix E is said to be orthogonal.

When G is not the identity matrix, then in the process of evaluating a scalar product of two vectors $\mathbf{x}$ and $\mathbf{y}$, given x and y, one must form either Gx or Gy. Since if one knows the vector

$$x' = Gx \tag{2.04.7}$$

one can always find x by solving the equations, knowing x' is (in principle) equivalent to knowing x. Hence x' is also a representation of $\mathbf{x}$, but of a different kind from x. It is customary to speak of x' as giving the covariant representation or as being a covariant vector, and of x as giving the contravariant representation or as being a contravariant vector. If y' is the covariant representation of $\mathbf{y}$, then

$$\mathbf{xy} = x^{\mathsf{T}}y' = x'^{\mathsf{T}}y = y^{\mathsf{T}}x' = y'^{\mathsf{T}}x.$$

In case $\mathbf{x}$ and $\mathbf{y}$ are the same vector, the scalar product is the square of the length:

$$\mathbf{xx} = |\mathbf{x}|^2 = x^{\mathsf{T}}Gx.$$

Since any numerical vector x represents some geometric vector $\mathbf{x}$, it follows that always

$$x^{\mathsf{T}}Gx \geq 0,$$

and the equality can hold only when x is the null vector, $x = 0$. By virtue of this property of the matrix G, it is said to be positive definite. The vectors $\mathbf{e}_i$ can always be referred to a unit-orthogonal system $\mathbf{f}$, and in this case (2.04.6) becomes

$$G = E^{\mathsf{T}}E,$$

since H is then the identity. It will be shown in §2.201 that any positive definite matrix can be expressed as a product of a matrix by its transpose. Hence any positive definite matrix represents a metric in some coordinate system.

There are always several geometric settings, any one of which could give rise to a given set of linear algebraic equations. Hence given the equations, we are at liberty to associate any geometric picture that seems convenient. This will be done from time to time in the presentation of the various methods for solving linear systems. The geometric vectors $\mathbf{e}$, $\mathbf{f}$, etc., seldom if ever need to be introduced explicitly. Given a numerical vector x, it is sufficient to know that, given a coordinate system $\mathbf{e}$, the numerical vector x defines a geometric vector $\mathbf{x}$ by the relation

$$\mathbf{x} = \mathbf{e}x.$$

Hence we shall often speak of x as though it were itself a geometric vector, and we shall refer to it simply as a vector, without qualification.

2.05. *Rank and Nullity; Adjoint and Reciprocal Matrices.* The outer product of two vectors has been interpreted geometrically as representing an oriented parallelogram; the outer product of r vectors represents an oriented parallelepiped of r dimensions. We shall take it as geometrically evident that the outer product of r vectors can vanish if and only if the vectors lie in a space of dimension less than r, or in other words, if they are linearly dependent.

A matrix is said to have rank r in case it has r linearly independent columns, but every $r + 1$ columns are linearly dependent. Hence every column is expressible as a linear combination of these r columns. The outer product of the vectors represented by these r columns must be non-null, so that at least one submatrix formed from these r columns must have a nonvanishing determinant. On the other hand, no submatrix of order $r + 1$ can have a nonvanishing determinant. Hence a matrix is of rank r if and only if the largest submatrix with nonvanishing determinant is of order r. By applying the above argument to the transpose of the matrix, everything that has been said of the columns applies equally to the rows.

A square matrix of order n and rank r is said to have nullity $n - r$. If we suppose the coordinate vectors $\mathbf{e}$ to be a unitary orthogonal system, the homogeneous equations

$$A^{\mathsf{T}}x = 0$$

are satisfied by any vector x orthogonal to all the columns of A. But if A has rank r, its columns determine an r-dimensional subspace of the n-dimensional vector space, and there is an $(n - r)$-dimensional subspace orthogonal to it. Hence the equations have $n - r$ linearly independent solutions.

A matrix is nonsingular in case its determinant is nonzero. If

$$\mathbf{x} = \Sigma\xi_i\mathbf{a}_i = \mathbf{a}x \tag{2.05.1}$$

and the vectors $\mathbf{a}_i$ are linearly independent, then

$$\begin{aligned}\xi_1[\mathbf{a}_1, \mathbf{a}_2, \ldots, \mathbf{a}_n] &= [\xi_1\mathbf{a}_1, \mathbf{a}_2, \ldots, \mathbf{a}_n] \\ &= [\xi_1\mathbf{a}_1 + \xi_2\mathbf{a}_2 + \cdots + \xi_n\mathbf{a}_n, \mathbf{a}_2, \ldots, \mathbf{a}_n] \\ &= [\mathbf{x}, \mathbf{a}_2, \ldots, \mathbf{a}_n].\end{aligned}$$

Hence if

$$\mathbf{a} = \mathbf{e}A,$$

these equations give

$$\xi_1|a_1 \quad a_2 \quad \ldots \quad a_n| = |x \quad a_2 \quad \ldots \quad a_n| \tag{2.05.2}$$

when we drop the outer product of the $\mathbf{e}_i$. Since the vectors $\mathbf{a}_i$ are linearly independent, the matrix A is nonsingular, and hence

$$\xi_1 = |x \quad a_2 \quad \ldots \quad a_n|/|a_1 \quad a_2 \quad \ldots \quad a_n|, \tag{2.05.3}$$

and, in general, each ξ_i is the quotient by $|A|$ of the determinant obtained from $|A|$ when x replaces a_i. This is Cramer's rule.

If we write

$$A = (\alpha_{ij}), \tag{2.05.4}$$

and if A_{ji} is the cofactor of α_{ij} in $|A|$, the matrix

$$\operatorname{adj}(A) = (A_{ji}) \tag{2.05.5}$$

of the cofactors is called the adjoint of A. The expansion rules illustrated in §2.03 for a determinant of order 3 can be expressed

(2.05.6) $$A \operatorname{adj}(A) = \operatorname{adj}(A)A = |A|I;$$

the product of a matrix by its adjoint in either order is a matrix with 0 everywhere except along the principal diagonal, and there every element has the value $|A|$.

Now let

(2.05.7) $$\alpha^{ji} = A_{ji}/|A|.$$

Then

$$(\alpha^{ji}) = |A|^{-1} \operatorname{adj}(A),$$
$$(\alpha_{ij})(\alpha^{ji}) = I,$$

so that

(2.05.8) $$A^{-1} = |A|^{-1} \operatorname{adj}(A).$$

This gives the explicit representation of the elements of the inverse matrix.

2.051. *Projection operators.* Let **a** represent a set of $m \leq n$ linearly independent vectors, and

(2.051.1) $$\mathbf{a} = \mathbf{e}A.$$

These vectors form a basis for a subspace of m dimensions. If $\mathbf{e}^\mathsf{T}\mathbf{e} = I$, then

$$\mathbf{a}^\mathsf{T}\mathbf{a} = A^\mathsf{T}A$$

defines the metric for that subspace. The matrix is nonsingular, since otherwise a non-null vector x would exist satisfying $A^\mathsf{T}Ax = 0$, and hence $x^\mathsf{T}A^\mathsf{T}Ax = 0$, and the non-null geometric vector $\mathbf{a}x$ would have length zero, which is impossible.

Hence the symmetric matrix

(2.051.2) $$P = A(A^\mathsf{T}A)^{-1}A^\mathsf{T}$$

exists, and is said to be idempotent since

$$P^2 = A(A^\mathsf{T}A)^{-1}A^\mathsf{T}A(A^\mathsf{T}A)^{-1}A^\mathsf{T} = A(A^\mathsf{T}A)^{-1}A^\mathsf{T} = P.$$

If $\mathbf{x} = \mathbf{e}x$ is any vector, then

$$\mathbf{e}Px = \mathbf{e}A(A^\mathsf{T}A)^{-1}A^\mathsf{T}x = \mathbf{a}(A^\mathsf{T}A)^{-1}A^\mathsf{T}x$$

is a vector in the space of **a**. If $\mathbf{y} = \mathbf{a}y = \mathbf{e}Ay$ is any vector in the space of **a**, then

$$\mathbf{e}P(Ay) = \mathbf{e}A(A^\mathsf{T}A)^{-1}A^\mathsf{T}(Ay) = \mathbf{e}Ay = \mathbf{y}.$$

Hence if x represents any vector, Px represents a vector in the space of A, and if $x = Ay$ represents a vector in the space of A, then Px represents the same vector. Thus the matrix P projects any vector into a particular subspace and leaves unchanged any vector already in that subspace. It is therefore called a projection operator.

The projection is, indeed, an orthogonal projection. For given any x, the projection is Px, and the residual is $x - Px$. But since P is symmetric, the projection and the residual have the scalar product

$$x^{\mathsf{T}}P(I - P)x = x^{\mathsf{T}}(P - P^2)x = 0,$$

since P is idempotent.

Any symmetric idempotent matrix P is a projection operator. For if P has rank m, we can find a matrix A of m linearly independent columns such that every column of P is a linear combination of columns of A. Hence, for some matrix B we can write

$$P = AB^{\mathsf{T}}.$$

We have only to show that

$$B = A(A^{\mathsf{T}}A)^{-1}.$$

Since P is idempotent,

$$P = P^2 = AB^{\mathsf{T}}AB^{\mathsf{T}} = AB^{\mathsf{T}}.$$

Since the columns of A are linearly independent, $B^{\mathsf{T}}AB^{\mathsf{T}} = B^{\mathsf{T}}$. The rank of $P = AB^{\mathsf{T}}$ cannot exceed the rank of B^{T}; hence B^{T} has rank at least m, and having only m rows, the rank is exactly m. Hence

$$B^{\mathsf{T}}A = I = A^{\mathsf{T}}B.$$

Since P is symmetric,

$$\begin{aligned} AB^{\mathsf{T}} &= BA^{\mathsf{T}}, \\ AB^{\mathsf{T}}A &= BA^{\mathsf{T}}A, \\ A &= BA^{\mathsf{T}}A. \end{aligned}$$

This is the desired result.

For an arbitrary metric, $\mathbf{e}^{\mathsf{T}}\mathbf{e} = G$, the orthogonal projection is represented by a matrix

$$P = A(A^{\mathsf{T}}GA)^{-1}A^{\mathsf{T}}G, \tag{2.051.3}$$

where the columns of A are contravariant vectors.

2.06. *Cayley-Hamilton Theorem; Canonical Form of a Matrix.* If with any vector x we associate its successive transforms

$$x_i = T^i x \qquad (i = 0, 1, 2, \ldots), \tag{2.06.1}$$

at most n vectors of the sequence $x_0, x_1, x_2, \ldots$ will be linearly independent. Hence for some $r \leq n$ there exist scalars $\gamma_0, \gamma_1, \ldots, \gamma_r$ not all null such that

$$(\gamma_0 I + \gamma_1 T + \cdots + \gamma_r T^r)x = 0. \tag{2.06.2}$$

The matrix

$$\psi(T) = \gamma_0 I + \gamma_1 T + \cdots + \gamma_r T^r \tag{2.06.3}$$

is a polynomial in T, and since

$$\psi(T)x = 0,$$

x is said to lie in the null space of $\psi(T)$. If both x and y are in the null space of $\psi(T)$, then $\alpha x + \beta y$ is also in this null space for any scalars α and β. The null space consists of the null vector only, unless $\psi(T)$ is a singular matrix. If it is nonsingular, we shall say it has no null space, disregarding the trivial case of the null vector alone.

If $\phi(T)$ and $\psi(T)$ are polynomials in T, and if there is a non-null vector x which lies in the null space of each, then the scalar polynomials $\phi(\lambda)$ and $\psi(\lambda)$ have a common divisor which is not constant. For if they do not, then by a classical theorem in algebra (proved below in §3.06), there exist polynomials $f(\lambda)$ and $g(\lambda)$ such that

$$f(\lambda)\phi(\lambda) + g(\lambda)\psi(\lambda) = 1 \tag{2.06.4}$$

identically. But then

$$f(T)\phi(T) + g(T)\psi(T) = I, \tag{2.06.5}$$

and hence

$$f(T)\phi(T)x + g(T)\psi(T)x = x. \tag{2.06.6}$$

But since x lies in the null spaces of both ϕ and ψ, the left member of this identity vanishes, and hence x is a null vector, contrary to supposition.

Next to a constant, whose null space is the null vector alone, the simplest type of polynomial is linear. Hence consider a polynomial $T - \lambda I$. It has a null space if and only if its determinant vanishes:

$$|T - \lambda I| = 0. \tag{2.06.7}$$

This determinant, when expanded, is a polynomial in λ of degree n,

$$\phi(\lambda) = |T - \lambda I| = (-\lambda)^n + \gamma_1(-\lambda)^{n-1} + \cdots + \gamma_n, \tag{2.06.8}$$

called the characteristic function of T. Equation (2.06.7) is called the characteristic equation of T. One verifies easily that

$$\gamma_1 = \Sigma\tau_{ii} = t(T), \qquad \gamma_n = |T|, \tag{2.06.9}$$

where $t(T)$ is called the trace of T and is equal to the sum of the diagonal elements.

Any λ satisfying (2.06.7) is called a proper value of T. If λ is any proper value, there is at least one vector in the null space of $T - \lambda I$, and any vector in the null space of $T - \lambda I$ is called a proper vector associated with the proper value λ. If λ is not a proper value, $T - \lambda I$ has no null space.

For any proper value λ, any vector in the null space of $T - \lambda I$ is also in the null space of $(T - \lambda I)^r$ for any positive integer r, but the converse is not true. A vector in the null space of $(T - \lambda I)^r$ for any positive integer r is a principal vector. If x is in the null space of $(T - \lambda I)^r$ but not in the null space of $(T - \lambda I)^{r-1}$, it is called a principal vector of grade r.

The characteristic function $\phi(\lambda)$ has the remarkable property that

$$\phi(T) = 0. \tag{2.06.10}$$

This is the Cayley-Hamilton theorem, which can be stated otherwise by saying that the null space of $\phi(T)$ is the entire space. This might be expected from the fact, shown above, that the null spaces of two polynomials in T have a non-null vector in common only if they have a common divisor. A proof of the Cayley-Hamilton theorem is as follows Since

$$\phi(T) - \phi(\lambda)I = [(-T)^n - (-\lambda)^n I] + \gamma_1[(-T)^{n-1} - (-\lambda)^{n-1}I] + \cdots + \gamma_{n-1}(-T + \lambda I),$$

therefore this difference is equal to a polynomial in T multiplied by $T - \lambda I$, and hence is said to be divisible by $T - \lambda I$. Also, by Eq. (2.05.6),

$$(T - \lambda I) \operatorname{adj} (T - \lambda I) = \phi(\lambda)I.$$

Hence $\phi(\lambda)I$ is also divisible by $T - \lambda I$. It follows, then, that $\phi(T)$ is divisible by $T - \lambda I$. However, $\phi(T)$ is independent of λ and must therefore vanish.

If F is any nonsingular matrix, then the matrices T and

$$T' = F^{-1}TF \tag{2.06.11}$$

are said to be similar. They represent the same transformation but in different coordinate systems. Since

$$F^{-1}(T - \lambda I)F = T' - \lambda I,$$

they have the same characteristic function and hence the same proper values. One proves inductively that for any positive integer r

$$T'^r = F^{-1}T^rF,$$

and hence that

$$\psi(T') = F^{-1}\psi(T)F \tag{2.06.12}$$

for any polynomial ψ.

It is reasonable to expect that a given transformation might be more simply represented in some coordinate systems than in others, and this will now be shown. Note first that the theorem expressed by (2.06.4) can be generalized as follows: If there is no nonconstant factor common to all polynomials $\phi_1(\lambda)$, $\phi_2(\lambda)$, . . . , $\phi_m(\lambda)$, then there exist polynomials $f_1(\lambda)$, $f_2(\lambda)$, . . . , $f_m(\lambda)$ such that

$$f_1(\lambda)\phi_1(\lambda) + \cdots + f_m(\lambda)\phi_m(\lambda) = 1 \tag{2.06.13}$$

identically. The proof is inductive and sufficiently indicated by taking $m = 3$. It may be that ϕ_1 and ϕ_2 have a common factor $d_{12}(\lambda)$, but if so, it is prime to ϕ_3. By applying (2.06.4) to $\phi = \phi_1/d_{12}$ and $\psi = \phi_2/d_{12}$, and then multiplying, we have for some g_1 and g_2

$$g_1\phi_1 + g_2\phi_2 = d_{12}.$$

Also for some g and h,

$$gd_{12} + h\phi_3 = 1.$$

Hence

$$gg_1\phi_1 + gg_2\phi_2 + h\phi_3 = 1.$$

Now let the characteristic function $\phi(\lambda)$ be factored completely:

$$\phi(\lambda) = (\lambda - \lambda_1)^{n_1}(\lambda - \lambda_2)^{n_2} \cdots (\lambda - \lambda_m)^{n_m},$$

where

$$n_1 + n_2 + \cdots + n_m = n,$$

and the λ_i are all different. Clearly the polynomials

$$\phi_i(\lambda) = (\lambda - \lambda_i)^{-n_i}\phi(\lambda)$$

satisfy the conditions of the theorem. Hence with suitable polynomials f_i, (2.06.13) can be satisfied, and hence

$$f_1(T)\phi_1(T) + \cdots + f_m(T)\phi_m(T) = I.$$

Hence for any vector x

$$f_1(T)\phi_1(T)x + \cdots + f_m(T)\phi_m(T)x = x.$$

But $\phi_i(T)x$, and hence also $f_i(T)\phi_i(T)x$, is a principal vector since

$$(T - \lambda_i I)^{n_i}\phi_i(T) = \phi(T) = 0.$$

Hence any vector is expressible as a sum of principal vectors.

Moreover, the representation of x as a sum of principal vectors is unique. For if not, the difference of distinct representations would express the null vector as a sum of principal vectors in the form

$$y_1 + y_2 + \cdots + y_m = 0,$$

where y_i is a principal vector associated with λ_i, and at least one y_i is non-null. Suppose $y_1 \neq 0$. Then

$$\phi_1(T)(y_1 + y_2 + \cdots + y_m) = 0.$$

But since $\phi_1(T)$ contains every factor $(\lambda - \lambda_i)^{n_i}$ except $(\lambda - \lambda_1)^{n_1}$, it follows that

$$\phi_1(T)(y_2 + \cdots + y_m) = 0.$$

Hence

$$\phi_1(T)y_1 = 0.$$

But

$$(T - \lambda_1 I)^{n_1} y_1 = 0,$$

whereas $(\lambda - \lambda_1)^{n_1}$ and $\phi_1(\lambda)$ have no common factor. Hence $y_1 = 0$, contrary to supposition.

Now for a new coordinate system choose a matrix

$$F = (F_1 F_2 \cdots F_m) \tag{2.06.14}$$

in which the columns of F_1 form a coordinate system for the null space of $(T - \lambda_1 I)^{n_1}$, the columns of F_2 form a coordinate system for the null space of $(T - \lambda_2 I)^{n_2}$, ... If x is any vector in the null space of $(T - \lambda_i I)^{n_i}$, so also is Tx since

$$(T - \lambda_i I)^{n_i} Tx = T(T - \lambda_i I)^{n_i} x = 0.$$

Hence any column of TF_i is expressible as a linear combination of columns of F_i, and therefore

$$TF = FT' \tag{2.06.15}$$

where T' has the form

$$T' = \begin{pmatrix} T_1 & 0 & 0 & \cdots \\ 0 & T_2 & 0 & \cdots \\ 0 & 0 & T_3 & \cdots \\ \cdots & \cdots & \cdots & \cdots \end{pmatrix}. \tag{2.06.16}$$

But F must be nonsingular, whence

$$F^{-1}TF = T', \tag{2.06.17}$$

and except for the trivial case $m = 1$, a partial simplification has been effected. We proceed next to specialize the choice of the columns within any F_i.

These are all principal vectors, and there is some $\nu_i \leq n_i$ such that every vector in the space of F_i is of grade ν_i or less. First select a maximal linearly independent set of proper vectors in the space of F_i (corresponding to the proper value λ_i). Adjoin to these a maximal linearly independent set of (principal) vectors of grade 2, . . . , and finally complete F_i by adjoining vectors of maximal grade ν_i. Thus F_i has the form

$$F_i = (F_{i1}, F_{i2}, \ldots, F_{i\nu_i}),$$

where all columns of F_i are linearly independent and where every column of F_{ij} is a principal vector of grade j.

Now the columns of $F_{i\nu_i}$ are of grade ν_i, while those of

$$(T - \lambda_i I)F_{i\nu_i}$$

are of grade $\nu_i - 1$. Furthermore, the columns of

$$(T - \lambda_i I)F_{i\nu_i}, F_{i\nu_i}$$

are linearly independent. If this were not the case, there would exist equal linear combinations $(T - \lambda_i I)x$ of the columns of the first submatrix and y of the columns of the second:

$$(2.06.18) \qquad (T - \lambda_i I)x = y.$$

But then

$$(T - \lambda_i I)^{\nu_i}x = (T - \lambda_i I)^{\nu_i - 1}y = 0$$

since all columns of F_i are of grade $\leq \nu_i$. But then y is of grade $\nu_i - 1$. Hence y, which is by definition a linear combination of columns of $F_{i\nu_i}$, is also a linear combination of the remaining columns of F_i, and this is another way of saying that the columns of F_i are linearly dependent. Since this is not the case, $y = 0$. Hence by (2.06.18) x is a proper vector, and hence both a linear combination of columns of F_{i1} and of $F_{i\nu_i}$. Hence $x = 0$.

The argument may be continued to show that the columns of

$$(2.06.19) \qquad ((T - \lambda_i I)^{\nu_i - 1}F_{i\nu_i}, \ . \ . \ . \ , (T - \lambda_i I)F_{i\nu_i}, F_{i\nu_j})$$

are linearly independent. If there are more columns in F_{i1} than in $(T - \lambda_i I)^{\nu_i - 1}F_{i\nu_i}$, we adjoin additional ones to form a new matrix F_{i1} and continue to form a new matrix F_i of which the matrix (2.06.19) is a submatrix. We then show that the columns of

$$((T - \lambda_i I)^{\nu_i - 2}F_{i\nu_i - 1}, \ . \ . \ . \ , F_{i\nu_i - 1}, F_{i\nu_i})$$

are linearly independent. Proceeding as before, we obtain finally a matrix F_i such that $(T - \lambda_i I)F_{i\nu_i}$ is a submatrix of $F_{i\nu_i - 1}$, . . . , and $(T - \lambda_i I)F_{i2}$ is a submatrix of F_{i1}.

Finally, the columns of F_i are rearranged as follows: For f_{i1} take any column of $F_{i\nu_i}$. Let

$$\begin{aligned} f_{i2} &= (T - \lambda_i I)f_{i1}, \\ f_{i3} &= (T - \lambda_i I)f_{i2}, \\ &\cdots\cdots\cdots \\ f_{i\nu_i} &= (T - \lambda_i I)f_{i,\nu_i - 1}. \end{aligned}$$

Then

$$0 = (T - \lambda_i I)f_{i\nu_i},$$

and every vector of the sequence is a vector in F_i. If there are other columns in $F_{i\nu_i}$, take one of them as f_{i,ν_i+1} and proceed as before. When all columns of $F_{i\nu_i}$ are exhausted, pass next to a column (if any) from F_{i,ν_i-1} which does not appear in one of the above chains; then pass to $F_{i,\nu_i-2}, \ldots$.

By so forming and rearranging the matrices F_i which make up F, we obtain a matrix F whose columns are grouped into sequences such that, when the double subscripts of the f's are replaced by single subscripts from 1 to n, we have either

$$Tf_j = \lambda_i f_j + f_{j+1}$$

or else

$$Tf_j = \lambda_i f_j$$

for some λ_i. Hence

$$TF = FT',$$

where now T' has the form

$$T' = \begin{pmatrix} T'_1 & 0 & 0 & \cdots \\ 0 & T'_2 & 0 & \cdots \\ 0 & 0 & T'_3 & \cdots \\ \cdots & \cdots & \cdots & \cdots \end{pmatrix} \tag{2.06.20}$$

and each T'_j is eit er a scalar λ_i or else has the form

$$T'_j = \lambda_i I + I_1, \tag{2.06.21}$$

$$I_1 = \begin{pmatrix} 0 & 0 & 0 & \cdots \\ 1 & 0 & 0 & \\ 0 & 1 & 0 & \\ \cdots & \cdots & \cdots & \cdots \end{pmatrix}. \tag{2.06.22}$$

The matrix I_1 is called the auxiliary unit matrix, and has units along the first subdiagonal and elsewhere has zeros. Note that

$$I_2 = I_1^2$$

has units along the second subdiagonal, and if I_1 is of order ν, then $I_1^\nu = 0$.

Every column of F is a principal vector of T. We could apply the above theorem to T^T and in the process obtain a matrix G every column of which is a principal vector of T^T. If f is a principal vector of T corresponding to the proper value λ, and if g is a principal vector of T^T corresponding to the proper value $\mu \neq \lambda$, then g and f satisfy

$$g^\mathsf{T}Tf = g^\mathsf{T}f = 0. \tag{2.06.23}$$

The proof can be made inductively. Suppose first that g and f are proper vectors: Then

$$g^\mathsf{T}T = \mu g^\mathsf{T}, \qquad Tf = \lambda f,$$

so that

$$g^\mathsf{T}Tf = \mu g^\mathsf{T}f = \lambda g^\mathsf{T}f.$$

Since $\lambda \neq \mu$, this proves the relation for that case. Next suppose g is a proper vector but f of grade 2. Hence

$$(T - \lambda I)f = f_1 \neq 0,$$

but

$$(T - \lambda I)f_1 = 0.$$

Hence f_1 is a proper vector. Now

$$g^{\mathsf{T}}f_1 = 0$$

as was just shown. Hence

$$g^{\mathsf{T}}Tf = \lambda g^{\mathsf{T}}f,$$

whereas

$$g^{\mathsf{T}}Tf = \mu g^{\mathsf{T}}f,$$

and again, since $\lambda \neq \mu$, the relation is proved. By continuing one proves the relation for proper vectors g and f of any grade.

If T is symmetric, $T^{\mathsf{T}} = T$, then any principal vector of T^{T} is also a principal vector of T. But for a symmetric matrix we now show that all principal vectors are proper vectors, and in the normalized form (2.06.20) of T all matrices T'_i are scalars.

This is clearly the case when the proper values are all distinct. In that case, in fact, every proper vector is orthogonal to every other proper vector, whence $F^{\mathsf{T}}F$ is a diagonal matrix, and by choosing every vector f to be of unit length, one has even

$$F^{\mathsf{T}}F = I \tag{2.06.24}$$

so that F is an orthogonal matrix. Suppose the proper values of T are not all distinct. One can, nevertheless, vary the elements of T slightly so that the matrix $T + \delta T$ is still symmetric and has all proper values distinct. Then $F + \delta F$ is an orthogonal matrix. As the elements of $T + \delta T$ vary continuously while the matrix remains symmetric, the columns $f + \delta f$ of $F + \delta F$ also vary continuously but remain mutually orthogonal and can be held at unit length. Hence these properties remain while δT vanishes. Hence for any symmetric matrix T there exists an orthogonal matrix F such that

$$F^{\mathsf{T}}TF = \Lambda, \tag{2.06.25}$$

where Λ is a diagonal matrix whose elements are the proper values of T.

2.07. *Analytic Functions of a Matrix; Convergence.* The relation (2.06.12), valid for any polynomial, is easily extended. Consider first any of the matrices T'_i of (2.06.20), neglecting the trivial case when T'_i is a scalar. Any power can be written

$$T'^r_i = \lambda_i^r I + r\lambda_i^{r-1} I_1 + \binom{r}{2} \lambda_i^{r-2} I_2 + \cdots, \tag{2.07.1}$$

and if T'_i is of order ν, there are at most ν terms for any integer r. If $|\lambda_i| < 1$, then for any fixed s

$$\lim_{r \to \infty} \binom{r}{s} \lambda_i^{r-s} = 0.$$

Hence in this case as r becomes infinite, $T_i'^r$ approaches the null matrix, in the sense that every one of its elements approaches zero. If for every proper value λ_i of T it is true that $|\lambda_i| < 1$, then also T'^r approaches the null vector as r becomes infinite. Since F and F^{-1} are fixed, this is true also of $F^{-1}T'^rF$, and hence of T^r. Hence if every proper value of T has modulus less than unity, then $T^r \to 0$ as r becomes infinite. This condition is necessary as well as sufficient.

Now consider any function $\psi(\lambda)$ analytic at the origin:

$$\psi(\lambda) = \psi_0 + \psi_0'\lambda + \tfrac{1}{2}\psi_0''\lambda^2 + \cdots .$$

If λ_i lies within the circle of convergence of $\psi(\lambda)$, then since any derivative of ψ is analytic within the same circle, we may write formally

$$\text{(2.07.2)} \quad \begin{aligned} \psi(T_i') &= \psi_0 I + \psi_0'(\lambda_i I + I_1) + \tfrac{1}{2}\psi_0''(\lambda_i^2 I + 2\lambda_i I_1 + I_2) + \cdots \\ &= \psi(\lambda_i)I + \psi'(\lambda_i)I_1 + \tfrac{1}{2}\psi''(\lambda_i')I_2 + \cdots , \end{aligned}$$

where at most ν terms are non-null. Hence for a matrix of the form of T_i' the analytic function $\psi(T_i')$ is defined by this relation. But then if every λ_i lies within the circle of convergence, we may take

$$\text{(2.07.3)} \qquad \psi(T') = \begin{pmatrix} \psi(T_1') & 0 & \cdots \\ 0 & \psi(T_2') & \cdots \\ \cdots & \cdots & \cdots \end{pmatrix},$$

and then we may further take

$$\text{(2.07.4)} \qquad \psi(T) = F\psi(T')F^{-1}.$$

Hence if $\psi(\lambda)$ is any function that is analytic in a circle about the origin which contains all proper values of T in its interior, then $\psi(T)$ is defined and, in fact, is given by the convergent power series

$$\text{(2.07.5)} \qquad \psi(T) = \psi_0 I + \psi_0' T + \tfrac{1}{2}\psi_0'' T^2 + \cdots .$$

2.08. *Measures of Magnitude.* Most iterative processes with matrices are equivalent to successive multiplications of a matrix by another matrix or by a vector, and hence involve the formation of successively higher powers of the first matrix. The success of the process depends upon the successive powers approaching the null matrix. An adequate criterion for this is given in terms of the proper values of the matrix, but the calculation of proper values is generally a most laborious procedure. Hence other criteria that can be more readily applied are much to be desired.

For this purpose certain measures of magnitude of a vector or matrix are now introduced, and some relations among them developed. These measures are of use also as an aid in assessing the several types of error that enter any numerical computation.

For any matrix A we define the bound $b(A)$, the norm $N(A)$, and the maximum $M(A)$, and for a vector x we define the bound $b(x)$ and the norm $N(x)$. It turns out that a natural definition of a maximum $M(x)$ for a vector x is identical with $N(x)$.

Taking first the vector x whose elements are ξ_i, we define

$$\begin{aligned} b(x) &= \max_i |\xi_i|, \\ N(x) &= [\Sigma \xi_i^2]^{1/2} = [x^\mathsf{T} x]^{1/2}. \end{aligned} \tag{2.08.1}$$

Thus $b(x)$ is the magnitude of the numerically largest element, while $N(x)$ is the ordinary geometric length defined by the metric I.

We define $b(A)$ and $N(A)$ analogously:

$$\begin{aligned} b(A) &= \max_{i,j} |\alpha_{ij}|, \\ N(A) &= [\Sigma \alpha_{ij}^2]^{1/2}. \end{aligned} \tag{2.08.2}$$

Clearly

$$b(A^\mathsf{T}) = b(A), \qquad N(A^\mathsf{T}) = N(A). \tag{2.08.3}$$

If we use the notion of a trace of a matrix

$$\mathrm{tr}\ (A) = \Sigma \alpha_{ii}, \tag{2.08.4}$$

then an equivalent expression for $N(A)$ is

$$N(A) = [\mathrm{tr}\ (A^\mathsf{T} A)]^{1/2} = [\mathrm{tr}\ (AA^\mathsf{T})]^{1/2}. \tag{2.08.5}$$

If a_i are the column vectors of A, and a_j' the row vectors, then

$$N^2(A) = \Sigma N^2(a_i) = \Sigma N^2(a_j'), \tag{2.08.6}$$

where the exponent applies to the functional value

$$N^2(A) = [N(A)]^2.$$

Hence if $a_i = x$ and all other $a_j = 0$,

$$N(A) = N(x).$$

A useful inequality is the Schwartz inequality which states that for any vectors x and y

$$|x^\mathsf{T} y| = |y^\mathsf{T} x| \le N(x)N(y). \tag{2.08.7}$$

Geometrically this means that a scalar product of two vectors does not exceed the product of the lengths of the vectors (in fact it is this product multiplied by the cosine of the included angle). This generalizes immediately to matrices

$$N(AB) \le N(A)N(B). \tag{2.08.8}$$

Another useful inequality is the triangular inequality

(2.08.9) $$N(x + y) \leq N(x) + N(y),$$

which says that one side of a triangle does not exceed the sum of the other two, and which also generalizes immediately

(2.08.10) $$N(A + B) \leq N(A) + N(B).$$

Also we have

(2.08.11) $$b(A + B) \leq b(A) + b(B).$$

But

(2.08.12) $$|x^{\mathsf{T}}y| \leq nb(x)b(y),$$

since in $x^{\mathsf{T}}y$ there are n terms each of which could have the maximum value $b(x)b(y)$. Hence for matrices

(2.08.13) $$b(AB) \leq nb(A)b(B).$$

If in (2.08.8) we take

$$b_1 = x, \qquad b_2 = b_3 = \cdots = b_n = 0,$$

then we have

(2.08.14) $$N(Ax) \leq N(A)N(x).$$

We now introduce the third measure:

(2.08.15) $$M(A) = \max_{x \neq 0} N(Ax)/N(x) = \max_{x,y \neq 0} |x^{\mathsf{T}}Ay|/[N(x)N(y)],$$

or equivalently

(2.08.16) $$M(A) = \max_{N(x) = 1} N(Ax) = \max_{N(x) = N(y) = 1} |x^{\mathsf{T}}Ay|.$$

It is obvious that (2.08.15) and (2.08.16) are equivalent. We show now that the two parts of (2.08.15) or of (2.08.16) are equivalent. Take $M(A)$ as defined by the first equality, and designate, for the moment, the last member by $M'(A)$. We wish to show that

$$M(A) = M'(A).$$

First we have

$$|x^{\mathsf{T}}Ay| \leq N(x)N(Ay)$$

by the Schwartz inequality (2.08.7). But

$$N(Ay) \leq M(A)N(y)$$

by definition of $M(A)$. Hence for any x and y

$$|x^{\mathsf{T}}Ay| \leq M(A)N(x)N(y),$$
$$|x^{\mathsf{T}}Ay|/[N(x)N(y)] \leq M(A).$$

Hence the maximum of the left member cannot exceed the right member, so that

$$M'(A) \leq M(A).$$

Now to prove that

$$M(A) \leq M'(A),$$

we take

$$x = Ay$$

and obtain

$$|x^{\mathsf{T}}Ay| = |y^{\mathsf{T}}A^{\mathsf{T}}Ay| = N^2(Ay).$$

Hence when x is so defined

$$\frac{|x^{\mathsf{T}}Ay|}{N(x)N(y)} = \frac{N^2(Ay)}{N(Ay)N(y)} = \frac{N(Ay)}{N(y)},$$

and this quantity has $M(A)$ for its maximum. Hence $M'(A)$ cannot be less than $M(A)$, and the theorem is proved.

Since

$$x^{\mathsf{T}}Ay = y^{\mathsf{T}}A^{\mathsf{T}}x,$$

it follows from the second definition of $M(A)$ that

(2.08.17) $$M(A^{\mathsf{T}}) = M(A).$$

Also

(2.08.18) $$M(A + B) \leq M(A) + M(B),$$

(2.08.19) $$M(AB) \leq M(A)M(B).$$

To prove the first of these, we have

$$N(Ax + Bx) \leq N(Ax) + N(Bx) \leq [M(A) + M(B)]N(x),$$

the first inequality being a consequence of the triangular inequality applied to the vectors Ax and Bx, and the second following from the definition of the maximum. Since, therefore,

$$N[(A + B)x]/N(x) \leq M(A) + M(B)$$

for all vectors $x \neq 0$, the maximum value of the left member cannot exceed the right.

For the second inequality we have

$$N[A(Bx)] \leq M(A)N(Bx) \leq M(A)M(B)N(x),$$

both inequalities being consequences of the definition of the maximum. If we divide again by $N(x)$, the conclusion follows as before.

We now establish the following relations among the functions N, b, and M of the same matrix:

(2.08.20) $$b(A) \leq M(A) \leq N(A) \leq nb(A),$$

(2.08.21) $$N(A) \leq n^{1/2}M(A).$$

To prove the first, we use the second definition of M in (2.08.16). If we take

$$x = e_i, \qquad y = e_j,$$

then

$$x^{\mathsf{T}} A y = \alpha_{ij}.$$

Any choice of unit vectors for x and y will give a number $x^{\mathsf{T}} A y$ which cannot exceed the maximum. Hence

$$|\alpha_{ij}| \leq M(A),$$

and since this holds for any i and j, it follows that

$$b(A) \leq M(A).$$

Next, since by (2.08.14)

$$N(Ax)/N(x) \leq N(A),$$

the maximum value of the left member cannot exceed $N(A)$, and this maximum is $M(A)$ by definition. Hence

$$M(A) \leq N(A).$$

To show that

$$N(A) \leq nb(A),$$

we go to the definition and write

$$N^2(A) = \Sigma \alpha_{ij}^2 \leq n^2 b^2(A),$$

since $b^2(A)$ is the greatest of the n^2 terms in the sum. When we take the square root, we have the desired result.

Before proving (2.08.21), we prove first a more general result:

$$\text{(2.08.22)} \qquad \begin{aligned} N(AB) &\leq N(A)M(B), \\ N(AB) &\leq M(A)N(B). \end{aligned}$$

If the columns of B are b_i, then the columns of AB are Ab_i. By (2.08.6) applied to the matrices B and AB,

$$N^2(AB) = \Sigma N^2(Ab_i) \leq M^2(A) \Sigma N^2(b_i) = M^2(A) N^2(B).$$

We get the second relation on taking square roots. The first follows after taking transposes from (2.08.3) and (2.08.17). Now (2.08.21) is an immediate consequence of the second of (2.08.22) when we take $B = I$, since

$$N(I) = n^{1/2}.$$

Of the three functions b, N, and M, the first is obtainable for any given matrix by inspection, and the second by direct computation. The third, however, is only obtainable in general from rather elaborate computations, though it generally yields the best estimates of error.

A few additional properties of these functions will be useful. First we have

(2.08.23) $$M(A^{\mathsf{T}}A) = M^2(A).$$

In view of (2.08.17) and (2.08.19), we know that the left member cannot exceed the right. On the other hand

$$N^2(Ax) = x^{\mathsf{T}}A^{\mathsf{T}}Ax \leq N(x)N(A^{\mathsf{T}}Ax) \leq M(A^{\mathsf{T}}A)N^2(x),$$

the first inequality being a consequence of the Schwartz inequality (2.08.7), while the second follows from the definition of M. Hence the right member of (2.08.23) cannot exceed the left, and the equality therefore follows.

An orthogonal matrix X is one whose transpose is its inverse:

$$X^{\mathsf{T}}X = XX^{\mathsf{T}} = I.$$

Multiplication by an orthogonal matrix does not affect the norm of a vector:

$$N^2(Xx) = x^{\mathsf{T}}X^{\mathsf{T}}Xx = x^{\mathsf{T}}Ix = x^{\mathsf{T}}x = N^2(x).$$

Hence for any matrix A if X is orthogonal,

(2.08.24) $$N(AX) = N(A), \qquad M(AX) = M(A).$$

By (2.06.25), for any symmetric matrix B there exists an orthogonal matrix X such that

(2.08.25) $$X^{\mathsf{T}}BX = \Lambda$$

is a diagonal matrix. The columns of X are the proper vectors of the matrix, and the diagonal elements of Λ the proper values. If

$$B = A^{\mathsf{T}}A,$$

then B is non-negative, and these proper values are all non-negative. We may suppose that they are arranged in order of decreasing magnitude,

(2.08.26) $$\lambda_1 \geq \lambda_2 \geq \cdots \geq \lambda_n \geq 0,$$

the last equality holding only when A is singular. Clearly

$$M(\Lambda) = e_1^{\mathsf{T}}\Lambda e_1,$$

and furthermore

$$M(B) = M(\Lambda).$$

Hence

(2.08.27) $$M(A) = \lambda_1^{1/2}.$$

Also

(2.08.28) $$N^2(A) = \Sigma\lambda_i.$$

To see this, we observe that by definition the proper values λ_i of a matrix B satisfy the algebraic equation

$$|B - \lambda I| = 0$$

and that the trace tr (B) is the sum of the proper values, while by definition

$$N^2(A) = \text{tr}\,(A^{\mathsf{T}}A) = \text{tr}\,(B).$$

These relations provide an alternative proof for (2.08.21) and the second inequality in (2.08.20).

By analogy with M, we define

$$m(A) = \min_{x \neq 0} N(Ax)/N(x). \tag{2.08.29}$$

Then

$$m(A) = \lambda_n^{1/2}. \tag{2.08.30}$$

Also if A is nonsingular,

$$M(A^{-1}) = \lambda_n^{-1/2}, \qquad m(A^{-1}) = \lambda_1^{-1/2}. \tag{2.08.31}$$

Hence for a nonsingular matrix

$$M(A^{-1})m(A) = 1. \tag{2.08.32}$$

These relations arise from the fact that, if B is nonsingular,

$$X^{\mathsf{T}}B^{-1}X = \Lambda^{-1},$$

which is a special case of the relation

$$X^{\mathsf{T}}B^{r}X = \Lambda^{r}, \tag{2.08.33}$$

where r is any integer, positive or negative.

We conclude this discussion by noting that, if x' is the vector whose elements are $|\xi_i|$ and if 1_n is the vector each of whose elements is unity, then from (2.08.7) it follows that

$$\Sigma|\xi_i| \leq n^{1/2}N(x), \tag{2.08.34}$$

where x has the elements ξ_i. This follows from the fact that

$$N(1_n) = n^{1/2}.$$

2.1. Iterative Methods. Generally speaking, an iterative method for solving an equation or set of equations is a rule for operating upon an approximate solution x_p in order to obtain an improved solution x_{p+1}, and such that the sequence $\{x_q\}$ so defined has the solution x as its limit. This is to be contrasted with a direct method which prescribes only a finite sequence of operations whose completion yields an exact solution. Since the exact operations must generally be replaced by pseudo operations, in which round-off errors enter, the exact solution is seldom attainable in practice, and one may wish to improve the result actually obtained by one or more iterations. Also since the "approximation" x_0 with which one may start an iteration does not necessarily need to be close, it is sometimes advantageous to omit the direct method altogether, start with an arbitrary x_0, perhaps $x_0 = 0$, and iterate until the approach is sufficiently close.

2.11. *Some Geometric Considerations.* A large class of iterative methods are based upon the following simple geometric notion: Take any vector b and a sequence of vectors $\{u_p\}$, and define the sequence $\{b_p\}$ by

$$b = b_0, \qquad b_{p-1} = b_p + \lambda_p u_p, \tag{2.11.1}$$

where the scalar λ_p is chosen so that b_p is orthogonal to u_p. Then if the vectors u_p "fill out" some n space, the vectors b_p approach as a limit a vector that is orthogonal to this n space. Without attempting a more precise definition of what is meant by "filling out," we can see that it must imply the following: If the vectors $\mathbf{e}_i$ represent any set of reference vectors for this n space, then however far out we may go in the sequence $\{u_p\}$, it must always be possible to find vectors $\mathbf{u}_q = \mathbf{e}u_q$ with a nonvanishing projection on any $\mathbf{e}_i$ and, in fact, with components that have some fixed positive lower bound. A possible choice for the vectors $\mathbf{u}_p$ would be the reference vectors $\mathbf{e}_i$ taken in order and then repeated. If e_i is the arithmetic vector associated with the geometric vector $\mathbf{e}_i$, we have then

$$u_{\nu n+i} = e_i,$$

where ν is any integer. It is easily verified that the vector e_i has 1 in the ith place and 0 in every other position.

The vectors $\mathbf{e}_i$ are entirely subject to our choice, and we may choose any convenient positive definite matrix H to represent the metric. Then the orthogonal projection upon u_p is represented by $u_p(u_p^\mathsf{T} H u_p)^{-1} u_p^\mathsf{T} H$ (cf. §2.051), whence

$$\lambda_p = u_p^\mathsf{T} H b_{p-1} / u_p^\mathsf{T} H u_p. \tag{2.11.2}$$

Now we write

$$Ax = y \tag{2.11.3}$$

as the equations to be solved. Let x_p represent any approximation to the solution x. If

$$As_p = r_p = y - Ax_p, \tag{2.11.4}$$

then either s_p or r_p can be taken as representing the deviation of the approximation from the true solution. Hence either s_p or r_p may be taken as b_p. We consider separately the case when A is itself positive definite and that when A is not itself positive definite.

2.111. *The matrix A is positive definite.* Let

$$r_p = b_p.$$

Then

$$y - Ax_p = b_p = b_{p-1} - \lambda_p u_p = y - Ax_{p-1} - \lambda_p u_p.$$

Hence by comparing first and last members in the equality

$$Ax_p = Ax_{p-1} + \lambda_p u_p,$$
$$x_p = x_{p-1} + \lambda_p A^{-1} u_p.$$

Therefore we take

$$u_p = Av_p$$

and obtain

$$(2.111.1) \qquad x_p = x_{p-1} + \lambda_p v_p.$$

Now

$$\lambda_p = u_p^{\mathsf{T}} H b_{p-1} / u_p^{\mathsf{T}} H u_p$$
$$= v_p^{\mathsf{T}} A H b_{p-1} / v_p^{\mathsf{T}} A H A v_p.$$

Consequently it is natural to take

$$H = A^{-1},$$

which gives

$$(2.111.2) \qquad \lambda_p = v_p^{\mathsf{T}} r_{p-1} / v_p^{\mathsf{T}} A v_p.$$

Alternatively we may take

$$s_p = b_p.$$

Then

$$y - Ax_p = Ab_p = A(b_{p-1} - \lambda_p u_p) = y - Ax_{p-1} - \lambda_p A u_p.$$

Hence

$$Ax_p = Ax_{p-1} + \lambda_p A u_p,$$

or

$$(2.111.3) \qquad x_p = x_{p-1} + \lambda_p u_p.$$

But

$$\lambda_p = u_p^{\mathsf{T}} H b_{p-1} / u_p^{\mathsf{T}} H u_p.$$

If we take

$$H = A,$$

we have

$$(2.111.4) \qquad \lambda_p = u_p^{\mathsf{T}} r_{p-1} / u_p^{\mathsf{T}} A u_p.$$

Equations (2.111.3) and (2.111.4) are identical with (2.111.1) and (2.111.2) except for the replacement of v_p by u_p.

In geometric terms, since A is taken to represent the metric, our equation (2.11.3) requires the contravariant representation x of a vector whose covariant representation y is known. We begin with an approximate representation x_0. Then

$$r_0 = y - Ax_0 = A(x - x_0)$$

is the covariant representation of the error $\mathbf{x} - \mathbf{x}_0$, and

$$\lambda_1 u_1 = \{u_1^{\mathsf{T}} r_0 / u_1^{\mathsf{T}} A u_1\} u_1$$

is the contravariant representation of the orthogonal component of $\mathbf{r}_0$ in the direction of $\mathbf{u}_1$. When this vector is added to x_0, the new residual $\mathbf{r}_1 = \mathbf{x} - \mathbf{x}_1$ is thus a leg of a right triangle of which $\mathbf{r}_0$ is the hypotenuse. Hence we have a better approximation provided only $\mathbf{u}_1$ and $\mathbf{r}_0$ were not orthogonal.

Any rule for selecting the u_p (or equivalently the v_p) at each step defines a particular iterative process. There are three in common use:

1. The *method of steepest descent* prescribes that we take

$$u_p = r_{p-1}.$$

To understand the reason for this selection, we note first that, since A is positive definite, by hypothesis, there exists therefore a matrix C for which

$$A = C^{\mathsf{T}} C.$$

The equations

$$Ax = y$$

are therefore equivalent to the equations

$$Cx = z,$$

where

$$C^{\mathsf{T}} z = y.$$

Define the function

$$\begin{aligned} g(x) &= (Cx - z)^{\mathsf{T}}(Cx - z) \\ &= x^{\mathsf{T}} C^{\mathsf{T}} C x - 2x^{\mathsf{T}} C^{\mathsf{T}} z + z^{\mathsf{T}} z \\ &= x^{\mathsf{T}} A x - 2x^{\mathsf{T}} y + z^{\mathsf{T}} z. \end{aligned}$$

This function is a sum of squares which has a minimum of zero for

$$x = C^{-1} z = A^{-1} y.$$

The function $g(x)$ is a function of the n variables $\xi_1, \ldots, \xi_n$. Its partial derivatives with respect to these variables are the coordinates of the gradient vector

$$g_x = 2(Ax - y),$$

and the gradient at the point x_{p-1} is $-2r_{p-1}$. Hence the function $g(x)$ evaluated at the point x_{p-1} is undergoing its most rapid variation in the direction r_{p-1}, and if we think of our problem as one of minimizing $g(x)$ by proceeding in successive steps, it is natural to take each step in the direction of most rapid decrease. If we define

$$\phi(\lambda) = f(x_{p-1} + \lambda u_p),$$

we shall find that ϕ, as a function of the single variable λ, takes on its minimum value at $\lambda = \lambda_p$ given by (2.111.4).

2. The *method of Seidel* takes

$$u_p = u_{\nu n+i} = e_i.$$

From (2.111.3) we see that x_p differs from x_{p-1} only in the ith element. Furthermore, since b_p and e_i are to be orthogonal, this means that the ith element of $r_p = As_p$ must equal zero, which means that the ith equation is satisfied exactly. Hence the ith element of x_p is chosen so that the ith equation will be satisfied when all other elements are the same as for x_{p-1}. While we may expect that this process will require more steps than does the method of steepest descent, the simplicity of each step is a great advantage, especially in using automatic machinery.

3. The *method of relaxation* always takes u_p to be some e_i, but the selection is made only at the time. Since the choice $u_p = e_i$ has the effect of eliminating the ith component of r_p, one chooses to eliminate the largest residual. However this is not necessarily the best choice. The effectiveness of the correction is measured by the magnitude of the correcting vector $\lambda_p u_p$, and this magnitude is

$$u_p^{\mathsf{T}} A s_{p-1} \{u_p^{\mathsf{T}} A u_p\}^{-1/2}.$$

Now when $u_p = e_i$, then $u_p^{\mathsf{T}} A s_{p-1}$ is the ith component of the residual r_{p-1}, but this is divided by the length of e_i which has the value of $\sqrt{\alpha_{ii}}$. Hence one should examine the quotients of the residual components divided by the corresponding $\sqrt{\alpha_{ii}}$ and eliminate the largest quotient.

This method clearly converges more rapidly than the Seidel method, which projects upon the same vectors but in a fixed sequence. Therefore for "hand" calculations it is to be preferred. For automatic machinery, however, the fixed sequence is almost certainly to be preferred.

2.112. *The matrix A is not necessarily positive definite.* This case can always be reduced to the preceding if we multiply throughout by A^{T}. However, this extra matrix multiplication is to be avoided if possible. With regard to the equations

$$Ax = y,$$

we may adopt either of two obvious geometric interpretations.

The simplest interpretation is that we wish to change the vector coordinates, as in Eq. (2.01.8), where y, taking the place of x', is known, and A, taking the place of E, is known. In the symbols used here, therefore, the columns of A are the numerical vectors which represent the $\mathbf{e}_i$ in the system $\mathbf{f}$, and the column vector y represents $\mathbf{x}$ in the same system. The vector y is to be expressed as a linear combination of the columns of A, which is another way of saying that the vector $\mathbf{x}$, whose representation is known in the $\mathbf{f}$ system, is to be resolved along the vectors $\mathbf{e}_i$.

The other interpretation comes from regarding each of the n equations as the equation of a hyperplane in n space. If a_i represents the ith row vector in A, then the ith equation is

$$a_i x = \eta_i,$$

and this equation is satisfied by any vector x leading from the origin of the point-coordinate system to a point in the hyperplane. If we divide through this equation by $N(a_i)$, the length of a_i, we obtain

$$[a_i/N(a_i)]x = \eta_i/N(a_i),$$

and since the vector multiplying x is a unit vector, the equation says that the projection of x upon the direction of a_i is of length $\eta_i/N(a_i)$, and hence the same for all points in the plane. Consequently the vector a_i is orthogonal to the plane, and the distance of the plane from the origin is $|\eta_i|/N(a_i)$.

In case we think of the underlying coordinate system $\mathbf{e}$ as nonorthogonal, the vectors a_i are taken to be covariant representations of the normals, and the vector x as the contravariant representation of the vector $\mathbf{x}$ drawn to the common intersection of the n planes.

These two geometric interpretations suggest different iterative schemes. We begin with the hyperplane interpretation.

2.1121. The Equations Represent a System of Hyperplanes. If v is any column vector, then

$$v^{\mathsf{T}}Ax = v^{\mathsf{T}}y \tag{2.1121.1}$$

is also the equation of a hyperplane passing through the point x. The normal, written as a column vector, is $A^{\mathsf{T}}v$. If x_p is any approximation to x, and s_p and r_p are defined as before,

$$As_p = r_p = y - Ax_p = A(x - x_p), \tag{2.1121.2}$$

then we may take

$$b_p = s_p = x - x_p = A^{-1}r_p, \tag{2.1121.3}$$

and project upon the vector $u_{p+1} = A^{\mathsf{T}}v_{p+1}$. This amounts to writing

$$Ax_p = y - Ab_p$$

so that as b_p vanishes, x_p approaches x. The basic sequence as defined by (2.11.1) and (2.11.4) takes the form

(2.1121.4) $$b_{p-1} = b_p + \lambda_p A^{\mathsf{T}} v_p,$$

(2.1121.5) $$\lambda_p = v_p^{\mathsf{T}} r_{p-1} / v_p^{\mathsf{T}} A A^{\mathsf{T}} v_p,$$

if the identity matrix is taken to define the metric. But then (2.1121.4) gives

(2.1121.6) $$x_p = x_{p-1} + \lambda_p A^{\mathsf{T}} v_p.$$

By analogy with the method of steepest descent as described for the positive definite case, we may define the non-negative function

$$g(x) = (Ax - y)^{\mathsf{T}}(Ax - y)$$

and find that at x_{p-1} its gradient lies in the direction $A^{\mathsf{T}} r_{p-1}$, whence the choice

$$v_p = r_{p-1}$$

is optimal from the point of view of most rapid minimization of $f(x)$.

If we take

$$v_p = v_{\nu n+i} = e_i,$$

then, as appears from (2.1121.4), we are projecting the vectors b_p in rotation upon the normals to the basic hyperplanes of the system. In this event x_p is caused to lie in the ith hyperplane so that the ith equation is satisfied, though in general all components of x_p will differ from those of x_{p-1}. The numerator in λ_p is simply the ith residual, *i.e.*, the ith element of r_{p-1}, while the denominator is the sum of squares of the ith row of A.

Hence if we follow the lead of the method of relaxation, we would choose v_p to be that e_j that would provide the maximal correction

$$\lambda_p A^{\mathsf{T}} e_j.$$

Hence to make the optimal choice, we should divide each residual by the square root of the sum of the squares of the corresponding row of A, and select the largest quotient. Presumably all these square roots will be used repeatedly and should be calculated in advance.

2.1122. The Equations Represent a Resolution of the Vector y along the Column Vectors of A. If x_p is any set of trial multipliers,

$$r_p = y - Ax_p$$

represents the deviation of the vector Ax_p from the required vector y. Take

$$b_p = r_p,$$

and let

$$u_p = Av_p$$

represent any linear combination of the columns of A. Then Eqs. (2.11.1) and (2.11.4) give

$$y - Ax_{p-1} = y - Ax_p + \lambda_p A v_p,$$

or

$$x_p = x_{p-1} + \lambda_p v_p, \tag{2.1122.1}$$

with

$$\lambda_p = v_p^{\mathsf{T}} A^{\mathsf{T}} r_{p-1} / v_p^{\mathsf{T}} A^{\mathsf{T}} A v_p. \tag{2.1122.2}$$

The method of steepest descent would take

$$v_p = A^{\mathsf{T}} r_{p-1},$$

a choice that complicates the denominator in λ_p excessively. In taking $v_p = e_i$ for some i, we alter only one element of x_{p-1} in obtaining x_p, but no element of r_p is made to vanish, so no one of the equations is necessarily satisfied exactly. To find the optimal e_i according to the principle of the method of relaxation, we observe that we wish to maximize the vector

$$\lambda_p u_p = \lambda_p A v_p$$

or

$$\lambda_p A e_i.$$

However, the numerators of the possible λ_p have the form

$$e_i^{\mathsf{T}} A^{\mathsf{T}} r_{p-1}$$

which is a complete scalar product of the ith column of A with the residual vector r_{p-1}. Taking this scalar product represents the greater portion of the labor involved in the complete projection, so that one would probably always take the vectors e_i in strict rotation.

2.113. *Some generalizations.* The methods described in §2.111 consisted in taking each residual $s_{p-1} = x - x_{p-1}$ from the true solution x, projecting orthogonally upon a vector u_p, and adding the projection to x_{p-1} to obtain an improved approximation x_p. The new residual s_p was orthogonal to the projection on u_p. Clearly if the projection is made on a linear space of two or more dimensions, the projection, *i.e.*, the correction, will be at least as large as the projection on any single direction in this space. Hence it is to be expected that the rate of convergence of the process would be more rapid if, instead of projecting each time upon a single vector u_p, we were to project upon a linear space of two or more dimensions. Such a space may be represented by a matrix U_p such that any vector in the space is a linear combination of columns of U_p.

The problem is now the following: Given any matrix U_p, we wish to project the residual s_{p-1} orthogonally upon the space U_p (that is, the space of linear combinations of its columns). The projection will be taken as a correction to be added to x_{p-1} to yield the improved approximation x_p. The orthogonal projection is represented by the matrix $U_p(U_p^\mathsf{T} A U_p)^{-1} U_p^\mathsf{T} A$, and we find now that

$$x_p = x_{p-1} + U_p(U_p^\mathsf{T} A U_p)^{-1} U_p^\mathsf{T} r_{p-1}. \tag{2.113.1}$$

The scalar λ_p is for present purposes to be replaced by the vector

$$(U_p^\mathsf{T} A U_p)^{-1} U_p^\mathsf{T} r_{p-1}.$$

Its expression is the same as for λ_p except that the matrix U_p replaces the vector u_p, and it is to be noted that the reciprocal matrix enters as a premultiplier. While the method does provide a larger correction, in general this advantage is offset by the necessity for calculating an inverse matrix whose order is equal to the dimensionality of the subspace upon which the projection is being made. Nevertheless in special cases this inversion may prove to be fairly simple.

If the columns of the matrix U_p are unit vectors e_i, then the matrix $U_p^\mathsf{T} A U_p$ is a principal submatrix of the matrix A. If, say, we take U_p to be the two-column matrix (e_i, e_j), then

$$U_p^\mathsf{T} A U_p = \begin{pmatrix} \alpha_{ii} & \alpha_{ij} \\ \alpha_{ji} & \alpha_{jj} \end{pmatrix}.$$

The correction is made so as to eliminate both the ith and the jth elements of r_p by appropriate selection of the ith and jth elements of x_p. Hence the ith and jth equations are solved simultaneously for these elements in terms of the current approximations to the other elements. This is to be contrasted with the method of §2.111 where the ith and jth equations are solved at different times and not simultaneously.

The iterations for the case when A is not a symmetric matrix can be generalized in like manner. When we interpret the equations as representing a system of hyperplanes and project upon a space determined by normals to these hyperplanes, we obtain

$$x_p = x_{p-1} + A^\mathsf{T} V_p(V_p^\mathsf{T} A A^\mathsf{T} V_p)^{-1} V_p r_{p-1}, \tag{2.113.2}$$

where V_p is an arbitrary matrix of linearly independent columns. When we interpret the equations as requiring the resolution of the vector y along the column vectors of A, we get

$$x_p = x_{p-1} + V_p(V_p^\mathsf{T} A^\mathsf{T} A V_p)^{-1} V_p^\mathsf{T} A^\mathsf{T} r_{p-1}. \tag{2.113.3}$$

2.12. *Some Analytic Considerations.* The iterative methods so far discussed have been suggested by geometry. Analytical considerations suggest several others.

2.121. *Cesari's method.* In the Seidel iteration x_1 differs from x_0 only in the first element, which is so chosen that the first equation is satisfied exactly. Next x_2 differs from x_1 only in the second element, which is so chosen that the second equation is satisfied exactly. In the nth step, x_n differs from x_{n-1} only in the nth element, which is chosen to satisfy the nth equation exactly. Next x_{n+1} is obtained from x_n by readjusting the first element to satisfy the first equation, and this begins a new cycle.

Let us write $A = A_1 + A_2$, where in A_1 all diagonal and subdiagonal elements are the same as those in A, while all supradiagonal elements are zero:

$$\begin{aligned} A_1 &= (\alpha'_{ij}), \qquad \alpha'_{ij} \begin{array}{l} = \alpha_{ij} \text{ when } i \geq j, \\ = 0 \text{ when } i < j; \end{array} \\ A_2 &= (\alpha''_{ij}), \qquad \alpha''_{ij} \begin{array}{l} = 0 \text{ when } i \geq j, \\ = \alpha_{ij} \text{ when } i < j. \end{array} \end{aligned} \tag{2.121.1}$$

Then we verify easily that

$$A_1 x_{(p+1)n} = y - A_2 x_{pn}.$$

Since the vectors x_{pn+i} for $0 < i < n$ need not enter explicitly, we may modify the notation by writing simply x_p for what had been designated x_{pn}, and the iteration is written

$$A_1 x_{p+1} = y - A_2 x_p, \qquad A = A_1 + A_2. \tag{2.121.2}$$

We know that this iteration converges when A is positive definite and A_1 and A_2 are given by (2.121.1). However (2.121.2) formally defines an iterative process whether or not A is positive definite, and whether or not A_1 and A_2 satisfy (2.121.1). It is required only that A_1 be nonsingular. Since (2.121.2) is equivalent to

$$x_p = A_1^{-1}(y - A_2 x_{p-1}),$$

and x satisfies

$$x = A_1^{-1}(y - A_2 x),$$

it follows that the residuals $s_p = x - x_p$ satisfy

$$s_p = -A_1^{-1} A_2 s_{p-1} = (-A_1^{-1} A_2)^p s_0.$$

Hence for the sequence (2.121.2) to converge for an arbitrary y, it is necessary and sufficient that the proper values of the matrix $-A_1^{-1}A_2$ should all be of modulus < 1. The characteristic equation for this matrix can be written

$$|\lambda A_1 + A_2| = 0. \tag{2.121.3}$$

A sufficient condition for convergence is that

$$M(A_1^{-1}A_2) < 1, \tag{2.121.4}$$

and hence, a fortiori, it is sufficient that

$$N(A_1^{-1}A_2) < 1, \tag{2.121.5}$$

the latter criterion being one that is fairly readily applied.

Cesari's derivation of the same sequence is interesting. Let

$$A_1 + A_2 = \gamma A, \tag{2.121.6}$$

where γ is any scalar and A_1 is nonsingular. Let the vector $x(\mu)$ as a function of the scalar μ be defined by

$$(A_1 + \mu A_2)x(\mu) = \gamma y + (\mu - 1)A_2 v, \tag{2.121.7}$$

with v an arbitrary vector. Then

$$x(1) = x.$$

Let

$$x(\mu) = x_0 + \mu x_0' + \mu^2 x_0''/2! + \cdots.$$

Then on substituting into (2.121.7) and grouping terms, one obtains

$$0 = (A_1x_0 - \gamma y + A_2v) + \mu(A_2x_0 + A_1x_0' - A_2v) + \mu^2(A_2x_0' + A_1x_0''/2!) \\ + \mu^3(A_2x_0''/2! + A_1x_0'''/3!) + \cdots.$$

If

$$x_p = x_0 + x_0' + \cdots + x_0^{(p)}/p!,$$

then

$$A_1x_0 = \gamma y - A_2v,$$
$$A_1x_{p+1} = \gamma y - A_2x_p \qquad (p \geq 1),$$

and we have again the recursion (2.121.2).

When A_1 is a diagonal matrix whose diagonal is the same as that of γA, we have a method discussed by von Mises and Geiringer. Since for such a choice of A_1 the diagonal elements of A_2 are all zeros, and since A_1 is a diagonal matrix, the diagonal elements of $A_1^{-1}A_2$ are also all zeros. It is no restriction to suppose that $A_1 = I$ and that $\gamma = 1$, for we may replace the original system by the equivalent one,

$$\gamma A_1^{-1}Ax = \gamma A_1^{-1}y.$$

Then (2.121.5) yields one of the criteria given by von Mises and Geiringer which they state in the form

$$\sum_{\substack{i \neq j \\ i,j}} \alpha_{ij}^2 < 1.$$

We obtain another of their criteria,

$$\sum_{i \neq j} |\alpha_{ij}| \leq \alpha < 1,$$

by noting that, if $\sigma_i^{(p)}$ are the elements of s_p, then $|\sigma_i^{(p+1)}| \le \sum_{j \ne i} |\alpha_{ij}| \cdot |\sigma_i^{(p)}|$, whence the criterion implies that $\Sigma|\sigma_i^{(p+1)}| \le \alpha \Sigma |\sigma_i^{(p)}|$.

Now suppose A is positive definite and write

$$\gamma(I + B) = A. \tag{2.121.8}$$

Thus we take $A_1 = I$, $A_2 = B$ for (2.121.2). Since

$$\lambda I - A = \gamma[(\lambda/\gamma - 1)I - B],$$

it follows that for each proper value λ_i of A there is a proper value $(\lambda_i - \gamma)/\gamma$ of B, and convergence of the process requires therefore that every $\lambda_i < 2\gamma$. If γ is so chosen, and if the proper values of λ are arranged in order of magnitude,

$$\lambda_1 \ge \lambda_2 \ge \cdot \cdot \cdot \ge \lambda_n,$$

then the proper values of B, in the same order, are

$$\lambda_1/\gamma - 1, \lambda_2/\gamma - 1, \ . \ . \ . \ , \lambda_n/\gamma - 1,$$

and at least $\lambda_n/\gamma - 1$ is negative. If γ is taken too near to $\lambda_1/2$, then $\lambda_1/\gamma - 1$ will be close to 1, and convergence will be slow; if γ is taken too large, $\lambda_n/\gamma - 1$ will be close to -1, and convergence again slow. The optimal choice is

$$\gamma = (\lambda_1 + \lambda_2)/2, \tag{2.121.9}$$

for then

$$\lambda_1/\gamma - 1 = -(\lambda_n/\gamma - 1) = (\lambda_1 - \lambda_n)/(\lambda_1 + \lambda_n), \tag{2.121.10}$$

the extreme proper values being equal in magnitude.

We may now ask, with Cesari, whether by any choice of a polynomial $f(\lambda)$, with $F(\lambda) = \lambda f(\lambda)$, the system

$$F(A)x = f(A)y,$$

equivalent to the original, yields a more rapidly convergent sequence.

The proper values of $F(A)$ are $\mu_i = F(\lambda_i)$. If μ' and μ'' are the largest and smallest of the μ_i, we wish to choose $F(\lambda)$ so that $(\mu' - \mu'')/(\mu' + \mu'')$ is as small as possible, as we see by (2.121.10). Hence we wish to choose $F(\lambda)$ to be positive over the range (λ_1, λ_n), and with the least possible variation.

The simplest case is that of a quadratic function F, and the optimal choice is then

$$F = \lambda(\alpha - \lambda), \qquad \alpha = (\lambda_1 + \lambda_n)/2.$$

Ordinarily one does not know the proper values in advance, though one might wish to estimate the two extreme ones required (*e.g.*, see Bargmann,

Montgomery, and von Neumann), or these might be required for other purposes.

2.122. *The method of Hotelling and Bodewig.* The iterations so far considered have begun with an arbitrary initial approximation x_0 (which might be $x_0 = 0$). Suppose, now, that by some process of operating upon y in the system

$$Ax = y, \tag{2.122.1}$$

perhaps by means of one of the direct methods of solution to be described below, one obtains a "solution" x_0 which, however, is inexact because it is infected by round-off. The operations performed upon the vector y are equivalent to the multiplication by an approximate inverse C:

$$x_0 = Cy. \tag{2.122.2}$$

Then by (2.11.4) the unknown residual s_0 satisfies the same system as does x, except that r_0 replaces y, and hence we might expect that Cr_0 is also an approximation to s_0. Hence we might suppose that

$$x_1 = x_0 + Cr_0 = C(2I - AC)y$$

is a closer approximation to x than is x_0. Otherwise stated, it would appear that, if C_0 is an approximation to A^{-1}, then

$$C_1 = C_0(2I - AC_0)$$

is a better one.

If C_0 is an approximation to A^{-1}, then AC_0 is an approximation to I. Let

$$B_p = I - AC_p, \qquad C_{p+1} = C_p(I + B_p). \tag{2.122.3}$$

Then

$$B_{p+1} = I - AC_{p+1} = I - AC_p(I + B_p) = B_p^2,$$

and therefore

$$B_p = B_0^{2^p}. \tag{2.122.4}$$

Hence if $M(B_0) < 1$, then $M(B_{p+1}) < M(B_p)$, and if $N(B_0) < 1$, then $N(B_{p+1}) < N(B_p)$, and in fact,

$$M(B_p) \leq [M(B_0)]^{2^p}, \qquad N(B_p) \leq [N(B_0)]^{2^p}. \tag{2.122.5}$$

If A is positive definite, then we can always transform the equations if necessary and secure that $M(A) < 1$. Then all proper values μ of $I - A$ satisfy $0 < \mu < 1$ so that $M(I - A) < 1$. In this case we may take

$$C_0 = I, \qquad B_0 = I - A,$$

whence

$$B_p = (I - A)^{2^p},$$

and convergence is assured.

To return to the general case, if C_0 is any approximate inverse, so that $M(B_0)$ is small, we have

$$A^{-1} = C_0(I - B_0)^{-1} = C_0(I + B_0 + B_0^2 + \cdots),$$

and the series converges. It is easily verified that

$$\begin{aligned}(I + B_0)(I + B_0^2) &= I + B_0 + B_0^2 + B_0^3,\\ (I + B_0)(I + B_0^2)(I + B_0^4) &= I + B_0 + \cdots + B_0^7,\\ \cdots\cdots\cdots\cdots\cdots\cdots\cdots\cdots\cdots\end{aligned}$$

Hence A^{-1} is expressible as the infinite product

$$A^{-1} = C_0(I + B_0)(I + B_0^2)(I + B_0^4) \cdots . \tag{2.122.6}$$

In applying this scheme, one retains the successively squared powers of B_0, adding I after a squaring, and multiplying by the previously obtained approximation. The identical argument carries through if we take

$$B_0' = I - C_0A,$$

and obtain

$$A^{-1} = \cdots (I + B_0'^4)(I + B_0'^2)(I + B_0')C_0. \tag{2.122.7}$$

This scheme is given by Hotelling and by Bodewig.

In solving a system of equations, it may be preferable to operate directly on the vector y and the successive remainders, as was originally suggested. In this case the matrix C may not be given explicitly. Let

$$\begin{aligned} v_0 &= x_0 = Cy, & r_0 &= y - Ax_0 = By,\\ v_{p+1} &= Cr_p, & r_{p+1} &= r_p - Av_{p+1} = Br_p.\end{aligned} \tag{2.122.8}$$

Then

$$\begin{aligned} x_p &= v_0 + v_1 + \cdots + v_p,\\ r_p &= y - Ax_p = B^{p+1}y.\end{aligned} \tag{2.122.9}$$

The procedure is to compute v_{p+1} from the last remainder r_p, and then compute the next remainder r_{p+1}.

2.13. *Some Estimates of Error*. The fact that an iterative process converges to a given limit does not of itself imply that the sequence obtained by a particular digital computation will approach this limit. If the machine operates with σ significant figures in the base β, we are by no means sure of σ significant figures in the result. At some stage the round-off errors introduced in the process being used will be of such magnitude that continuation of the process is unprofitable. However another, perhaps more slowly convergent, process might permit further improvement. In any case it is important to be able to estimate both the residual errors and the generated errors. In presenting these estimates, it will be supposed that the equations to be solved are themselves exact. The extent to which an error in the original coefficients affects the solution will be discussed in a later section.

2.131. *Residual errors.* We first consider residual or truncation errors neglecting any effects of round-off. If $M(H) < 1$, then from the identity

$$\begin{aligned}(I - H)^{-1} &= I + H(I - H)^{-1} \\ &= I + H(I + H + H^2 + \cdots)\end{aligned}$$

it follows that

$$M[(I - H)^{-1}] \leq 1 + M(H)M[(I - H)^{-1}].$$

Hence

$$M[(I - H)^{-1}] \leq 1/[1 - M(H)]. \tag{2.131.1}$$

Thus in certain cases a bound for the maximum value of a reciprocal can be obtained from the matrix itself. From the same identity, since $N(I) \leq n^{1/2}$, if $N(H) < 1$, then

$$N[(I - H)^{-1}] \leq n^{1/2}/[1 - N(H)], \tag{2.131.2}$$

and if $nb(H) < 1$, then

$$b[(I - H)^{-1}] \leq 1/[1 - nb(H)]. \tag{2.131.3}$$

The last two inequalities are generally less sharp but more easily applied.

Now consider the sequences C_p and B_p defined by (2.122.3). Since

$$A^{-1} = C_0(I - B_0)^{-1},$$

if we set $H = B_0$, then

$$M(A^{-1}) \leq M(C_0)/[1 - M(B_0)], \tag{2.131.4}$$

provided the denominator is positive. To establish the analogous inequalities using N and b, we note that

$$A^{-1} = C_0 + C_0B_0 + C_0B_0^2 + \cdots ,$$

whence

$$\begin{aligned}N(A^{-1}) &\leq N(C_0) + N(C_0)N(B_0) + N(C_0)N^2(B_0) + \cdots \\ &= N(C_0)/[1 - N(B_0)],\end{aligned} \tag{2.131.5}$$

and

$$\begin{aligned}b(A^{-1}) &\leq b(C_0) + nb(C_0)b(B_0) + n^2b(C_0)b^2(B_0) + \cdots \\ &= b(C_0)/[1 - nb(B_0)],\end{aligned} \tag{2.131.6}$$

again provided the denominators are positive.

Now from (2.122.4) and (2.122.3) we can deduce that

$$C_p = A^{-1}(I - B_0^{2^p}),$$

or

$$A^{-1} - C_p = A^{-1}B_0^{2^p}. \tag{2.131.7}$$

Hence given the hypotheses on $M(B_0)$, $N(B_0)$, or $nb(B_0)$, as the case may be, we have

$$M(A^{-1} - C_p) \leq M(A^{-1})M^{2^p}(B_0),$$

whence by (2.131.4)

$$M(A^{-1} - C_p) \leq M(C_0)M^{2^p}(B_0)/[1 - M(B_0)], \tag{2.131.8}$$

and analogously

$$N(A^{-1} - C_p) \leq N(C_0)N^{2^p}(B_0)/[1 - N(B_0)], \tag{2.131.9}$$

$$b(A^{-1} - C_p) \leq n^{2^p}b(C_0)b^{2^p}(B_0)/[1 - nb(B_0)]. \tag{2.131.10}$$

We can write (2.131.7) in the form

$$A^{-1} - C_p = C_0(I - B_0)^{-1}B_0^{2^p}.$$

If

$$x_p = C_py, \qquad s_p = x - x_p,$$

then

$$s_p = C_0(I - B_0)^{-1}B_0^{2^p}y.$$

Hence we have, for example,

$$N(s_p) \leq M(C_0)M^{2^p}(B_0)N(y)/[1 - M(B_0)].$$

If e_i is the ith unit vector, $e_i^\mathsf{T}s_p$ is the ith element in s_p. Hence we may write, with de la Garza,

$$|e_i^\mathsf{T}s_p| \leq N(e_i^\mathsf{T}C_0)M^{2^p}(B_0)N(y)/[1 - M(B_0)], \tag{2.131.11}$$

$$|e_i^\mathsf{T}s_p| \leq N(e_i^\mathsf{T}C_0)N^{2^p}(B_0)N(y)/[1 - N(B_0)]. \tag{2.131.12}$$

For the Seidel iteration we have

$$A = A_1 + A_2,$$
$$A^{-1} = (I - H)^{-1}A_1^{-1}, \qquad H = -A_1^{-1}A_2.$$

Let

$$C_p = (I + H + H^2 + \cdots + H^{p-1})A_1^{-1}.$$

Then

$$A^{-1} - C_p = H^p(I - H)^{-1}A_1^{-1},$$

and

$$M(A^{-1} - C_p) \leq M^p(H)M(A_1^{-1})/[1 - M(H)], \tag{2.131.13}$$

$$N(A^{-1} - C_p) \leq N^p(H)N(A_1^{-1})/[1 - M(H)], \tag{2.131.14}$$

$$b(A^{-1} - C_p) \leq n^{p+1}b^p(H)b(A_1^{-1})/[1 - nb(H)]. \tag{2.131.15}$$

If we take $x_0 = 0$ in the Seidel iteration, then $x_p = C_py$, and the deviation is $x - x_p = (A^{-1} - C_p)y$. In general, however, for an arbitrary x_0 if $r_p = y - Ax_p$, then

$$r_p = A_1H^pA_1^{-1}r_0,$$

and

$$N(r_p) \leq M(A_1)M(A_1^{-1})M^p(H)N(r_0).$$

Since

$$x_{p+1} - x_p = H^p A_1^{-1} y,$$
$$x - x_p = (I - H)^{-1} H^p A_1^{-1} y,$$

then

$$s_p = x - x_p = (I - H)^{-1}(x_{p+1} - x_p).$$

Hence

(2.131.16)
$$N(s_p) \leq N(x_{p+1} - x_p)/[1 - M(H)] \leq N(x_{p+1} - x_p)/[1 - N(H)].$$

Also

$$s_p = (I - H)^{-1} H(x_p - x_{p-1}),$$

(2.131.17)
$$N(s_p) \leq M(H)N(x_p - x_{p-1})/[1 - M(H)] \leq N(H)N(x_p - x_{p-1})/[1 - N(H)].$$

These inequalities provide estimates of the error in terms of the magnitude of a particular correction.

2.132. *Generated errors.* If A is symmetric, let μ and u be the numerically smallest proper value and an associated proper vector, respectively. If x_0 is any approximation to $x = A^{-1}y$, then $Ax_0 = y - r_0$ while

$$A(x_0 + u) = y - r_0 + \mu u.$$

Hence if μ is very small, a large component in x_0 along u would appear in r_0 as only a small component in the same direction. Another way of saying this is to say that a putative solution x_0 might yield a residual r_0 that would be regarded as negligibly small even when x_0 has a large erroneous component along u.

In general, for any matrix if x_0 and x_1 are two putative solutions, then

(2.132.1)
$$x_1 - x_2 = A^{-1}(r_2 - r_1),$$
$$N(x_1 - x_2) \leq N(r_2 - r_1)M(A^{-1}) = N(r_2 - r_1)/m(A),$$

and if $m(A)$ is small, then a large difference $x_1 - x_2$ could result in only a small $r_2 - r_1$, possibly less than the maximum round-off error. In fact, if ϵ is the limit of the round-off error, then $\epsilon/m(A)$ represents the limit of detectable accuracy in the solution.

There is no a priori assurance, however, that any particular method of solution will give a result that is even that close. We therefore consider this question for some of the iterative methods described above. It will be assumed, for definiteness, that the operations are fixed-point with maximal round-off ϵ, all numbers being scaled to magnitude less than unity, and that in the multiplication of vectors and matrices it is possible to accumulate complete products and round off the sum. If each product is rounded, then generally in the estimates given below the factor ϵ must be multiplied by n, the order of the matrix.

In any iterative process which utilizes one approximation to obtain one that is theoretically closer, the given approximation actually utilized in the computation, however it may have been obtained, is digital. To the digital approximation one applies certain pseudo operations to obtain another digital approximation. Two partially distinct questions arise: Given a digital approximation and a particular method of iteration, can we be sure that the next iteration will give improvement? Given two digital approximations, however obtained, when can we be sure that one is better than the other? These are questions relating to both the generated and the residual errors, since for iterative methods they merge together.

Basic to the discussion is the fact that, when a product Ax_0, say, of a digital matrix by a digital vector, is rounded off by rounding only the accumulated sums and not the separate products of the element, then the resulting digital vector, which will be designated $(Ax_0)^*$, satisfies

$$\begin{aligned} b[Ax_0 - (Ax_0)^*] &\leq \epsilon, \\ N[Ax_0 - (Ax_0)^*] &\leq n^{1/2}\epsilon. \end{aligned} \tag{2.132.2}$$

An additional factor n would appear if each product of elements were rounded. Likewise, for the multiplication of two digital matrices the digital product satisfies

$$\begin{aligned} b[AC - (AC)^*] &\leq \epsilon, \\ N[AC - (AC)^*] &\leq n\epsilon. \end{aligned} \tag{2.132.3}$$

2.1321. The Seidel Process. We consider only a positive definite matrix A. The process is based upon the fact that the vector x which satisfies the equations is the vector x which minimizes the function

$$x^{\mathsf{T}}(Ax - 2y) = -x^{\mathsf{T}}(y + r).$$

Hence it maximizes $x^{\mathsf{T}}(y + r)$. Hence, given two approximate solutions x_0 and x_1, we shall say that x_1 is a better approximation than x_0 provided

$$x_1^{\mathsf{T}}(y + r_1) > x_0^{\mathsf{T}}(y + r_0), \tag{2.1321.1}$$

and if the two quantities are equal, the two approximations are equally good. However for making the test in a particular instance, there will be available only the vectors

$$r_p^* = y - (Ax_p)^*, \qquad p = 0, 1.$$

By (2.132.2),

$$N(r_p - r_p^*) = N[(Ax_p)^* - Ax_p] \leq n^{1/2}\epsilon,$$

and therefore

$$|x_p^{\mathsf{T}}(r_p - r_p^*)| \leq N(x_p)N(r_p - r_p^*) \leq n^{1/2}\epsilon N(x_p).$$

Also

$$x_p^{\mathsf{T}}(y + r_p) = x_p^{\mathsf{T}}(y + r_p^*) + x_p^{\mathsf{T}}(r_p - r_p^*).$$

Hence

$$x_0^{\mathsf{T}}(y + r_0) - x_1^{\mathsf{T}}(y + r_1) = x_0^{\mathsf{T}}(y + r_0^*) - x_1^{\mathsf{T}}(y + r_1^*) + x_0^{\mathsf{T}}(r_0 - r_0^*) - x_1^{\mathsf{T}}(r_1 - r_1^*),$$

and (2.1321.1) will certainly be satisfied if

$$(2.1321.2) \qquad x_1^{\mathsf{T}}(y + r_1^*) - x_0^{\mathsf{T}}(y + r_0^*) > n^{1/2}\epsilon[N(x_0) + N(x_1)].$$

Since we can also say that

$$|x_p^{\mathsf{T}}(r_p - r_p^*)| \leq nb(x_p)b(r_p - r_p^*) \leq n\epsilon b(x_p),$$

therefore (2.1321.1) is also implied by

$$(2.1321.3) \qquad x_1^{\mathsf{T}}(y + r_1^*) - x_0^{\mathsf{T}}(y + r_0^*) > n\epsilon[b(x_0) + b(x_1)].$$

This requirement is somewhat more stringent.

Now consider a particular approximation x_0 and the digital approximation that would be obtained from x_0 following a single projection. Can we be assured that the digital result of making the projection will be a better approximation than x_0? If the projection is made on e_i, we wish to know whether

$$(x_0 + \lambda^* e_i)^{\mathsf{T}}\{y + [y - A(x_0 + \lambda^* e_i)]\} > x_0^{\mathsf{T}}(y + r_0),$$

where

$$\lambda^* = e_i^{\mathsf{T}} r_0^* \div \alpha_{ii}.$$

We suppose every $\alpha_{ii} = 1$. This does not violate the requirement that all stored quantities be in magnitude less than unity since the α_{ii} need not be stored explicitly in this case. Hence

$$\lambda^* = e_i^{\mathsf{T}} r_0^*, \qquad \lambda = e_i^{\mathsf{T}} r_0.$$

The above inequality reduces to

$$2\lambda^*\lambda > \lambda^{*2}.$$

For $\lambda^* > 0$, this is equivalent to

$$\lambda^* > 2(\lambda^* - \lambda),$$

and for $\lambda^* < 0$ it is equivalent to

$$\lambda^* < 2(\lambda^* - \lambda).$$

Since $|\lambda^* - \lambda| < \epsilon$, either condition is assured by

$$(2.1321.4) \qquad |\lambda^*| = |e_i^{\mathsf{T}} r_0^*| > 2\epsilon.$$

If $|\lambda^*| < 2\epsilon$, then $\lambda^* = e_i^{\mathsf{T}} r_0^* = 0$, and no change is made in x_0; if $|\lambda^*| = 2\epsilon$, then at least the modified vector is not worse than x_0. Hence in spite of the round-off, no step in the process can yield a poorer approximation, and in general any step will yield a better one until ultimately some $r_p^* = 0$.

2.1322. Iteration with an Approximate Inverse. Next consider an arbitrary nonsingular matrix A with a given approximate inverse C and a given approximation x_0 to $x = A^{-1}y$. Elements of the inverse will be out of range if the elements of A are digital. Hence C must be stored in the form $\beta^{-\gamma}C$, where γ is some positive integer, and the elements of this matrix will be assumed digital. Also γ will be supposed large enough so that all prescribed operations yield numbers in range. We suppose y scaled so that x and any approximation are in range. Hence x_0 is supposed digital.

As a measure of magnitude of a vector r, we use $b(r)$, and associate with it a measure of magnitude of a matrix A, denoted $c(A)$, and defined by

$$c(A) = \max_i \sum_j |\alpha_{ij}|.$$

One verifies easily that for any two matrices A and B

$$c(AB) \leq c(A)c(B),$$
$$c(A + B) \leq c(A) + c(B).$$

If we form a matrix having any vector r in the first column and zero elsewhere, and apply the first of these inequalities, we conclude that

$$b(Ar) \leq c(A)b(r).$$

Moreover

$$c(A) = \max_{r \neq 0} b(Ar)/b(r),$$

though this property will not be required.

If x_0 and x_1 are any two digital approximations to x, we ask first under what conditions we can be assured that $b(r_1) < b(r_0)$. Since we calculate

$$r_p^* = y - (Ax_p)^*, \qquad p = 1, 2,$$

we can be assured of the relation only when

$$b(r_1^*) + b(r_1 - r_1^*) < b(r_0^*) - b(r_0 - r_0^*).$$

But each element of r_p^* can be in error by an amount ϵ (if individual products are rounded, it is $n\epsilon$), whence the condition is

$$b(r_1^*) < b(r_0^*) - 2\epsilon.$$

When the equality holds, then at worst $b(r_1) = b(r_0)$.

Now suppose we have the approximation x_0 and wish to decide whether to attempt to improve the approximation by forming $x_0 + Cr_0$. Are we assured of obtaining a better approximation? Actually we form a digital vector

$$x_1 = x_0 + \beta^{\gamma}(\beta^{-\gamma}Cr_0^*)^*,$$

and the question is whether this is a better approximation than x_0. We have identically

(2.1322.1) $$r_1 = r_0 - r_0^* + B^*r_0^* + (B - B^*)r_0^* + \beta^\gamma A[\beta^{-\gamma}Cr_0^* - (\beta^{-\gamma}Cr_0^*)^*],$$

where

$$B = I - AC, \qquad B^* = \beta^\gamma\{\beta^{-\gamma}I - [A(\beta^{-\gamma}C)]^*\}.$$

Hence an improvement is assured if $b(r_0)$ exceeds the b function of the right member of (2.1322.1), and this is certainly true when $b(r_0^*) - b(r_0 - r_0^*)$ exceeds the same quantity. Hence the condition is

(2.1322.2) $$b(r_0^*) > \{2b(r_0 - r_0^*) + \beta^\gamma c(A)b[\beta^{-\gamma}Cr_0^* - (\beta^{-\gamma}Cr_0^*)^*]\}/[1 - c(B^*) - c(B - B^*)].$$

In this relation $c(A)$, $c(B^*)$, and $b(r_0^*)$ can be evaluated directly while the other quantities are limited by the computational routine.

By the contemplated routine of rounding off the accumulation of products, each element of $(\beta^{-\gamma}Cr_0^*)^*$ can differ from $\beta^{-\gamma}Cr_0^*$ by as much as ϵ, whence

$$b[\beta^{-\gamma}Cr_0^* - (\beta^{-\gamma}Cr_0^*)^*] \leq \epsilon.$$

As for B^*, each element in $[A(\beta^{-\gamma}C)]^*$ can be in error by ϵ, and n terms contribute to the C function. Hence

$$c(B - B^*) \leq n\epsilon\beta^\gamma.$$

The condition required is therefore

(2.1322.3) $$b(r_0^*) > \epsilon[2 + \beta^\gamma c(A)]/[1 - n\epsilon\beta^\gamma - c(B^*)].$$

The dominant term in the numerator is probably $\beta^\gamma c(A)$.

A slight modification of the routine can improve the situation by "damping out" the term $\beta^\gamma c(A)$. Since r_0^* is presumably small, it should be possible to scale it up by a factor β^δ, forming $(\beta^{\delta-\gamma}Cr_0^*)^*$. If this is done, one has

$$b[\beta^{\delta-\gamma}Cr_0^* - (\beta^{\delta-\gamma}Cr_0^*)^*] \leq \epsilon,$$

and in place of (2.1322.3),

(2.1322.4) $$b(r_0^*) > \epsilon[2 + \beta^{\gamma-\delta}c(A)]/[1 - n\epsilon\beta^\gamma - c(B^*)].$$

On consecutive iterations as the residual diminishes, δ can be increased, possibly even until the term $\beta^{\gamma-\delta}c(A)$ becomes negligible. In the denominator, $c(B^*)$ can be reduced, if desired, by iterating to improve the approximate inverse. But the term $n\epsilon\beta^\gamma$ is not at our disposal.

By a further modification of the routine, the factor $\beta^{-\delta}$ can be brought before the entire numerator. If the products required in forming Ax_0 can be accumulated before rounding, the accumulation can also be subtracted from y and the result multiplied by β^δ before rounding. This gives

$$b(r_0 - r_0^*) \leq \beta^{-\delta}\epsilon.$$

If this is done, the condition

$$b(r_0^*) > \epsilon\beta^{-\delta}[2 + \beta^\gamma c(A)]/[1 - n\epsilon\beta^\gamma - c(B^*)] \tag{2.1322.5}$$

is sufficient to ensure improvement. Since the stored quantities are elements of $\beta^\delta r_0^*$, the elements of r_0^* can be made actually less than the maximal round-off.

2.2. Direct Methods. A direct method for solving an equation or system of equations is any finite sequence of arithmetic operations that will result in an exact solution. Since the operations one actually performs are generally pseudo operations, the direct methods do not generally in practice yield exact results. Nevertheless, the results may be as accurate as one requires, or it may be advantageous to apply first a direct method after which the solution may be improved by the application of one of the iterative methods.

Certainly all (correct) direct methods are equivalent in the sense that they all yield in principle the same exact solution (when the matrix is nonsingular and the solution unique). Nevertheless the methods differ in the total number of operations (additions and subtractions, multiplications, divisions, recordings) that they require and in the order in which these take place. As a consequence they differ also as to the opportunities for making blunders and as to the magnitude of the generated error.

Most direct methods involve obtaining, at one stage or another, a system of equations of such a type that one equation contains only one unknown, a second equation contains only this unknown and one other, a third only these and one other, etc. The procedure for solving such a system is quite obvious. The matrix of such a system is said to be triangular (or semidiagonal), since either every element below the principal diagonal, or else every element above the principal diagonal, has the value zero. A matrix of the first of these types is upper triangular; one of the second is lower triangular. If it happens, in addition, that every diagonal element is equal to 1, then the matrix is unit upper triangular or unit lower triangular, as the case may be. We shall say that a matrix M is of triangular type, upper or lower, if it can be partitioned into one of the two forms

$$M = \begin{pmatrix} M_{11} & M_{12} \\ 0 & M_{22} \end{pmatrix}, \qquad M = \begin{pmatrix} M_{11} & 0 \\ M_{21} & M_{22} \end{pmatrix}, \tag{2.2.1}$$

where M_{11} and M_{22} are both square matrices, as is M itself. We shall consider such matrices briefly.

2.201. *Matrices of triangular type.* If M is of upper triangular type, then M^T is of lower triangular type. Hence it is sufficient to consider just one of these types. One verifies directly that, if M and N are both of upper triangular type when similarly partitioned (*i.e.*, corresponding submatrices have the same dimensions), then the product MN is of upper triangular type when similarly partitioned. If, further, M_{11} and M_{22} are nonsingular, then M^{-1} exists, and is of upper triangular type when similarly partitioned. In fact,

$$M^{-1} = \begin{pmatrix} M_{11}^{-1} & -M_{11}^{-1}M_{12}M_{22}^{-1} \\ 0 & M_{22}^{-1} \end{pmatrix}. \tag{2.201.1}$$

Hence if M_{11} is upper triangular and M_{22} a scalar, then M itself is upper triangular and not merely of upper triangular type. In this case (2.201.1) provides a stepwise procedure for inverting a triangular matrix, if M_{22} is a scalar while M_{11} is a matrix which has been inverted.

If a matrix A is partitioned as

$$A = \begin{pmatrix} A_{11} & A_{12} \\ A_{21} & A_{22} \end{pmatrix} \tag{2.201.2}$$

with A_{11} nonsingular, then A can be expressed in many ways as the product of matrices of triangular type in the form

$$\begin{pmatrix} A_{11} & A_{12} \\ A_{21} & A_{22} \end{pmatrix} = \begin{pmatrix} N_{11} & 0 \\ N_{21} & N_{22} \end{pmatrix}\begin{pmatrix} M_{11} & M_{12} \\ 0 & M_{22} \end{pmatrix}. \tag{2.201.3}$$

In fact, for an arbitrary nonsingular N_{11} (of the dimensions of A_{11}), M_{11}, M_{12}, and N_{21} are uniquely determined independently of A_{22} and of the selection of N_{22} and M_{22}. For one verifies directly that

$$M_{11} = N_{11}^{-1}A_{11}, \qquad M_{12} = N_{11}^{-1}A_{12}, \qquad N_{21} = A_{21}A_{11}^{-1}N_{11}, \tag{2.201.4}$$

while N_{22} and M_{22} are restricted only by the relation

$$N_{22}M_{22} = A_{22} - A_{21}A_{11}^{-1}A_{12}.$$

This being the case, we can give an inductive algorithm for a factorization of A into the product of a unit lower triangular matrix L and an upper triangular matrix W. That such a factorization exists and is unique when A is of second order and $A_{11} \neq 0$ follows from the above by taking $N_{11} = N_{22} = 1$. For purposes of the induction suppose that N_{11} above was unit lower triangular and M_{11} upper triangular. Then M_{12} and N_{21} are uniquely determined by (2.201.4). We change the notation and partition further, writing

$$\begin{pmatrix} A_{11} & A_{12} & A_{13} \\ A_{21} & A_{22} & A_{23} \\ A_{31} & A_{32} & A_{33} \end{pmatrix} = \begin{pmatrix} L_{11} & 0 & 0 \\ L_{21} & L_{22} & 0 \\ L_{31} & L_{32} & L_{33} \end{pmatrix}\begin{pmatrix} W_{11} & W_{12} & W_{13} \\ 0 & W_{22} & W_{23} \\ 0 & 0 & W_{33} \end{pmatrix}, \tag{2.201.5}$$

where $L_{11} = N_{11}$, $W_{11} = M_{11}$, A_{11} is the same as above, but the submatrices previously designated A_{22}, A_{21}, and A_{12} are now further partitioned, as are the matrices N_{21} and M_{12}. When the necessary inverses exist, these last matrices or their own submatrices are determined uniquely by Eqs. (2.201.4) which now have the form

$$\begin{aligned} W_{12} &= L_{11}^{-1}A_{12}, & W_{13} &= L_{11}^{-1}A_{13}, \\ L_{21} &= A_{21}W_{11}^{-1}, & L_{31} &= A_{31}W_{11}^{-1}. \end{aligned}$$

Four conditions remain to be satisfied. Of these, three give

$$\begin{aligned} W_{22} &= L_{22}^{-1}(A_{22} - L_{21}W_{12}), \\ W_{23} &= L_{22}^{-1}(A_{23} - L_{21}W_{13}), \\ L_{32} &= (A_{32} - L_{31}W_{12})W_{22}^{-1}. \end{aligned} \tag{2.201.6}$$

Hence W_{22}, W_{23}, and L_{32} can be determined uniquely from A and from the portions of L and W already determined, provided only that $A_{22} - L_{21}W_{12}$ is nonsingular, and independently of the choice of the matrices L_{33} and W_{33}. The last condition merely specifies the product $L_{33}W_{33}$. Hence for the inductive algorithm take $L_{22} = 1$ and determine the scalar W_{22} and the vectors W_{23} and L_{32}. Now the matrices

$$\begin{pmatrix} L_{11} & 0 \\ L_{21} & L_{22} \end{pmatrix}, \qquad \begin{pmatrix} W_{11} & W_{12} \\ 0 & W_{22} \end{pmatrix}$$

are unit lower and upper triangular matrices, respectively, to be designated L_{11} and W_{11} for the next steps. The process fails if at some stage $W_{22} = 0$. If A is nonsingular, it is always possible to rearrange rows and columns and continue.

By applying the process to A^{T}, we note that we could equally well make W a unit upper triangular matrix and L lower triangular, and again the factorization is unique.

When A is symmetric and L is unit lower triangular, let D^2 represent the diagonal matrix of elements of the principal diagonal of W. Then

$$D^{-2}W = L^{\mathsf{T}},$$

and the factorization can be written

$$A = LD^2L^{\mathsf{T}}.$$

If we write

$$LD = K,$$

then

$$A = KK^{\mathsf{T}}.$$

If A is not positive definite, then D^2 will have negative elements, and hence D will have pure imaginary elements. However this presents no real computational difficulty.

We conclude this section by noting that, if B is an arbitrary matrix of $m \leq n$ linearly independent columns, there exists a unique unit upper triangular matrix V of order m such that the columns of

$$R = BV^{-1} \tag{2.201.7}$$

are mutually orthogonal with respect to the matrix G. This is to say that the matrix

$$R^{\mathsf{T}}GR = D \tag{2.201.8}$$

is a diagonal matrix. To make the proof inductive, and exhibit the algorithm, suppose this has been accomplished for the matrix B with $m < n$, and suppose that the vector b is independent of the columns of B. We wish then to select vectors r and v so that

$$(B, b) = (R, r)\begin{pmatrix} V & v \\ 0 & 1 \end{pmatrix} \tag{2.201.9}$$

and

$$r^{\mathsf{T}}GR = 0.$$

These conditions are satisfied by taking

$$v = D^{-1}R^{\mathsf{T}}Gb, \qquad r = b - Rv. \tag{2.201.10}$$

Geometrically, this process amounts to resolving b into a component in the space of the columns of B (or of R) and a component orthogonal to this space, the latter component becoming the vector r.

Most of the classical direct methods for solving systems of linear equations can now be deduced almost immediately.

2.21. *Methods of Elimination.* In elementary algebra one learns to solve a system of equations by "the method of elimination by addition and subtraction." In this method an equation is selected in which the coefficient of the first unknown ξ_1 is non-null, and one adds an appropriate multiple of this equation to each of the others in turn so that ξ_1 is eliminated from these equations. The resulting $n - 1$ equations, together with the one used for the elimination, constitute a new system equivalent to the first system, and the $n - 1$ equations contain only the $n - 1$ unknowns $\xi_2, \ldots, \xi_n$. The same process applied to these yields $n - 2$ equations containing only the $n - 2$ unknowns $\xi_3, \ldots, \xi_n$. Eventually one obtains a single equation in ξ_n alone. The final solution is now obtained by "back substitution," in which the value of ξ_n obtained from the last equation is substituted into the preceding which can then be solved for ξ_{n-1}, these are then substituted into the one before, etc. Thus the elimination phase followed by the back-substitution phase yields the final solution.

In the elimination phase, the operation of eliminating each of the unknowns is equivalent to the operation of multiplying the system by a particular unit lower triangular matrix—a matrix, in fact, whose off-diagonal non-null elements are all in the same column. The product of all these unit lower triangular matrices is again a unit lower triangular matrix, and hence the entire process of elimination (as opposed to that of back substitution) is equivalent to that of multiplying the system by a suitably chosen unit lower triangular matrix. Since the matrix of the resulting system is clearly upper triangular, these considerations constitute another proof of the possibility of factorizing A into a unit lower triangular matrix and an upper triangular matrix.

For the system

$$Ax = y,$$

after eliminating any one of the variables, the effect to that point is that of having selected a unit lower triangular matrix of the form

$$\begin{pmatrix} L_{11} & 0 \\ L_{12} & I_{22} \end{pmatrix}$$

where L_{11} is itself unit lower triangular, in such a way that A is factored

$$A = \begin{pmatrix} A_{11} & A_{12} \\ A_{21} & A_{22} \end{pmatrix} = \begin{pmatrix} L_{11} & 0 \\ L_{21} & I_{22} \end{pmatrix} \begin{pmatrix} W_{11} & W_{12} \\ 0 & M_{22} \end{pmatrix}, \tag{2.21.1}$$

with W_{11} upper triangular but M_{22} not. Hence

$$M_{22} = A_{22} - A_{21}A_{11}^{-1}A_{12}. \tag{2.21.2}$$

The original system has at this stage been replaced by the system

$$\begin{pmatrix} W_{11} & W_{12} \\ 0 & M_{22} \end{pmatrix} \begin{pmatrix} x_1 \\ x_2 \end{pmatrix} = \begin{pmatrix} z_1 \\ z_2 \end{pmatrix} \tag{2.21.3}$$

where

$$\begin{pmatrix} L_{11} & 0 \\ L_{21} & I_{22} \end{pmatrix} \begin{pmatrix} z_1 \\ z_2 \end{pmatrix} = \begin{pmatrix} y_1 \\ y_2 \end{pmatrix} = y. \tag{2.21.4}$$

The matrices L_{11} and L_{21} are not themselves written down. The partial system

$$M_{22}x_2 = z_2$$

represents those equations from which further elimination remains to be done, and this can be treated independently of the other equations of the system, which fact explains why it is unnecessary to obtain the L matrices explicitly.

If the upper left-hand element of M_{22} vanishes, this cannot be used in the next step of the elimination, and it is not advantageous to use it when it is small. Hence rows or columns, or both, in M_{22} must be rearranged to bring to this position an element that is sufficiently large. Corresponding changes must be made in the notation.

Crout's method differs in that the L matrices are written down explicitly at each stage while M_{22} is not. It utilizes the inductive algorithm given in (2.201.6) where, as we have seen, each successive column of L and row of W can be obtained from those previously computed together with the corresponding row and column of A. In order to compute the vector z as one goes along, one writes the augmented matrix (A, y). Then the partitioning of (2.201.5) is extended to the following:

$$(2.21.5)\quad \begin{pmatrix} A_{11} & A_{12} & A_{13} & y_1 \\ A_{21} & A_{22} & A_{23} & y_2 \\ A_{31} & A_{32} & A_{33} & y_3 \end{pmatrix} = \begin{pmatrix} L_{11} & 0 & 0 \\ L_{21} & L_{22} & 0 \\ L_{31} & L_{32} & L_{33} \end{pmatrix} \begin{pmatrix} W_{11} & W_{12} & W_{13} & z_1 \\ 0 & W_{22} & W_{23} & z_2 \\ 0 & 0 & W_{33} & z_3 \end{pmatrix}$$

and supposing L_{11}, L_{21}, L_{31}, W_{11}, W_{12}, W_{13}, z_1 already determined, L_{22} is prescribed (in practice $L_{22} = 1$), and L_{32}, W_{22}, W_{23}, z_2 are to be determined at this step. Equations (2.201.6) give L_{32}, W_{22}, and W_{23}, while z_2 is given by

$$(2.21.6)\qquad z_2 = L_{22}^{-1}(y_2 - L_{21}z_1).$$

While in practice one takes $L_{22} = 1$, this equation and Eqs. (2.201.6) are perfectly general. Since neither L_{33}, W_{33}, nor z_3 occurs in any of these relations, one can, with Crout, write the two matrices $L - I$ and (W, z) in the same rectangular array, filling out in sequence the first row, the first column, the second row, the second column, etc. When this array is filled out, the elements along and to the right of the principal diagonal are the coefficients and the constants in the triangular equations $Wx = z$.

In case one has two or more sets of equations with the same matrix A, then the vectors y and z may be replaced by the matrices Y and Z in (2.21.5) and (2.21.6). Alternatively one may solve one of these systems, after which, with L and W already known, the elements of any other column z in Z are obtained sequentially from the corresponding column y in Y by using (2.21.6), remembering that at the start there is no partial vector z_1 so that one has simply $\zeta_1 = \eta_1$. In particular, if a single system is solved by this method, and a result x_0 is obtained which is only approximate because of round-off errors, we have seen that the error vector $x - x_0$ satisfies a system with the same matrix A, so that (2.21.6) can be applied with $y - Ax_0$ replacing y.

Another modification of the method of elimination is that of Jordan. It is clear that, after ξ_1 has been eliminated from equations 2, . . . , n, and while the new second equation is being used to eliminate ξ_2 from equations 3, . . . , n, this can also be used to eliminate ξ_2 from the first equation. Next the third equation can be used to eliminate ξ_3 from what are now equations 1 and 2, as well as from equations 4, . . . , n. By proceeding thus, one obtains an equivalent system of the form $Dx = w$, where D is diagonal. This amounts to multiplying the original system $Ax = y$ sequentially by matrices each of which differs from the identity only in a single column. However this column will have non-null elements both above and below the diagonal.

Crout's method provides a routine for triangular factorization which minimizes the number of recordings and also the space required for the recordings. This is very desirable, whether the computations are by automatic machinery or not. For machine computation it has the disadvantage of requiring products such as $L_{31}W_{12}$ involving elements from a column of L and from a row of W. Jordan's method permits a similar economy of recording without requiring operations upon columns.

To see this we note first that, if J is a matrix satisfying

$$J(A, I) = (I, J),$$

then certainly $J = A^{-1}$. In words, if it is possible to find a sequence of row operations which, when performed upon the matrix (A, I), reduces it to a matrix (I, J), then J is the required inverse of A. Jordan's method forms (I, J) stepwise from (A, I) by multiplying on the left by matrices of the form $I + J_i$, where J_i differs from the null matrix in the ith column only. The process will be complicated somewhat in "positioning for size," but we neglect this here and assume that the process can be carried out in natural order from the first column to the last. Then the sequence starts with the operation

$$(I + J_1)(A, I) = (A_1, K_1)$$

and continues with

$$(I + J_2)(A_1, K_1) = (A_2, K_2),$$

and generally

$$(I + J_i)(A_{i-1}, K_{i-1}) = (A_i, K_i).$$

We now observe that in A_i the first i columns are the same as those of I, and in K_i the last $n - i$ columns are the same as those of I. Thus one needs to record only the first i columns of K_i and the last $n - i$ columns of A_i. The ith column of J_i is to be selected so that the ith column of the product $(I + J_i)A_{i-1}$ is equal to e_i, and need not be recorded. Instead one records the ith column of $I + J_i$, which becomes the ith column of K_i. If the ith column of A_{i-1} is $\alpha_1, \ldots, \alpha_n$, and the ith column of $I + J_i$ is $\phi_1, \ldots, \phi_n$, then $\phi_i = \alpha_i^{-1}$, and $\phi_j = -\alpha_j/\alpha_i$.

Hence in forming the composite matrix of nontrivial columns of A_i

and of K_i, one first forms the ith row from the ith row of the previous composite. For this one divides every element but the ith (which is α_i) by α_i, recording the quotient, and in the ith place records α_i^{-1}. To obtain the jth row ($j \neq i$), one increases each element except the ith by $-\alpha_j$ times the corresponding element in the new ith row. In the ith place one records $\phi_j = -\alpha_j/\alpha_i = 0 - \alpha_j/\alpha_i$.

Clearly if one operates in this fashion upon the matrix (A, I, y), then one comes out with (I, A^{-1}, x). Thus in using automatic machinery if $n(n + 1)$ places are reserved in the memory for (A^{-1}, x), then these places are to be filled first by the matrix (A, y) arranged by rows. Each multiplication by an $I + J_i$ requires first an operation upon the elements of the ith row, followed by an operation upon the elements of the old jth and the new ith row.

2.22. *Methods of Orthogonalization.* Let

$$A = RV, \qquad R^{\mathsf{T}}R = D^2 \tag{2.22.1}$$

where V is unit upper triangular and D^2 is diagonal. We have seen in §2.201 that such matrices exist. The general metric G of §2.201 is here taken to be I. The matrices V and R can be computed sequentially by applying Eqs. (2.201.9) and (2.201.10) with appropriate modification of notation. Then the equations $Ax = y$ can be written $RVx = y$ so that $D^2Vx = R^{\mathsf{T}}y$, and

$$x = V^{-1}D^{-2}R^{\mathsf{T}}y. \tag{2.22.2}$$

Since D is diagonal and V unit upper triangular, their inversion is straightforward. This is Schmidt's method.

In the least-squares problem one has a matrix B, with $m < n$ rows, and a vector y, and one seeks a vector x of m elements such that

$$Bx = y + d, \qquad d^{\mathsf{T}}B = 0.$$

Geometrically the vector y is to be projected orthogonally upon the space of the columns of B, and the components x of the projection are required. Since

$$B^{\mathsf{T}}Bx = B^{\mathsf{T}}y,$$

these equations yield the required x. If, however,

$$B = RV, \qquad R^{\mathsf{T}}R = D^2,$$

then

$$d^{\mathsf{T}}R = 0,$$

whence

$$D^2Vx = R^{\mathsf{T}}y,$$

and

$$x = V^{-1}D^{-2}R^{\mathsf{T}}y.$$

The orthogonalization process can therefore be applied directly to the matrix B, and $B^{\mathsf{T}}B$ is not required explicitly.

To return to the system $Ax = y$, we may equally well write

$$A = US, \qquad SS^{\mathsf{T}} = D^2, \tag{2.22.3}$$

where U is unit lower triangular. Let w satisfy

$$x = S^{\mathsf{T}}w.$$

Then the equations can be written

$$USS^{\mathsf{T}}w = UD^2w = y$$

so that

$$w = D^{-2}U^{-1}y,$$

and therefore

$$x = S^{\mathsf{T}}D^{-2}U^{-1}y. \tag{2.22.4}$$

In this method the rows of S are orthogonal combinations of the rows of A, and since $x^{\mathsf{T}} = w^{\mathsf{T}}S$, the vector x^{T} is expressed as a linear combination of these orthogonal row vectors.

If A is positive definite, we could attempt to use A, or possibly A^{-1}, as the metric to obtain an orthogonalized set of vectors along which x might be resolved easily, and in fact from §2.201 we can form matrices R and V such that

$$I = RV, \qquad R^{\mathsf{T}}AR = D^2,$$

so that if

$$x = Rw,$$

then

$$\begin{aligned} R^{\mathsf{T}}ARw &= R^{\mathsf{T}}y, \\ w &= D^{-2}R^{\mathsf{T}}y, \\ x &= RD^{-2}R^{\mathsf{T}}y. \end{aligned}$$

Indeed, $R = V^{-1}$, and it is therefore unit upper triangular. Hence the relation $R^{\mathsf{T}}AR = D^2$ is equivalent to $A = V^{\mathsf{T}}D^2V$, which is the triangular resolution already obtained but arrived at in a different way.

An orthogonalization process of somewhat different type has been devised by Stiefel and Hestenes, independently. The process leads to a fairly simple iteration, which, however, terminates in n steps to yield the exact solution apart from round-off. Since the n steps yield progressively better approximations to the true solution, the process can be continued beyond n steps for reduction of the round-off error.

The first step, as applied to a positive definite matrix A, is the same as in the method of steepest descent, in that one starts with an arbitrary initial approximation x_0 and improves it by adding a multiple of the residual r_0. Thereafter, however, instead of adding to each x_i a multiple of r_i, one adds a vector so chosen that r_{i+1} is orthogonal, with respect to the metric I, to all preceding r_i. If this can be accomplished, then for some $m \leq n$, $r_m = 0$, and hence $Ax_m = y$. For if all the vectors $r_0, r_1, \ldots, r_{n-1}$ are non-null, then being mutually orthogonal they are linearly independent, and only the null vector is orthogonal to all of them.

Geometrically the method has other points of interest. We have already noted that the solution x of the equations $Ax = y$ minimizes the function

$$2f(x) = x^{\mathsf{T}}Ax - 2x^{\mathsf{T}}y. \tag{2.22.5}$$

In fact it represents the common center of the hyperdimensional ellipsoids

$$f(x) = \text{const.} \tag{2.22.6}$$

This fact provides the usual approach to the method of steepest descent. Also at x_0 the function $f(x)$ is varying most rapidly in the direction of r_0, which is the gradient at x_0 of the function $-f(x)$. Hence one takes

$$x_1 = x_0 + \alpha_0 r_0,$$

where α_0 minimizes the function $f(x_0 + \alpha r_0)$ of the single variable α. It can be shown that the point x_1 is the mid-point of the chord through x_0 in the direction r_0 of that particular ellipsoid $f(x) = f(x_0)$ which passes through x_0. It is easy to write the equation of the diametral plane which bisects all chords in the direction r_0. This is a diametral plane of the ellipsoid $f(x) = f(x_1)$ through x_1, as well as a diametral plane of

$$f(x) = f(x_0),$$

and it intersects the original ellipsoids in hyperdimensional ellipsoids whose dimensionality is one less than that of the original ellipsoids. Stiefel and Hestenes now improve the approximation x_1 by adjoining a vector in the direction of the gradient to the lower dimensional ellipsoid, which is the orthogonal projection upon the diametral plane of the gradient r_1 to the n-dimensional ellipsoid. One proceeds then to get ellipsoids of progressively lower dimensionality until one finally reaches the center itself.

The success of the method depends upon a theorem (Lanczos) which will have application also in other connections. Beginning with a vector r_0, suppose one seeks to orthogonalize the vectors $r_0, Ar_0, A^2r_0, \ldots$ by selecting a set of vectors $r_0, r_1, r_2, \ldots$ in such a way that r_{i+1} is a linear combination of the vectors $r_0, Ar_0, \ldots, A^ir_0$, and r_{i+1} is orthogonal to $r_0, r_1, \ldots, r_i$. At most n such vectors will be non-null, and we have already shown how any set of linearly independent vectors can be orthogonalized. We can then express r_{i+1} as a linear combination of the vectors $r_0, r_1, \ldots, r_i$ and of Ar_i:

$$r_{i+1} = \rho_{i+1,0}r_0 + \rho_{i+1,1}r_1 + \cdots + \rho_{i+1,i}r_i + \rho_i Ar_i.$$

For $i = 0$ the statement is trivial, for it merely says that r_1 is a linear combination of r_0 and Ar_0. Suppose that all vectors $r_0, r_1, \ldots, r_i$ are non-null and that the statement holds for them. Then $\rho_i \neq 0$, since otherwise the mutually orthogonal vectors $r_0, \ldots, r_{i+1}$ would be linearly dependent. Since r_i is a linear combination of $r_0, Ar_0, \ldots, A^{i-1}r_0$, therefore Ar_i is a linear combination of $Ar_0, A^2r_0, \ldots, A^ir_0$. Hence r_{i+1} is expressed as a linear combination of $r_0, Ar_0, \ldots, A^ir_0$.

The theorem in question states that

$$\rho_{i+1,0} = \rho_{i+1,1} = \cdots = \rho_{i+1,i-2} = 0.$$

For suppose the resolution made. Then since $\rho_i \neq 0$, each Ar_i is a linear combination of $r_0, r_1, \ldots, r_{i+1}$. Hence Ar_i is orthogonal to every r_j for $j > i + 1$. Hence

$$r_j^\mathsf{T} Ar_i = 0, \qquad |j - i| > 1.$$

Hence Ar_i is a linear combination of only r_{i-1}, r_i, and r_{i+1}, which proves the theorem.

After simplifying the notation, we can therefore set

$$r_{i+1} = \gamma_i(r_i - \beta_{i-1}r_{i-1} - \alpha_i Ar_i). \tag{2.22.7}$$

We would like to arrange it so that these vectors r_i are residuals $y - Ax_i$. Then we require that

$$y - Ax_{i+1} = \gamma_i[y - Ax_i - \beta_{i-1}(y - Ax_{i-1}) - \alpha_i Ar_i].$$

If we impose the condition that

$$\gamma_i(1 - \beta_{i-1}) = 1, \tag{2.22.8}$$

then we can achieve this by taking

$$x_{i+1} = \gamma_i(x_i - \beta_{i-1}x_{i-1} + \alpha_i r_i). \tag{2.22.9}$$

For $i = 0$ we have

$$\begin{aligned} r_1 &= r_0 - \alpha_0 Ar_0, \\ x_1 &= x_0 + \alpha_0 r_0. \end{aligned} \tag{2.22.10}$$

Let

$$\rho_i = r_i^\mathsf{T} r_i, \qquad \sigma_i = r_i^\mathsf{T} Ar_i. \tag{2.22.11}$$

When we apply the orthogonality criterion, we find first

$$\alpha_i = \rho_i/\sigma_i. \tag{2.22.12}$$

Also

$$\rho_{i+1} = -\alpha_i\gamma_i r_{i+1}^\mathsf{T} Ar_i,$$

so that by reducing subscripts

$$r_i^\mathsf{T} Ar_{i-1} = -\rho_i/(\alpha_{i-1}\gamma_{i-1}). \tag{2.22.13}$$

Hence, since

$$0 = \beta_{i-1}\rho_{i-1} + \alpha_i r_{i-1}^{\mathsf{T}} A r_i,$$

therefore

$$\beta_{i-1} = \alpha_i \rho_i / (\alpha_{i-1}\rho_{i-1}\gamma_{i-1}). \tag{2.22.14}$$

Hence, beginning with α_0, and $\gamma_0 = 1$, we can find r_1, then α_1, β_0, γ_1, and hence r_2, and so on sequentially.

From (2.22.7) we can write

$$\begin{aligned} r_{i+1} - r_i &= \gamma_i[\beta_{i-1}(r_i - r_{i-1}) - \alpha_i A r_i], \\ x_{i+1} - x_i &= \gamma_i[\beta_{i-1}(x_i - x_{i-1}) + \alpha_i r_i]. \end{aligned}$$

Therefore

$$x_{i+1} = x_i + \lambda_i z_i, \tag{2.22.15}$$

and hence

$$r_{i+1} = r_i - \lambda_i A z_i, \tag{2.22.16}$$

$$z_{i+1} = r_{i+1} + \mu_i z_i, \tag{2.22.17}$$

where

$$\lambda_i = \alpha_i \gamma_i, \qquad \mu_i = \gamma_{i+1}\beta_i\lambda_i/\lambda_{i+1} = \rho_{i+1}/\rho_i. \tag{2.22.18}$$

It can be shown inductively that

$$z_i^{\mathsf{T}} A z_j = 0, \qquad i \neq j. \tag{2.22.19}$$

To begin with, from

$$\begin{aligned} z_0 &= r_0, \\ r_1 &= r_0 - \lambda_0 A z_0, \\ z_1 &= r_1 + \mu_0 z_0, \end{aligned}$$

we have

$$z_0^{\mathsf{T}} A z_1 = z_0^{\mathsf{T}} A r_1 + \mu_0 z_0^{\mathsf{T}} A z_0$$

and

$$r_1^{\mathsf{T}} r_1 = -\lambda_0 r_1^{\mathsf{T}} A z_0.$$

Hence, elimination of $r_1^{\mathsf{T}} A z_0$ gives with (2.22.11)

$$z_0^{\mathsf{T}} A z_1 = -\rho_1/\lambda_0 + \mu_0\sigma_0,$$

and this is seen to vanish from (2.22.18), (2.22.14), and (2.22.12). Now suppose (2.22.19) verified for all $j < i \leq k$. From (2.22.16) we have

$$\begin{aligned} r_j^{\mathsf{T}} A z_i &= 0, \qquad j \leq i - 1, j > i + 1, \\ \rho_{i+1} &= -\lambda_i r_{i+1}^{\mathsf{T}} A z_i, \\ \rho_i &= \lambda_i r_i^{\mathsf{T}} A z_i. \end{aligned} \tag{2.22.20}$$

Hence, from (2.22.17) with $i = k$ we have the required relation verified when $i = k + 1$ and $j \leq k - 1$.

Again, from (2.22.17) with $i = k$

$$z_k^{\mathsf{T}} A z_{k+1} = z_k^{\mathsf{T}} A r_{k+1} + \mu_k z_k^{\mathsf{T}} A z_k.$$

But from (2.22.17) with $i = k - 1$ and from (2.22.20)

$$z_k^{\mathsf{T}} A z_k = z_k^{\mathsf{T}} A r_k = \rho_k/\lambda_k, \tag{2.22.21}$$

whence, again with (2.22.20),

$$z_k^{\mathsf{T}} A z_{k+1} = -\rho_{k+1}/\lambda_k + \mu_k \rho_k/\lambda_k = 0.$$

This completes the proof of (2.22.19).

If we set

$$\tau_i = z_i^{\mathsf{T}} A z_i, \tag{2.22.22}$$

then from (2.22.16) and (2.22.17) we can calculate sequentially $z_0 = r_0$, λ_0, r_1, μ_0, z_1, λ_1, r_2, μ_1, z_2, . . . from the formulas

$$\lambda_i = \rho_i/\tau_i, \qquad \mu_i = \rho_{i+1}/\rho_i. \tag{2.22.23}$$

These equations are obtained by making use of (2.22.20) and (2.22.21). From these it is clear that

$$\lambda_i > 0, \qquad \mu_i > 0.$$

Hence, from (2.22.18) $\gamma_i > 0$ and $\beta_i > 0$. Hence

$$1 > \beta_i > 0.$$

We can now relate the method to the minimizing problem. The ellipsoid

$$f(x) = f(x_0)$$

passes through the point x_0; as λ varies, the points $x_0 + \lambda u$ lie on the secant through x_0 in the direction u, and this secant intersects the ellipsoid for $\lambda = 0$ and again for

$$\lambda = \lambda' = 2u^{\mathsf{T}} r_0 / u^{\mathsf{T}} A u.$$

To see this we have only to solve the equation $f(x_0 + \lambda u) = f(x_0)$ for λ. If we take $u = r_0$, then $\lambda' = 2\lambda_0$. Hence x_1 is the mid-point of the chord in the direction r_0.

Now the plane $r_0^{\mathsf{T}} A(x - x_1) = 0$ passes through x_1 and also through the point $x = A^{-1}y$, for by direct substitution the left member of this equation becomes $r_0^{\mathsf{T}} r_1$ which vanishes because of orthogonality. This plane is a diametral plane of the ellipsoid $f(x) = f(x_1)$; it intersects this latter ellipsoid in an ellipsoid of lower dimensionality. Instead of choosing x_2 to lie on the gradient to $f(x) = f(x_1)$, as is done by the method of steepest descent, the method of Hestenes and Stiefel now takes x_2 to lie on the orthogonal projection of the gradient in this hyperplane, or, what amounts to the same, along the gradient to the section of the ellipsoid which lies in the hyperplane. At the next step a diametral space of dimension $n - 2$ is formed, and x_3 is taken in the gradient to the section of the ellipsoid $f(x) = f(x_2)$ by this $(n - 2)$ space. Ultimately a diametral line is obtained, and x_n is the center itself. With the formulas already given these statements can be proved in detail, but the proof will be omitted here.

2.23. *Escalator Methods.* Various schemes have been proposed for utilizing a known solution of a subsystem as a step in solving the complete system. Let A be partitioned into submatrices,

$$A = \begin{pmatrix} A_{11} & A_{12} \\ A_{21} & A_{22} \end{pmatrix}, \tag{2.23.1}$$

and suppose the inverse A_{11}^{-1} is given or has been previously obtained. If

$$A^{-1} = C = \begin{pmatrix} C_{11} & C_{12} \\ C_{21} & C_{22} \end{pmatrix}, \tag{2.23.2}$$

then

$$AC = \begin{pmatrix} I_{11} & 0_{12} \\ 0_{21} & I_{22} \end{pmatrix},$$

where the I_{ii} and the 0_{ij} are the identity and null matrices of dimensions that correspond to the partitioning. Hence if we multiply out, we obtain

$$\begin{aligned} A_{11}C_{11} + A_{12}C_{21} &= I_{11}, \\ A_{11}C_{12} + A_{12}C_{22} &= 0_{12}, \\ A_{21}C_{11} + A_{22}C_{21} &= 0_{21}, \\ A_{21}C_{12} + A_{22}C_{22} &= I_{22}. \end{aligned} \tag{2.23.3}$$

The following solution of the system can be verified:

$$\begin{aligned} C_{22} &= (A_{22} - A_{21}A_{11}^{-1}A_{12})^{-1}, \\ C_{12} &= -A_{11}^{-1}A_{12}C_{22}, \\ C_{21} &= -C_{22}A_{21}A_{11}^{-1}, \\ C_{11} &= A_{11}^{-1}(I_{11} - A_{12}C_{21}). \end{aligned} \tag{2.23.4}$$

If A_{22} is a scalar, A_{12} a column vector, and A_{21} a row vector, then C_{22} is a scalar, C_{12} a column vector, and C_{21} a row vector, and the inverse required for C_{22} is trivial. The matrices are to be obtained in the order given, and it is to be noted that the product $A_{11}^{-1}A_{12}$ occurs three times, and can be calculated at the outset. If A is symmetric, then

$$C_{21} = C_{12}^{\mathsf{T}}.$$

In any event the matrix C_{22} is of lower dimension than C, and the required inversion more easily performed. It is therefore feasible to invert in sequence the matrices

$$(\alpha_{11}), \begin{pmatrix} \alpha_{11} & \alpha_{12} \\ \alpha_{21} & \alpha_{22} \end{pmatrix}, \begin{pmatrix} \alpha_{11} & \alpha_{12} & \alpha_{13} \\ \alpha_{21} & \alpha_{22} & \alpha_{23} \\ \alpha_{31} & \alpha_{32} & \alpha_{33} \end{pmatrix}, \ldots,$$

each matrix in the sequence taking the place of the A_{11} in the inversion of the next.

In the following section it will be shown how from a known inverse A^{-1} one can find the inverse of a matrix A' which differs from A in only a single element or in one or more rows and columns. It is clearly possible to start from any matrix whose inverse is known, say the identity I, and by modifying a row or a column at a time, obtain finally the inverse required. However, these formulas have importance for their own sake, and will be considered independently.

2.24. *Inverting Modified Matrices.* The following formulas can be verified directly:

$$(2.24.1) \qquad (A + USV^{\mathsf{T}})^{-1} = A^{-1} - A^{-1}US(S + SV^{\mathsf{T}}A^{-1}US)^{-1}SV^{\mathsf{T}}A^{-1},$$

$$(2.24.2) \qquad (A + US^{-1}V^{\mathsf{T}})^{-1} = A^{-1} - A^{-1}U(S + V^{\mathsf{T}}A^{-1}U)^{-1}V^{\mathsf{T}}A^{-1},$$

provided the indicated inverses exist and the dimensions are properly matched. Thus A and S are square matrices, U and V rectangular. In particular, if U and V are column vectors u and v, and if the scalar $S = 1$, then

$$(2.24.3) \qquad (A + uv^{\mathsf{T}})^{-1} = A^{-1} - (A^{-1}u)(v^{\mathsf{T}}A^{-1})/(1 + v^{\mathsf{T}}A^{-1}u).$$

If $u = e_i$, then the ith row of uv^{T} is v^{T}, and every other row is null; if $v = e_i$, then the ith column of uv^{T} is u, and every other column is null; if $u = \sigma e_i$, where σ is some scalar, and $v = e_j$, then the element in the ith row and jth column of uv^{T} is σ, and every other element is zero. In the last instance, $v^{\mathsf{T}}A^{-1}u$ is $\sigma(\alpha^{-1})_{ij}$, where $(\alpha^{-1})_{ij}$ is the indicated element of A^{-1}. We have then the interesting corollary that the matrix $A + uv^{\mathsf{T}}$ becomes singular when $\sigma = -1/(\alpha^{-1})_{ij}$.

2.25. *Matrices with Complex Elements.* If the coefficients of a system of linear equations are complex, then the matrix can be written in the form $A + iB$, where A and B have only real elements. In general we may expect the solution to have complex elements. Hence the equations can be written in the form

$$(A + iB)(x + iy) = c + id,$$

where the vectors x, y, c, and d are all real. However, this is equivalent to

$$(Ax - By) + i(Ay + Bx) = c + id,$$

and since the real parts and the pure imaginary parts must be separately equal, this is equivalent to the real system of order $2n$:

$$\begin{aligned} Ax - By &= c, \\ Bx + Ay &= d, \end{aligned}$$

or

$$\begin{pmatrix} A & -B \\ B & A \end{pmatrix} \begin{pmatrix} x \\ y \end{pmatrix} = \begin{pmatrix} c \\ d \end{pmatrix}.$$

Thus the complex system of order n is equivalent to a real system of order $2n$, since these steps can be reversed. The complex matrix $A + iB$ is singular if and only if the system with $c + id = 0$ has a nontrivial solution, and this occurs if and only if the real matrix of order $2n$ is singular.

A complex matrix is called Hermitian in case A is symmetric and B skew-symmetric, *i.e.*, in case

$$A^{\mathsf{T}} = A, \qquad B^{\mathsf{T}} = -B.$$

But then the real matrix of order $2n$ can be written

$$\begin{pmatrix} A & B^{\mathsf{T}} \\ B & A \end{pmatrix},$$

and it is symmetric. Hence the complex matrix is Hermitian if and only if the corresponding real matrix is symmetric. A Hermitian matrix is positive definite if and only if for every non-null complex vector $x + iy$ it is true that

$$(x^{\mathsf{T}} - iy^{\mathsf{T}})(A + iB)(x + iy) > 0.$$

This implies that the quantity is real. But if we evaluate the quantity on the left, we obtain

$$x^{\mathsf{T}}(Ax - By) + y^{\mathsf{T}}(Bx + Ay) + i[x^{\mathsf{T}}(Bx + Ay) - y^{\mathsf{T}}(Ax - By)].$$

Since A is symmetric and B skew-symmetric, the quantity within brackets vanishes, and the quantity in question is certainly real whenever the matrix is Hermitian. As for the rest, we have

$$x^{\mathsf{T}}(Ax - By) + y^{\mathsf{T}}(Bx + Ay) = (x^{\mathsf{T}}, y^{\mathsf{T}}) \begin{pmatrix} A & -B \\ B & A \end{pmatrix} \begin{pmatrix} x \\ y \end{pmatrix}.$$

If this is positive for every choice of x and y, then the real matrix of order $2n$ is positive definite. Hence a Hermitian matrix of order n is positive definite if and only if the corresponding real matrix of order $2n$ is positive definite.

Throughout the discussion of methods of inverting matrices and solving systems of equations, we have tacitly assumed that all quantities were real, though in fact many of the processes were equally applicable to the complex case when appropriate changes are made in the wording. However, rather than complicate the exposition, we have preferred to treat only the real case and reduce the complex case to the real.

2.3. Some Comparative Evaluations. For either inverting a matrix or for solving a system of equations, there is no single method that is clearly best for all matrices or all systems. Some matrices may have many null elements, especially those which result from the finite difference approximation to a differential equation. Analysis of the system may show that a method that would be highly inefficient in general would work admirably well for the particular case. On the other hand, if one has many systems to solve, all differing among themselves, it may be more efficient to use the same scheme for all of them than it would be to analyze each system as it arises before deciding upon how to proceed. This is especially true if one is using automatic computing machinery for which the arrangement of the program is a major task.

When computing machinery is used, the method must be adapted to the machine. Generally speaking, the number of multiplications and divisions and the number of recordings of intermediate results together provide a rough over-all estimate of the efficiency of a computational scheme. It is possible to estimate these numbers as functions of n, the order of the system, but the functions may be discontinuous. That is to say, if n is small enough so that all quantities, initial, intermediate, and final, can be retained in the internal memory of the machine, the functions will be of one form. But if n is so large that the auxiliary storage must be utilized, then transfers must be made between the internal memory and the external, and additional operations may be required. In fact, it may be necessary to use an entirely different computational scheme.

2.31. *Operational Counts.* The possible occurrence of null elements will be ignored. With this understanding we consider the number of operations required in the application of some of the methods discussed above.

2.311. *The method of Seidel and the method of relaxation.* The equations can be written in scalar form

$$\alpha_{ii}\xi_i = \eta_i - \sum_{j \neq i} \alpha_{ij}\xi_j,$$

$$\xi_i = \eta_i/\alpha_{ii} - \sum_{j \neq i} (\alpha_{ij}/\alpha_{ii})\xi_j.$$

The $n(n+1)/2$ divisions η_i/α_{ii} and α_{ij}/α_{ii} (for a symmetric matrix) can always be done in advance. Thereafter each correction of a single ξ_i requires $n - 1$ multiplications and a single recording provided the products can be accumulated. For a complete Seidel cycle this is $n(n-1)$ products and n recordings. The number of cycles required, however, depends upon the system, the starting values, and the required accuracy.

2.312. *The method of steepest descent.* One requires at each step the product Ax_{p-1}, or n^2 products, the n^2 products Ar_p, the n products $r_p^{\mathsf{T}}(Ar_p)$, the n products $r_p^{\mathsf{T}}r_p$, the quotient $r_p^{\mathsf{T}}r_p/r_p^{\mathsf{T}}Ar_p = \lambda_p$, and the n products $\lambda_p r_p$, as well as the various sums and differences. Counting multiplications and divisions as equivalent, this is $2n^2 + 2n + 1$ product operations at each step. This is somewhat more than two complete Seidel cycles performed on a prepared system in which the indicated quotients have been taken in advance. As for recordings, one requires at least the n elements r_p, the n elements Ar_p, the products $r_p^{\mathsf{T}}r_p$ and $r_p^{\mathsf{T}}Ar_p$, the quotient λ_p, and the n elements x_p, or altogether $3n + 3$ quantities, as compared with n for a complete Seidel cycle.

2.313. *The Stiefel-Hestenes method.* The general step requires n^2 products for Ax_i and n^2 for Ar_i, n for $r_i^{\mathsf{T}}r_i$ and n more for $r_i^{\mathsf{T}}Ar_i$, and individual products and quotients for α_i and β_i. In principle the iteration terminates in n steps with something over $2n^3$ product operations. Each additional projection for reducing round-off requires something over $2n^2$ product operations. In recordings, each step requires at least the n elements of x_i, the n elements of r_i, the n elements of Ar_i, and a few scalars. This is something over $3n$ recordings per step or $3n^2$ for a complete cycle. In addition r_{i-1} must be carried over from the previous step.

2.314. *The Crout factorization.* One is to write

$$(A, y) = L(W, z),$$

where L is unit lower triangular and W upper triangular, then solve the triangular system

$$Wx = z.$$

Consider Eqs. (2.201.6) supposing W_{11} and L_{11} to be of order i, W_{22} a scalar, and $L_{22} = 1$. Then W_{22} requires i products, as does each element in W_{23}, making $i(n-i)$ products. Each of the $n - i - 1$ elements of L_{32} requires i products in $L_{31}W_{12}$ and the quotient of the result by W_{22}, making $(i+1)(n-i-1)$ product operations. Finally, from Eq. (2.21.6) one requires i products for z_2. This is $n(2i+1) - 2i^2 - i - 1$ product operations in all. Summing from $i = 0$ to $i = n - 1$, we get a total of $n(n^2-1)/3$ products to be formed. Solving the triangular system requires a quotient for ξ_n, a product and a quotient for ξ_{n-1}, . . . , or altogether $n(n+1)/2$ product operations. Altogether we have a total of $n(2n+1)(n+1)/6$ product operations, or something over $n^3/3$.

The recordings required are the triangular matrices L and W and the vectors z and x, or $n^2 + 2n$ quantities altogether.

To iterate the process in order to reduce round-off, formation of $r_0 = y - Ax_0$ requires n^2 products; $L^{-1}r_0$ requires $n(n-1)/2$ (all multiplications, no divisions since L is unit lower triangular); and $W^{-1}L^{-1}r_0$ requires $n(n+1)/2$, or altogether $2n^2$ products and at least $3n$ recordings. These give the corrections to x_0, so that an additional n recordings of x_1 itself are required. If the matrix is symmetric, the operations are reduced by nearly one-half.

2.315. *Orthogonalization.* In forming $RV = A$, $R^{\mathsf{T}}R = D^2$ as in §2.22, suppose i columns of A have been orthogonalized. As in (2.201.9) and (2.201.10), i elements of the next column of v are to be found, each requiring n products and a division. Then the next column of R requires $n(n+i)$ products, and n more are required for the next element of D^2. Hence to orthogonalize the columns of A requires a total of

$$n(4n^2 + n - 1)/2,$$

or approximately $2n^3$ products. Beyond this one requires $R^{\mathsf{T}}y$ with n^2 products; n more products in multiplying this by D^{-2}; and $n(n-1)/2$ more in solving the triangular system $Vx = D^{-2}R^{\mathsf{T}}y$. Altogether it amounts to $2n^2(n+1)$ products. For recordings we require at least the n^2 elements of R; $n(n-1)/2$ elements of V; n elements of D^2; and n elements each of $R^{\mathsf{T}}y$, of $D^{-2}R^{\mathsf{T}}y$, and of x. This makes $n(3n+7)/2$ recordings.

2.316. *Inverting a modified matrix.* In Eq. (2.24.3), if $u = e_i$, and A^{-1} is given, then the inversion of $A + uv^{\mathsf{T}}$ requires n^2 multiplications for $v^{\mathsf{T}}A^{-1}$; n quotients of $1 + v^{\mathsf{T}}A^{-1}u$ into the vector $v^{\mathsf{T}}A^{-1}$ (or into $A^{-1}u$); n^2 products for multiplying the column vector by the row vector. Hence there are $2n^2 + n$ product operations for modifying the inverse when a single column of the matrix A is modified. If one builds up the inverse by modifying a column at a time, then in the worst case $n^2(2n+1)$ products are required. However, if one starts with the identity, then in the first step, since

$$(I + uv^{\mathsf{T}})^{-1} = I - uv^{\mathsf{T}}/(1 + v^{\mathsf{T}}u),$$

only n quotients are needed and no other products. The new inverse differs from I in only the ith row, so that many zeros remain if n is large. If the programing takes advantage of the presence of the zeros, the number of products is reduced considerably.

Once the inverse is taken, if a set of equations are to be solved, an additional n^2 products are needed.

2.4. Bibliographic Notes. Most of the methods described here are old, and have been independently discovered several times. A series of papers by Bodewig (1947, 1947–1948) compares various methods, including some not described here, with operational counts and other points of comparisons, and it contains a list of sources. Forsythe (1952) has compiled an extensive bibliography with a classification of methods. A set of mimeographed notes (anonymous) was distributed by the Institute for Numerical Analysis, and interpreted several of the standard iterative methods as methods of successive projection, much as is done here. This includes the iterative resolution of y along the columns of A, which is attributed to C. B. Tompkins. The same method had been given by A. de la Garza in a personal communication to the author in 1949.

Large systems of linear inequalities have important military and economic applications. Agmon (1951) has developed a method of relaxation for such systems, which reduces to the ordinary method of relaxation when the inequalities become equations.

Conditions for convergence of iterations are given by von Mises and Pollaczek-Geiringer (1929), Stein (1951*b*, 1952), Collatz (1950), Reich (1949), Ivanov (1939), and Plunkett (1950).

On the orthogonalizations of residuals see Lanczos (1950, 1951). For other discussions of the method of Lanczos, Stiefel, and Hestenes see Hestenes and Stein (1951), Stiefel (1952), Hestenes and Stiefel (1952).

Crout (1941) pointed out the possibility of economizing on the recordings in the triangular factorization. The author is indebted to James Alexander and Jean Hall for pointing out the possibility of a similar economy in recording in the use of Jordan's method. Turing (1948) discusses round-off primarily with reference to these two methods and refers to the former as the "unsymmetric Choleski method." The formulas apply to the assessment of an inverse already obtained. On the other hand, von Neumann and Goldstine (1947) obtain a priori estimates, but in terms of largest and smallest proper values. They assume the method of elimination to be applied to the system with positive definite matrix A or to the system which has been premultiplied by A^T to make the matrix positive definite. See also Bargmann, Montgomery, and von Neumann (1946).

Dwyer (1951) devotes some little space to a discussion of errors and gives detailed computational layouts. Hotelling (1943) deals with a variety of topics, including errors and techniques. Lonseth (1947) gives the essential formulas for propagated error.

Sherman and Morrison (1949, 1950), Woodbury (1950), and Bartlett (1951) give formulas for the inverse of a modified matrix, and Sherman applies these to inversion in general.

A number of detailed techniques appear in current publications by the International Business Machines Corporation, especially in the reports of the several Endicott symposia.

In the older literature, reference should be made especially to Aitken (1932*b*, 1936–1937*a*).

An interesting and valuable discussion of measures of magnitude is given by Fadeeva (1950, in Benster's translation). In particular Fadeeva suggests the association of the measures designated here as $c(A)$ and $b(x)$.

The use of Chebyshev polynomials for accelerating convergence is described by Grossman (1950) and Gavurin (1950).

On general theory the literature is abundant. Muir (1906, 1911, 1920, 1923) is almost inexhaustible on special identities and special forms, and many of the results are summarized in Muir and Metzler (1930). Frazer, Duncan, and Collar (1946) emphasize computational methods. MacDuffee (1943) is especially good on the normal forms and the characteristic equation.

CHAPTER 3

NONLINEAR EQUATIONS AND SYSTEMS

3. Nonlinear Equations and Systems

In the present chapter matrices and vectors will occur only incidentally. Consequently the convention followed in the last chapter of representing scalars only by Greek letters will be dropped here. The objective of this chapter is to develop methods for the numerical approximation to the solutions of nonlinear equations and systems of equations. With systems of nonlinear equations, the procedure is generally to obtain a sequence of systems of linear equations whose solutions converge to the required values, or else a sequence of equations in a single unknown.

A major objective in the classical theory of equations is the expression in closed form of the solutions of an equation and the determination of conditions under which such expressions exist. Aside from the fact that only a limited class of equations satisfy these conditions, the closed expressions themselves are generally quite unmanageable computationally. Thus it is easy to write the formula for the real solution of $x^9 = 9$, but if one needs the solution numerically to very many decimals, it is most easily obtained by solving the equation numerically by one of the methods which will be described. Nevertheless, certain principles from the theory of equations will be required, and we begin by developing them.

To begin with, we shall be concerned with an algebraic equation, which is one that can be written

$$P(x) = 0, \tag{3.0.1}$$

where

$$P(x) \equiv a_0 x^n + a_1 x^{n-1} + \cdots + a_{n-1} x + a_n, \tag{3.0.2}$$

a polynomial of degree n. Ordinarily we shall suppose that $a_0 \neq 0$, since otherwise the polynomial or the equation would be of some lower degree. This being the case, we can always, if we wish, write

$$p(x) \equiv a_0^{-1} P(x) \equiv x^n + \alpha_1 x^{n-1} + \cdots + \alpha_n, \tag{3.0.3}$$

and the equation

$$p(x) = 0 \tag{3.0.4}$$

is equivalent to the original one.

3.01. *The Remainder Theorem; the Symmetric Functions.* A basic theorem is the remainder theorem, which states that, if the polynomial $P(x)$ is divided by $x - r$, where r is any number, then the remainder is the number $P(r)$. Thus suppose

$$P(x) \equiv (x - r)Q(x) + R, \tag{3.01.1}$$

where $Q(x)$ is the quotient of degree $n - 1$, and R is the constant remainder. This is an identity, valid for all values of x. Hence in particular it is valid for $x = r$, which leads to

$$P(r) = R. \tag{3.01.2}$$

A corollary is the factor theorem, which states that, if r is a zero of the polynomial $P(x)$, that is, a root of Eq. (3.0.1), then $x - r$ divides $P(x)$ exactly. Conversely, if $x - r$ divides $P(x)$, then r is a zero of $P(x)$. For by (3.01.2) if r is a zero of $P(x)$, then $R = 0$, and by (3.01.1) the division by $x - r$ is exact. The converse is obvious.

The fundamental theorem of algebra states that every algebraic equation has a root. The proof is a bit long and will not be given here. But it follows from that and the factor theorem that an algebraic equation of degree n has exactly n roots (which, however, are not necessarily distinct). For by the fundamental theorem (3.0.1) has a root, which we may call x_1. By the factor theorem we can write

$$P(x) \equiv (x - x_1)Q_1(x).$$

But $Q_1 = 0$ is an algebraic equation of degree $n - 1$; it has a root, say x_2, and hence

$$Q_1(x) \equiv (x - x_2)Q_2(x).$$

Eventually we get

$$Q_{n-1}(x) \equiv (x - x_n)Q_n,$$

where Q_{n-1} is linear and Q_n a constant. But then

$$P(x) \equiv (x - x_1)(x - x_2) \cdots (x - x_n)Q_n, \tag{3.01.3}$$

and not only x_1 but also $x_2, \ldots, x_n$ are roots of $P = 0$. But there can be no others. For if x_{n+1} were different from $x_1, \ldots, x_n$, and also a root of $P = 0$, then it would be true that

$$0 = P(x_{n+1}) = (x_{n+1} - x_1) \cdots (x_{n+1} - x_n)Q_n.$$

But if $x_{n+1} - x_i \neq 0$ for $i = 1, \ldots, n$, then $Q_n = 0$ and $P \equiv 0$ identically. Hence there are exactly n roots, and the theorem is proved. As a partial restatement we can say that, if a polynomial of degree not greater than n is known to vanish for $n + 1$ distinct values of x, then this polynomial vanishes identically.

Now consider the factorization (3.01.3). If we multiply out on the right, the polynomial we obtain must be precisely the polynomial P. But the coefficient of x^n on the right is Q_n, so therefore $Q_n = a_0$. By examining the other coefficients in turn, we find

$$\begin{aligned} a_1 &= -a_0 \sum_i x_i, \\ a_2 &= a_0 \sum_{i<j} x_i x_j, \\ a_3 &= -a_0 \sum_{i<j<k} x_i x_j x_k, \\ &\cdots\cdots\cdots\cdots \\ a_n &= (-1)^n a_0 x_1 x_2 \cdots x_n. \end{aligned} \tag{3.01.4}$$

Thus $(-1)^h a_h/a_0$ is the sum of the $\binom{n}{h}$ products of the r's taken h at a time. These sums are called the elementary symmetric functions of the roots. They are symmetric because interchanging any pair of the roots leaves the value of the function unchanged. It is a theorem that any rational symmetric function is expressible as a rational function of the elementary symmetric functions. The general theorem will not be required here, but special cases will appear. Consequently we introduce the notation σ_h for the elementary symmetric function of degree h:

$$a_h = (-1)^h a_0 \sigma_h = a_0 \alpha_h. \tag{3.01.5}$$

Of particular importance will be the sums of powers:

$$s_h = \sum_i x_i^h, \tag{3.01.6}$$

where h is any integer, positive, negative, or zero. For $h = 0$ we have $s_0 = n$. Expressions for these in terms of the elementary symmetric functions will be given later.

We conclude this section by noting that

$$\begin{aligned} x^n P(1/x) &\equiv a_n x^n + x_{n-1} x^{n-1} + \cdots + a_0 \\ &\equiv a_0(1 - x x_1)(1 - x x_2) \cdots (1 - x x_n). \end{aligned} \tag{3.01.7}$$

Hence the equation

$$a_n x^n + \cdots + a_0 = 0 \tag{3.01.8}$$

has the n roots $x_1^{-1}, \ldots, x_n^{-1}$, provided every $x_i \neq 0$. We call it the reciprocal equation. Then

$$a_{n-1} = -a_n \sum_i x_i^{-1},$$

(3.01.9)
$$a_{n-2} = a_n \sum_{i<j} x_i^{-1}x_j^{-1},$$
$$\cdots\cdots\cdots\cdots$$

3.02. *The Derivative Equations.* The derivative equations are those formed from the derivatives of P:

(3.02.1)
$$\begin{aligned} P'(x) &= 0, \\ P''(x) &= 0, \\ &\cdots\cdots \end{aligned}$$

If P is a real polynomial, *i.e.*, if all its coefficients are real, then the real roots of the derivative equations have important relations to the real roots of the original.

If we set $x = z + r$, where r is any real number, forming $P(z + r)$, expand each power of $z + r$, and collect like powers of z, the result is a polynomial in z of degree n. If in this polynomial we now replace z by $x - r$ but without expanding powers of $x - r$, we obtain an expression of the form

$$P(x) \equiv c_n + c_{n-1}(x - r) + c_{n-2}(x - r)^2 + \cdots + c_0(x - r)^n,$$

where the c's are the constant coefficients of the several powers of z in $P(z + r)$, and in particular, $c_0 = a_0$. This is an identity which therefore holds when we differentiate on both sides and continues to hold when we give to x any fixed value. In particular if we set $x = r$, we find (as in the proof of the remainder theorem) that

$$P(r) = c_n,$$

and if we first differentiate i times and then set $x = r$,

$$P^{(i)}(r) = i!c_{n-i}.$$

Hence

(3.02.2)
$$P(x) \equiv P(r) + (x - r)P'(r) + (x - r)^2P''(r)/2! + \cdots + (x - r)^nP^{(n)}(r)/n!.$$

This is Taylor's series for polynomials.

If $P(r) = 0$ but $P'(r) \neq 0$, then $x - r$ is a factor of $P(x)$, but $(x - r)^2$ is not. Hence r is a simple root of $P = 0$. But if $P(r) = P'(r) = 0$, then r is at least a double root. Hence a root of $P = 0$ is a multiple root if and only if it is also a root of $P' = 0$. In fact, r is a root of multiplicity m if and only if

$$0 = P(r) = P'(r) = \cdots = P^{(m-1)}(r) \neq P^{(m)}(r).$$

In that case (3.02.2) becomes

$$P(x) \equiv (x - r)^m P^{(m)}(r)/m! + \cdots .$$

Moreover, r is a root of multiplicity $m - 1$ of $P' = 0$, of multiplicity $m - 2$ of $P'' = 0, \ldots .$

If P is a real polynomial, then between consecutive real roots of $P = 0$ there is an odd number of roots of $P' = 0$. In particular, there is at least one. This is Rolle's theorem. For suppose x_1 is a root of multiplicity m_1, x_2 of multiplicity m_2. Then we can write

$$P(x) \equiv (x - x_1)^{m_1}(x - x_2)^{m_2}Q(x),$$

where Q does not vanish at x_1 or x_2 or anywhere between. Since Q is a polynomial, it must retain the same sign throughout the interval. Now

$$P'(x) = (x - x_1)^{m_1-1}(x - x_2)^{m_2-1}q(x),$$

where

$$q(x) = m_1(x - x_2)Q + m_2(x - x_1)Q + (x - x_1)(x - x_2)Q'.$$

Hence

$$\begin{aligned} q(x_1) &= m_1(x_1 - x_2)Q(x_1), \\ q(x_2) &= m_2(x_2 - x_1)Q(x_2). \end{aligned}$$

But $m_1Q(x_1)$ and $m_2Q(x_2)$ have the same sign, whereas

$$x_1 - x_2 = -(x_2 - x_1).$$

Hence $q(x_1)$ and $q(x_2)$ have opposite signs, and $q(x)$ must vanish an odd number of times between x_1 and x_2. Hence the same is true of $P'(x)$.

If we differentiate the factored form (3.01.3) of $P(x)$, we obtain for P' a sum of products of $n - 1$ factors each. In fact, each product can be written as $P(x)/(x - x_i)$ for some i. Hence

$$P'(x) \equiv P(x) \sum_i (x - x_i)^{-1}, \tag{3.02.3}$$

or

$$P'(x)/P(x) \equiv \sum_i (x - x_i)^{-1} \equiv p'(x)/p(x). \tag{3.02.4}$$

But for x sufficiently large

$$(x - x_i)^{-1} \equiv x^{-1} + x_i x^{-2} + x_i^2 x^{-3} + \cdots$$

and

$$\Sigma(x - x_i)^{-1} \equiv nx^{-1} + s_1x^{-2} + s_2x^{-3} + \cdots .$$

Since

$$\begin{aligned} p(x) &\equiv x^n - \sigma_1x^{n-1} + \sigma_2x^{n-2} - \cdots , \\ p'(x) &\equiv nx^{n-1} - (n - 1)\sigma_1x^{n-2} + (n - 2)\sigma_2x^{n-3} + \cdots . \end{aligned}$$

Hence if we multiply $p(x)$ by $\Sigma(x - x_i)^{-1}$ and equate the coefficients to those of $p'(x)$, we get the relations

$$
\begin{aligned}
s_1 - \sigma_1 &= 0, \\
s_2 - s_1\sigma_1 + 2\sigma_2 &= 0, \\
s_3 - s_2\sigma_1 + s_1\sigma_2 - 3\sigma_3 &= 0, \\
\cdots\cdots\cdots\cdots\cdots\cdots \\
s_n - s_{n-1}\sigma_1 + \cdots + (-1)^n n\sigma_n &= 0, \\
s_{n+p} - s_{n+p-1}\sigma_1 + \cdots + (-1)^n s_p\sigma_n &= 0.
\end{aligned}
\tag{3.02.5}
$$

These are Newton's identities, expressing recursively the sums of powers of the roots as polynomials in the elementary symmetric functions, and hence as rational functions of the coefficients. If one applies the same relations to the reciprocal equation (3.01.8), one obtains the sums of the powers with negative exponents.

If we set

$$Q(z) \equiv \Pi(1 - x_i z) \equiv 1 - \sigma_1 z + \sigma_2 z^2 - \cdots \pm \sigma_n z^n, \tag{3.02.6}$$

and expand

$$1/Q(z) \equiv \Pi(1 - x_i z)^{-1} \equiv 1 + S_1 z + S_2 z^2 + \cdots, \tag{3.02.7}$$

the coefficients S_p of this expansion are symmetric functions of the roots:

$$
\begin{aligned}
S_1 &= x_1 + x_2 + \cdots + x_n, \\
S_2 &= x_1^2 + x_1 x_2 + \cdots, \\
S_3 &= x_1^3 + x_1^2 x_2 + x_1 x_2 x_3 + \cdots,
\end{aligned}
\tag{3.02.8}
$$

the so-called "complete" symmetric functions. Since

$$(1 - \sigma_1 z + \sigma_2 z^2 - \cdots)(1 + S_1 z + S_2 z^2 + \cdots) = 1,$$

on comparing coefficients, one obtains

$$
\begin{aligned}
S_1 - \sigma_1 &= 0, \\
S_2 - \sigma_1 S_1 + \sigma_2 &= 0, \\
S_3 - \sigma_1 S_2 + \sigma_2 S_1 - \sigma_3 &= 0, \\
\cdots\cdots\cdots\cdots\cdots\cdots
\end{aligned}
\tag{3.02.9}
$$

Of the three sets of symmetric functions, each set can be expressed in terms of the others by means of these equations.

3.03. *Vandermonde Determinants.* It is easy to write in determinantal form an equation having specified roots. To illustrate for the case $n = 3$, if x_1, x_2, and x_3 are all distinct, the equation

$$
\begin{vmatrix}
1 & 1 & 1 & 1 \\
x_1 & x_2 & x_3 & x \\
x_1^2 & x_2^2 & x_3^2 & x^2 \\
x_1^3 & x_2^3 & x_3^3 & x^3
\end{vmatrix} = 0
\tag{3.03.1}
$$

has these and only these roots. The determinant, therefore, whose expansion is a cubic polynomial in x is equal to some constant times $(x - x_1)(x - x_2)(x - x_3)$ by application of the factor theorem. If we were to regard x_3 as the variable instead of x, and apply the factor theorem again, it appears that $(x_3 - x_2)$ and $(x_3 - x_1)$ are also factors of the expansion of the determinant. Likewise, regarding x_2 as the variable, $(x_2 - x_1)$ appears as an additional factor. Hence the determinant is equal to the product $(x_3 - x_2)(x_3 - x_1)(x_2 - x_1)(x - x_3)(x - x_2)(x - x_1)$, possibly multiplied by some factor as yet undetermined. However, the determinant is a cubic polynomial in x, and so has no other factors containing x; it is also a cubic in x_3, and so can have no other factors containing x_3; nor by the same rule can it have other factors containing x_2 or x_1. Hence any factor not yet found is a constant, independent of x or any of the x_i. But the expansion of the determinant contains the term $x_2x_3^2x^3$ once from the principal diagonal, and the expansion of the product contains this product also. Hence there is no other factor, and

$$\begin{vmatrix} 1 & 1 & 1 & 1 \\ x_1 & x_2 & x_3 & x \\ x_1^2 & x_2^2 & x_3^2 & x^2 \\ x_1^3 & x_2^3 & x_3^3 & x^3 \end{vmatrix} \equiv (x_3 - x_2)(x_3 - x_1)(x_2 - x_1)(x - x_3)(x - x_2)(x - x_1).$$

The coefficient of x^3 is

$$\begin{vmatrix} 1 & 1 & 1 \\ x_1 & x_2 & x_3 \\ x_1^2 & x_2^2 & x_3^2 \end{vmatrix} = (x_3 - x_2)(x_3 - x_1)(x_2 - x_1).$$

Such a determinant is called an alternant, or an elementary Vandermonde determinant. The negative of the coefficient of x^2 is

$$\begin{vmatrix} 1 & 1 & 1 \\ x_1 & x_2 & x_3 \\ x_1^3 & x_2^3 & x_3^3 \end{vmatrix} = \sigma_1 \begin{vmatrix} 1 & 1 & 1 \\ x_1 & x_2 & x_3 \\ x_1^2 & x_2^2 & x_3^2 \end{vmatrix}.$$

Again, the coefficient of x is

$$\begin{vmatrix} 1 & 1 & 1 \\ x_1^2 & x_2^2 & x_3^2 \\ x_1^3 & x_2^3 & x_3^3 \end{vmatrix} = \sigma_2 \begin{vmatrix} 1 & 1 & 1 \\ x_1 & x_2 & x_3 \\ x_1^2 & x_2^2 & x_3^2 \end{vmatrix},$$

and the negative of the constant term is

$$\begin{vmatrix} x_1 & x_2 & x_3 \\ x_1^2 & x_2^2 & x_3^2 \\ x_1^3 & x_2^3 & x_3^3 \end{vmatrix} = \sigma_3 \begin{vmatrix} 1 & 1 & 1 \\ x_1 & x_2 & x_3 \\ x_1^2 & x_2^2 & x_3^2 \end{vmatrix}.$$

The equation whose roots are x_1, $x_2 = x_1$ and $x_3 \neq x_1$ is

$$\begin{vmatrix} 1 & 0 & 1 & 1 \\ x_1 & 1 & x_3 & x \\ x_1^2 & 2x_1 & x_3^2 & x^2 \\ x_1^3 & 3x_1^2 & x_3^3 & x^3 \end{vmatrix} = 0.$$

For if we call the determinant $P(x)$, then $P(x_1) = P'(x_1) = 0$, and $P(x_3) = 0$. Likewise the equation whose roots are x_1, $x_2 = x_1$ and $x_3 = x_1$ is

$$\begin{vmatrix} 1 & 0 & 0 & 1 \\ x_1 & 1 & 0 & x \\ x_1^2 & 2x_1 & 1 & x^2 \\ x_1^3 & 3x_1^2 & 3x_1 & x^3 \end{vmatrix} = 0.$$

These representations are easily generalized, but the notation becomes cumbersome.

3.04. *Synthetic Division.* We consider now some further useful consequences of the remainder theorem. First we observe that $P(r)$, which is the remainder after dividing $P(x)$ by $x - r$, is most readily evaluated by evaluating sequentially

$$\begin{gathered} a_0r + a_1, \\ (a_0r + a_1)r + a_2, \\ [(a_0r + a_1)r + a_2]r + a_3, \\ \cdots\cdots\cdots\cdots\cdots \end{gathered}$$

with $P(r)$ obtained as the final step. This process can be systematized by writing the system

$$\begin{array}{cccccc|l} a_0 & a_1 & a_2 & \cdots & a_n & & r \\ & b_0r & b_1r & \cdots & b_{n-1}r & & \\ \hline b_0 & b_1 & b_2 & \cdots & R & & \end{array}$$

where $b_0 = a_0$, and in general every number along the bottom row is the sum of the two above it. The r is written in the upper right-hand box merely as a convenient reminder.

Having written this, we now observe that the b's are the coefficients of the quotient

$$Q(x) = b_0x^{n-1} + b_1x^{n-2} + \cdots + b_{n-1}.$$

One way to see this is to note that, when in ordinary long division we divide $P(x)$ by $x - r$, the b_1 is exactly the remainder we get after dividing $a_0x + a$ by $x - r$, the b_2 is the remainder after dividing $a_0x^2 + a_1x + a_2$, and so on sequentially. We have written above merely a scheme for evaluating these remainders in sequence.

If the coefficients of the equation $P(x) = 0$ are all integers, it is possible to obtain all its rational roots by inspection and a few synthetic divisions. This is a help even when the rational roots are of no interest for themselves, since for every known rational root the degree of the equation can be lowered by one. If we examine the scheme for synthetic division, we can see that, if r is an integer and the a's are all integers, then the b's are all integers. If r is a root, then $R = 0$, so that $a_n = -b_{n-1}r$. Hence r is a factor of a_n. Thus if the equation has any integral root, the root is a divisor of the constant term. More generally, if $P(x) = 0$ is a polynomial equation with integral coefficients, and if p/q is a rational root in lowest terms, then p is a divisor of the constant term, and q is a divisor of the leading coefficient.

For suppose r is a fraction p/q in lowest terms. If b_0r is a fraction, say with denominator s, then b_1 is a mixed number whose fractional term has the denominator s. But s cannot divide p since p/q is in lowest terms. Hence b_1r is certainly fractional. By continuing to the end, we conclude that p/q cannot be a root if q does not divide a_0. If we apply the argument to the reciprocal equation (3.01.8), we conclude that p must divide a_n. Since there are only a finite number of possible choices for p and q, these can be examined one by one.

In some of the numerical methods of evaluating roots of polynomial equations, and for other purposes too, one often starts with a polynomial $P(x)$, replaces x by $z + r$, and wishes to evaluate the coefficients of the polynomial $P(z + r)$ as in §3.02. For example, r might be a close approximation to a desired root of $P(x) = 0$, and we wish to replace the equation by one in z for which the desired root z is as small as possible. This is done in both Horner's method and in Newton's method, which will be described later.

As in deriving (3.02.2), we write

$$P(x) = c_0(x - r)^n + c_1(x - r)^{n-1} + \cdots + c_{n-1}(x - r) + c_n,$$

where $c_n = P(r)$, so that c_n is the remainder after dividing $P(x)$ by $x - r$. Again if we write

$$\begin{aligned} P(x) &= (x - r)Q(x) + c_n \\ &= (x - r)[c_0(x - r)^{n-1} + \cdots + c_{n-1}] + c_n, \end{aligned}$$

it is clear further that c_{n-1} is the remainder after dividing the quotient by $x - r$, ... Hence we extend our synthetic division scheme as follows:

$$
\begin{array}{cccccc|l}
a_0 & a_1 & \cdots & a_{n-2} & a_{n-1} & a_n & r \\
 & b_0 r & & b_{n-3} r & b_{n-2} r & b_{n-1} r & \\
\hline
b_0 & b_1 & \cdots & b_{n-2} & b_{n-1} & c_n & \\
 & b'_0 r & & b'_{n-3} r & b'_{n-2} r & & \\
\hline
b'_0 & b'_1 & \cdots & b'_{n-2} & c_{n-1} & & \\
 & b''_0 r & \cdots & b''_{n-3} r & & & \\
\hline
b''_0 & b''_1 & \cdots & c_{n-2} & & &
\end{array}
$$

At each division we cut off the final remainder and repeat the synthetic division with the preceding coefficients. This is sometimes called reducing the roots of the equation, since every root z_i of the equation $P(z + r) = 0$ is r less than a corresponding root x_i of $P(x) = 0$.

In solving equations by Newton's or Horner's method, it is first necessary to localize the roots roughly, and the first step in this is to obtain upper and lower bounds for all the real roots. If in the process of reducing the roots by a positive r the b's and the c of any line are all positive, as well as all the c's previously calculated, then necessarily all succeeding b's and c's will be positive. Hence the transformed equation will have only positive coefficients and hence can have no positive real roots. Hence the original equation can have no real roots exceeding r. Hence any positive number r is an upper bound to the real roots of an algebraic equation if in any line of the scheme for reducing the roots by r all numbers are positive along with the c's already calculated.

3.05. *Sturm Functions; Isolation of Roots.* The condition just given is sufficient for assuring us that r is an upper bound to the roots of $P = 0$, but it is not necessary. In particular if all coefficients of P are positive, the equation can have no positive roots. This again is a sufficient but not a necessary condition. A condition that is both necessary and sufficient will be derived in this section. In fact, we shall be able to tell exactly how many real roots lie in any interval. However, since it is somewhat laborious, some other weaker, but simpler, criteria will be given first.

Suppose r is an m-fold root so that

$$P(x) \equiv (x - r)^m P^{(m)}(r)/m! + \cdots .$$

Since $P^{(m)}(r) \neq 0$, there is some interval $(r - \epsilon,\ r + \epsilon)$ sufficiently small so that $P^{(m)}(x)$ is non-null throughout the interval, and $P^{(m-1)}(x)$, $\ldots$, $P'(x)$, $P(x)$ are non-null except at r. Suppose $P^{(m)}(r) > 0$. Then $P^{(m-1)}(x)$ is increasing throughout the interval, and so it must be negative at $r - \epsilon$, positive at $r + \epsilon$. Hence $P^{(m-2)}(x)$ is decreasing, and hence positive, at $r - \epsilon$; increasing, and hence again positive, at $r + \epsilon$. By extending the argument, it appears that the signs at $r - \epsilon$ and at $r + \epsilon$ can be tabulated as follows:

	...	$P^{(m-3)}$	$P^{(m-2)}$	$P^{(m-1)}$	$P^{(m)}$
$r-\epsilon$	...	$-$	$+$	$-$	$+$
$r+\epsilon$	...	$+$	$+$	$+$	$+$

If $P^{(m)}(r) < 0$, then every sign is reversed. If we count the variations in sign in the two sequences, we find that at $r - \epsilon$ there are m variations, for P and P' have opposite signs and present one variation, P' and P'' have opposite signs and present another, ... On the other hand, at $r + \epsilon$ all signs are alike, and there are no variations. Hence the sequence $P, P', P'', \ldots, P^{(m)}$ loses m variations in sign as one passes over an m-fold root of $P = 0$ in the direction of increasing x.

Next, suppose r is an m-fold root of $P^{(h)} = 0$ but not a root of $P^{(h-1)} = 0$ (and it may or may not be a root of $P = 0$). Then $P^{(h+m)}$ remains non-null and of fixed sign, say > 0, throughout some interval $(r - \epsilon, r + \epsilon)$, and from $P^{(h)}$ to $P^{(h+m)}$, m variations in sign are lost. However, we must consider the possible variations $P^{(h-1)}$, $P^{(h)}$, for the sign of $P^{(h)}$ may change, whereas that of $P^{(h-1)}$ does not. But if m is even, the sign of $P^{(h)}$ does not change, so from $P^{(h-1)}$ to $P^{(h+m)}$ there is still a loss of just m variations. If m is odd, $P^{(h)}$ does change, so that from $P^{(h-1)}$ to $P^{(h+m)}$ there is a loss of either $m + 1$ or of only $m - 1$ variations. In either event it is a non-negative even number.

By considering every point r on an interval (a, b), at which P or any of its derivatives may vanish, we conclude that, if V_a and V_b are the numbers of variations in sign at a and b, respectively, displayed by the sequence $P, P', P'', \ldots$, then $V_a - V_b$ exceeds by a non-negative even integer the number of roots of $P = 0$ on the interval.

This is Budan's theorem. In particular, $V_\infty = 0$, while

$$P^{(m)}(0) = m!a_{n-m}.$$

Hence the number of variations in sign in the coefficients is V_0, and this exceeds by a non-negative even integer the number of positive real roots. This is Descartes's rule of signs. In counting the variations, vanishing derivatives are ignored.

This is sometimes sufficient to give all the necessary information. Thus if there is a single variation or none, there will be one root or none; if an odd number, there is at least one.

Exact information, however, is always given by Sturm's theorem. In the following sequence set

$$P_0 = P, \qquad P_1 = P'$$

for uniformity. Divide P_0 by P_1 and denote the remainder with its sign changed by P_2. Divide P_1 by P_2 and denote that remainder with its sign changed by P_3, ... The polynomials $P_0, P_1, P_2, \ldots$ are of progressively lower degree, and the sequence must therefore terminate:

$$
\begin{aligned}
P_0 &= Q_1P_1 - P_2, \\
P_1 &= Q_2P_2 - P_3, \\
&\cdots\cdots\cdots \\
P_{m-2} &= Q_{m-1}P_{m-1} - P_m, \\
P_{m-1} &= Q_mP_m.
\end{aligned} \tag{3.05.1}
$$

(It is understood that any constant $\neq 0$ divides any polynomial exactly.)

Now P_m is the highest common factor of P_0 and P_1. For by the last equation, P_m divides P_{m-1}; by the one before, since P_m divides both itself and P_{m-1}, it divides P_{m-2}; by the one before that, it divides also P_{m-3}, ... Conversely, if p is any polynomial which divides both P_0 and P_1, it therefore divides P_2 by the first equation; by the second, it divides P_3, . . . , and by the next to last, it divides P_m. Thus the statement is established.

It follows that, if $P = 0$ has any multiple roots, they are roots of $P_m = 0$, and all roots of $P_m = 0$ are multiple roots of $P = 0$. Hence we can find all multiple roots by solving an equation of degree lower than the original, remove them from P, and continue with an equation of degree $< n$ whose roots are all simple.

We now suppose this to have been done in advance, that $P = 0$ has only simple roots, and that therefore P_m is a constant $\neq 0$. Then consider the variations in sign presented by the sequence P_i. Suppose $P_i(r) = 0$, where $0 < i < m$. Then $P_{i+1}(r) \neq 0$. For if

$$P_i(r) = P_{i+1}(r) = 0,$$

then P_i and P_{i+1} have a common divisor $x - r$. Since

$$P_{i-1} = Q_iP_i - P_{i+1}, \tag{3.05.2}$$

it follows that P_{i-1} has also $x - r$ as a divisor, and by continuing, we conclude that also P_1 and P_0 have the divisor $x - r$. But then r is a multiple root of $P = 0$, whereas there are no multiple roots. Hence if $P_i(r) = 0$, then $P_{i+1}(r) \neq 0$ and also $P_{i-1}(r) \neq 0$.

Consider again (3.05.2). At $x = r$, $P_{i-1} = -P_{i+1}$. Hence P_{i-1} and P_{i+1} have opposite signs at r and also throughout some small interval $(r - \epsilon, r + \epsilon)$. Hence whatever the signs of P_i at $r - \epsilon$ and $r + \epsilon$, these three polynomials present the same number of variations at $r - \epsilon$ as at $r + \epsilon$.

Now suppose $P_0(r) = 0$. Then $P_1(r) \neq 0$, and $P_1(x)$ keeps a fixed sign in some interval $(r - \epsilon, r + \epsilon)$. If $P_1 > 0$ on this interval, then $P_0(r - \epsilon) < 0 < P_0(r + \epsilon)$, while if $P_1 < 0$, $P_0(r - \epsilon) > 0 > P_0(r + \epsilon)$. In either case P_0 and P_1 present one variation at $r - \epsilon$ and none at $r + \epsilon$.

Hence if V_a and V_b are now the numbers of variations in sign at a and b of the sequence $P_0, P_1, P_2, \ldots, P_m$, then $V_a \geq V_b$ if $a < b$, and $V_a - V_b$

is exactly equal to the number of roots of $P = 0$ on the interval from a to b.

We have proved the theorem only for the case that all roots are simple roots. By modifying the argument slightly, it can be shown that it is true also when there are multiple roots, provided each multiple root is counted only once and not with its multiplicity. This is in contrast to Budan's theorem where a root of multiplicity m was to be counted m times.

In the practical application of Sturm's theorem, if for any $i \leq m$, P_i can be recognized to have no real zeros, then it is unnecessary to continue. Moreover, the sequence can be modified to

$$\begin{aligned} c_0P_0 &= Q_1P_1 - P_2, \\ c_1P_1 &= Q_2P_2 - P_3, \\ &\cdots\cdots\cdots \end{aligned}$$

where $c_0, c_1, \ldots$ are positive constants. Thus if the coefficients of P are integers, one can keep all coefficients of all the P_i integers and, moreover, remove any common numerical factors that may appear. A convenient algorithm is the following: Let $b_0, b_1, b_2, \ldots$ represent the coefficients of P_1 after removal of any common factor, and write the table

$$\begin{matrix} a_0 & a_1 & a_2 & a_3 & \cdots \\ b_0 & b_1 & b_2 & b_3 & \cdots \\ c'_0 & c'_1 & c'_2 & c'_3 & \cdots \end{matrix}$$

where

$$c'_p = -\begin{vmatrix} a_0 & a_{p+1} \\ b_0 & b_{p+1} \end{vmatrix}, \qquad p = 0, 1, \ldots .$$

Obtain, next, the sequence

$$c_0 \quad c_1 \quad c_2 \quad c_3 \quad \cdots$$

from the sequences b and c' by

$$c_p = \begin{vmatrix} b_0 & b_{p+1} \\ c'_0 & c'_{p+1} \end{vmatrix}.$$

Then these are the coefficients of P_2.

3.06. *The Highest Common Factor.* It is now easy to prove a theorem utilized in §2.06. Let P_0 and P_1 be any two polynomials with highest common factor D. There exist polynomials q_0 and q_1 such that

$$P_0q_0 + P_1q_1 \equiv D. \tag{3.06.1}$$

Suppose the degree of P_1 is not greater than that of P_0. In deriving relations (3.05.1), we were supposing that $P_1 = P'_0$, but this supposition is not at all necessary for those relations. The assumption was used only in the proof of Sturm's theorem. Hence for our arbitrary polynomials we can form the successive quotients and remainders (this is called the Euclidean algorithm) and obtain finally their highest common factor $P_m = D$. Write these in the form

$$\begin{aligned} P_2 &= Q_1P_1 - P_0, \\ -Q_2P_2 + P_3 &= -P_1, \\ P_2 - Q_3P_3 + P_4 &= 0, \\ &\cdots\cdots\cdots \\ P_{m-2} - Q_{m-1}P_{m-1} + P_m &= 0, \end{aligned}$$

and regard them as equations in the unknowns $P_2, P_3, \ldots, P_m$, the coefficients Q being supposed known. The matrix of coefficients is unit lower triangular and has determinant 1. Hence P_m is itself expressible as a determinant, in which P_1 and P_0 occur linearly, in the last column only. Hence the expansion of the determinant has indeed the form of the left number of (3.06.1), where q_0 and q_1 are polynomials, expressible in terms of the Q's.

3.07. *Power Series and Analytic Functions.* A few basic theorems on series, and in particular on power series, will be stated here for future reference. Proofs, when not given, can be found in most calculus texts. Consider first a series of any type

$$b_0 + b_1 + b_2 + \cdots \tag{3.07.1}$$

where the b_i are real or complex numbers. Let

$$s_n = b_0 + b_1 + \cdots + b_n \tag{3.07.2}$$

represent the sum of the first $n + 1$ terms. The series (3.07.1) converges to the limit s, provided $\lim_{n\to\infty} s_n = s$, that is, provided for any positive ϵ there exists an N such that $|s_n - s| < \epsilon$ whenever $n > N$. A theorem of Cauchy states that the series (3.07.1) converges if and only if

$$\lim_{n\to\infty} |s_{n+p} - s_n| = 0$$

for every positive integer p. In particular $s_{n+1} - s_n = b_{n+1}$, so that the theorem implies, with $p = 1$, that the individual terms in the series approach zero in absolute value.

If a new series is formed by dropping any finite number of terms from the original, or by introducing any finite number of terms into the original, the two series converge or diverge together.

The series is said to converge absolutely in case the series

$$|b_0| + |b_1| + |b_2| + \cdots$$

of moduli converges. If the series converges absolutely, then it converges since

$$|b_{n+1} + b_{n+2} + \cdots + b_{n+p}| \le |b_{n+1}| + \cdots + |b_{n+p}|.$$

If a series is absolutely convergent, its terms can be rearranged or associated in any fashion without affecting the fact of convergence or the limit to which it converges. This is not true of series which do not converge absolutely. Also if for every n it is true that $|b'_n| \leq |b_n|$, then the series

$$b'_0 + b'_1 + b'_2 + \cdots$$

converges absolutely if (3.07.1) converges absolutely. We shall say then that the series (3.07.1) dominates the other.

Since

$$1 + x + x^2 + \cdots + x^n = (1 - x^{n+1})/(1 - x)$$

identically, it follows that the geometric series, obtained by setting $b_i = \gamma x^i$, converges absolutely for any γ and for any x satisfying $|x| < 1$. In fact, for this series

$$(1 - x)^{-1} - s_n = x^{n+1}(1 - x)^{-1},$$

and when $|x| < 1$, this has the limit zero.

If for a real positive $\beta < 1$ it is true that $\lim_{n\to\infty} |b_{n+1}|/|b_n| = \beta$, then (3.07.1) converges absolutely. For select any positive $\epsilon < 1 - \beta$. Then there is an N such that for $n > N$

$$|b_{n+1}|/|b_n| < \beta + \epsilon.$$

Hence for any p

$$|b_{n+p}|/|b_n| < (\beta + \epsilon)^p.$$

Hence the series

$$|b_n| + |b_{n+1}| + \cdots = |b_n|(1 + |b_{n+1}|/|b_n| + |b_{n+2}|/|b_n| + \cdots)$$

converges since it is dominated by the terms of the geometric series. Hence (3.07.1) converges absolutely.

On the other hand, if for a real positive $\beta > 1$ it is true that

$$\lim_{n\to\infty} |b_{n+1}|/|b_n| = \beta,$$

then the series diverges.

If for some positive $\beta < 1$ it is true that $\lim_{n\to\infty} |b_n|^{1/n} = \beta$, then the series converges absolutely, but if $\beta > 1$, it diverges.

Any convergent series has a term of maximum modulus. For since the sequence of terms b_n has the limit zero, for any ϵ there is a term b_N such that all subsequent terms are less than ϵ in modulus. Choose ϵ less than the modulus of some term in the series. Among the $N + 1$ terms $b_0, \ldots, b_N$, there is one whose modulus is not exceeded by that of any other of these terms, nor is it exceeded by the modulus of any b_n for $n > N$. Hence this is a term of maximum modulus.

When the terms b_i of the series (3.07.1) are functions of x, the limit, when it exists, is also a function of x, and we may write

$$f(x) = b_0(x) + b_1(x) + b_2(x) + \cdots . \tag{3.07.3}$$

The series converges at x to the limit $f(x)$ in case for any ϵ there exists an N such that $|f(x) - s_n(x)| < \epsilon$ whenever $n > N$. Clearly N depends upon ϵ, and the smaller one requires ϵ to be, the larger N must be made. In general N will depend also on x. But if for every x in some region the series converges, and if, moreover, for every ϵ there is an N independent of x in that region, then the series is said to be uniformly convergent in the region.

If every $b_i(x)$ is continuous, and the series is uniformly convergent in some region, then $f(x)$ is continuous in that region. To show that $f(x)$ is continuous at x_0 in that region, one must show that for any positive ϵ there is a δ such that $|f(x) - f(x_0)| < \epsilon$ whenever $|x - x_0| < \delta$. Any finite sum $s_n(x)$ is continuous, whence there exists a δ such that $|s_n(x) - s_n(x_0)| < \epsilon/3$ whenever $|x - x_0| < \delta$. Let N be chosen so that $|f(x) - s_n(x)| < \epsilon/3$ for all x in the region whenever $n > N$. Hence

$$|f(x) - f(x_0)| \le |s_n(x) - s_n(x_0)| + |f(x) - s_n(x)| + |f(x_0) - s_n(x_0)| < \epsilon.$$

In case

$$f(x) = a_0 + a_1 x + a_2 x^2 + \cdots , \tag{3.07.4}$$

and the series converges for any x_0, then the series converges absolutely and uniformly for all x satisfying $|x| < |x_0|$. For the series $f(x_0)$ has a maximal term. Let this be γ. Hence for every i

$$|a_i x_0^i| \le \gamma. \tag{3.07.5}$$

But if $|x| < |x_0|$, then since

$$|a_i x^i| \le \gamma |x/x_0|^i,$$

the geometric series $\gamma \Sigma |x/x_0|^i$ converges and dominates the series for $f(x)$. Also any N that is effective for the series (3.07.4) when $x = x_0$ is a fortiori effective when $|x| < |x_0|$, whence the series is uniformly convergent for $|x| < |x_0|$. Since every term is a continuous function of x, therefore $f(x)$ is a continuous function of x.

If the series (3.07.4) diverges for any $x = x_0$, then it diverges for all x of greater moduli. For if $|x_1| > |x_0|$, and the series converged at x_1, then by the theorem just proved the series would converge at x_0, contrary to hypothesis. Hence if the series converges for any $x \neq 0$, it either converges for all x, or else there is some circle about the origin such that the series converges throughout the interior and diverges at every point outside the circle. The behavior at points on the circle can only be determined by further study. This circle is the circle of convergence of the power series.

From (3.07.5) it follows that, if the series (3.07.4) converges for $x = x_0$, then for every i

$$|a_i| \leq \gamma/|x_0^i|, \tag{3.07.6}$$

where γ is the modulus of the term of maximum modulus.

We have seen that $f(x)$ defined by (3.07.4) is continuous throughout its circle of convergence. It is also differentiable throughout the same circle, and

$$f'(x) = a_1 + 2a_2x + 3a_3x^2 + \cdots . \tag{3.07.7}$$

To prove this, let $r = |x| < R$, where R is the radius of the circle of convergence; let $\alpha_i = |a_i|$; and let

$$F(r) = \alpha_0 + \alpha_1 r + \alpha_2 r^2 + \cdots .$$

If $0 < \delta < R - r$, then the series $F(r)$ and also the series $F(r + \delta)$ both converge. Hence the series

$$\delta^{-1}[F(r + \delta) - F(r)] = \alpha_1 + \alpha_2(2r + \delta) + \alpha_3(3r^2 + 3r\delta + \delta^2) + \cdots$$

converges. If $|h| \leq \delta$,

$$h^{-1}[f(x + h) - f(x)] = a_1 + a_2(2x + h) + a_3(3x^2 + 3xh + h^2) + \cdots ,$$

and this last series is dominated by the previous one. Hence as a series in functions of h for a fixed x the latter series converges uniformly and hence defines a continuous function of h. But for $h = 0$ we have the series (3.07.7).

Thus a function defined by a power series (3.07.4) is continuous and differentiable throughout its circle of convergence, and the series (3.07.7) for its derivative converges in the same circle. But the same can therefore be said of f', so that f'' exists, as does f''', . . . , and for each the series converges in the same circle. The function f is said to be analytic in the circle within which its power series converges. Since

$$f^{(n)}(0) = n!a_n,$$

the series can be written in the form

$$f(x) = f(0) + xf'(0) + x^2f''(0)/2! + \cdots , \tag{3.07.8}$$

and this is its Maclaurin expansion. By a change of origin one can also write the more general Taylor expansion

$$f(x) = f(r) + (x - r)f'(r) + (x - r)^2f''(r)/2! + \cdots \tag{3.07.9}$$

already given for the polynomials.

The series

$$F(x) = a_0x + a_1x^2/2 + a_2x^3/3 + \cdots \tag{3.07.10}$$

formed by integrating each term from 0 to x has a radius of convergence at least as large as that of f. For if $x \neq 0$. the series $x^{-1}F(x)$ is dominated by $f(x)$. However, since $f = F'$, the radius of convergence of F can be no greater than that of f, as we have seen.

If $f(x)$ is an analytic function, the equation

$$f(x) = 0 \tag{3.07.11}$$

will be said to have a root r of multiplicity m in case $(x - r)^{-m}f(x)$ is analytic at r but $(x - r)^{-m-1}f(x)$ is not. But from the expansion (3.07.9) it appears that this will be so if and only if

$$0 = f(r) = f'(r) = \cdots = f^{(m-1)}(r) \neq f^{(m)}(r).$$

Hence, in that case

$$f(x) = (x - r)^m f^{(m)}(r)/m! + \cdots.$$

In particular, the equation

$$f(x)/f'(x) = 0$$

will have only simple roots, if any.

Budan's theorem holds in the case of an analytic function provided, for some k, $f^{(k)}(x)$ keeps the same sign throughout the interval from a to b.

3.08. *König's Theorem.* Consider any function

$$f(z) = a_0 + a_1 z + a_2 z^2 + \cdots \tag{3.08.1}$$

for which the expansion converges in some circle about the origin. Suppose that within this circle $f(z)$ has one and only one zero α, which is simple. Let $g(z)$ be analytic throughout the circle and $g(\alpha) \neq 0$. Then the expansion

$$g/f = h(z) = h_0 + h_1 z + h_2 z^2 + \cdots \tag{3.08.2}$$

converges for all $|z| < |\alpha|$, while the expansion

$$(\alpha - z)h(z) = F(z) = k_0 + k_1 z + k_2 z^2 + \cdots + k_\nu z^\nu + \cdots \tag{3.08.3}$$

converges throughout the circle. Then for $|z| < |\alpha|$

$$(\alpha - z)(h_0 + h_1 z + \cdots) = k_0 + k_1 z + \cdots \tag{3.08.4}$$

so that

$$\begin{aligned} \alpha h_0 &= k_0, \\ -h_0 + \alpha h_1 &= k_1, \\ &\cdots\cdots\cdots \\ -h_{\nu-1} + \alpha h_\nu &= k_\nu. \end{aligned} \tag{3.08.5}$$

On multiplying these equations by $1, \alpha, \alpha^2, \ldots$ and adding, one obtains

$$\alpha^{\nu+1} h_\nu = k_0 + k_1 \alpha + \cdots + k_\nu \alpha^\nu.$$

Let

$$F_\nu(z) \equiv k_0 + k_1 z + \cdots + k_\nu z^\nu \equiv F(z) - R_{\nu+1}(z). \tag{3.08.6}$$

Then

$$h_\nu = \alpha^{-\nu-1}F_\nu(\alpha) = \alpha^{-\nu-1}[F(\alpha) - R_{\nu+1}(\alpha)], \tag{3.08.7}$$

and

$$h_\nu/h_{\nu+1} = \alpha F_\nu(\alpha)/F_{\nu+1}(\alpha). \tag{3.08.8}$$

However, F is analytic at α, and the series (3.08.3) converges for $z = \alpha$. Let ρ' be the radius of convergence of this series and let ρ satisfy $|\alpha| < \rho < \rho'$. Then the series converges for $z = \rho$. If γ is the modulus of the term of maximum modulus of the series $F(\rho)$, then

$$|k_\nu| \leq \gamma/\rho^\nu.$$

By (3.08.8)

$$\alpha - h_\nu/h_{\nu+1} = k_{\nu+1}\alpha^{\nu+2}/F_{\nu+1}(\alpha).$$

Hence

$$|\alpha - h_\nu/h_{\nu+1}| \leq \mu|\alpha^{\nu+1}/\rho^{\nu+1}|, \tag{3.08.9}$$

where μ is some positive quantity depending upon the value of γ, α, and $F(\alpha)$. Since $|\alpha/\rho| < 1$, this proves that

$$\lim_{\nu\to\infty} h_\nu/h_{\nu+1} = \alpha$$

and shows, moreover, that the convergence is geometric with a ratio $|\alpha/\rho|$.

König's theorem has an important extension to the case in which $f(z)$ has exactly n simple zeros within some circle about the origin. The extended theorem and its proof are sufficiently well illustrated by the case $n = 2$, and this will now be given. Let the zeros be α_1 and α_2, and suppose that within some small circle about the origin $f(z)$ has these and no other zeros. Take g and h as before but with $g(\alpha_1) \neq 0$ and $g(\alpha_2) \neq 0$. Let

$$P(z) \equiv b_0(1 - z/\alpha_1)(1 - z/\alpha_2) \equiv b_0 + b_1z + b_2z^2, \tag{3.08.10}$$

and now take

$$P(z)h(z) \equiv F(z) \equiv k_0 + k_1z + k_2z^2 + \cdots. \tag{3.08.11}$$

Then $F(z)$ is analytic throughout the circle. We set

$$(b_0 + b_1z + b_2z^2)(h_0 + h_1z + \cdots) \equiv k_0 + k_1z + \cdots$$

so that

$$\begin{aligned} b_0h_0 &= k_0,\\ b_1h_0 + b_0h_1 &= k_1,\\ b_2h_0 + b_1h_1 + b_0h_2 &= k_2,\\ b_2h_1 + b_1h_2 + b_0h_3 &= k_3,\\ &\cdots\cdots\cdots\\ b_2h_{\nu-2} + b_1h_{\nu-1} + b_0h_\nu &= k_\nu,\\ &\cdots\cdots\cdots \end{aligned} \tag{3.08.12}$$

If we multiply these equations by 1, α_i, α_i^2, . . . , α_i^ν, and add, we obtain

$$(\alpha_i^{\nu-1}b_0 + \alpha_i^\nu b_1)h_{\nu-1} + \alpha_i^\nu b_0 h_\nu = F_\nu(\alpha_i),$$

where $F_\nu(z)$ is defined as in (3.08.6). Likewise

$$(\alpha_i^\nu b_0 + \alpha_i^{\nu+1}b_1)h_\nu + \alpha_i^{\nu+1}b_0 h_{\nu+1} = F_{\nu+1}(\alpha_i),$$

and

$$(\alpha_i^{\nu+1}b_0 + \alpha_i^{\nu+2}b_1)h_{\nu+1} + \alpha_i^{\nu+2}b_0 h_{\nu+2} = F_{\nu+2}(\alpha_i).$$

These equations hold for $i = 1$ or $i = 2$. Multiply the first of these equations through by α_i^2, and the second by α_i. The equations then show that the determinant

$$\begin{vmatrix} h_{\nu-1} & h_\nu & \alpha_i^2 F_\nu(\alpha_i) \\ h_\nu & h_{\nu+1} & \alpha_i F_{\nu+1}(\alpha_i) \\ h_{\nu+1} & h_{\nu+2} & F_{\nu+2}(\alpha_i) \end{vmatrix} = 0,$$

since they express the linear dependence of the three columns.

Now if ρ' is the radius of convergence of (3.08.11), ρ satisfies $|\alpha_i| < \rho < \rho'$ for $i = 1$ and 2, and γ is the modulus of the term of maximum modulus of the series $F(\rho)$, then

$$\begin{aligned} |R_{\nu+1}(\alpha_i)| &\le \gamma|\alpha_i^{\nu+1}/\rho^{\nu+1}|(1 + |\alpha_i/\rho| + \cdots) \\ &= \gamma|\alpha_i^{\nu+1}/\rho^{\nu+1}|/(1 - |\alpha_i/\rho|). \end{aligned}$$

If $|\alpha_2| \ge |\alpha_1|$, then

$$|R_{\nu+1}(\alpha_i)| \le \gamma|\alpha_2^{\nu+1}/\rho^{\nu+1}|/(1 - |\alpha_2/\rho|),$$

and, a fortiori,

$$\begin{aligned} |R_{\nu+2}(\alpha_i)| &\le \gamma|\alpha_2^{\nu+1}/\rho^{\nu+1}|/(1 - |\alpha_2/\rho|), \\ |R_{\nu+3}(\alpha_i)| &\le \gamma|\alpha_2^{\nu+1}/\rho^{\nu+1}|/(1 - |\alpha_2/\rho|). \end{aligned}$$

Let $\alpha = \alpha_2$ in case $|F(\alpha_1)| \ge |F(\alpha_2)|$, otherwise $\alpha = \alpha_1$, and set

$$\begin{aligned} M &= \gamma/[|F(\alpha)|(1 - |\alpha_2/\rho|)], \\ \mu &= |\alpha_2/\rho|. \end{aligned}$$

Then if the equation

$$\begin{vmatrix} h_\nu & h_{\nu-1} & z^2 \\ h_{\nu+1} & h_\nu & z \\ h_{\nu+2} & h_{\nu+1} & 1 \end{vmatrix} = 0 \tag{3.08.13}$$

is expanded to the form

$$c_0^{(\nu)}z^2 + c_1^{(\nu)}z + c_2^{(\nu)} = 0,$$

where the coefficients represent the cofactors of the powers of z in the determinant (3.08.13), then each of these coefficients differs by a factor of less than $M\mu^{\nu+1}$ from the coefficients of a quadratic

$$c_0z^2 + c_1z + c_2 = 0$$

satisfied by α_1 and α_2. Hence in the limit for large ν the quadratic (3.08.13) is satisfied by the two roots α_1 and α_2 of smallest modulus of $f = 0$, if such exist within the circle of convergence of the expansion of f.

In the general case of n roots the equations of the sequence are

$$\begin{vmatrix} h_\nu & h_{\nu-1} & \cdots & h_{\nu-n+1} & z^n \\ h_{\nu+1} & h_\nu & \cdots & h_{\nu-n+2} & z^{n-1} \\ \cdots & \cdots & \cdots & \cdots & \cdots \\ h_{\nu+n} & h_{\nu+n-1} & \cdots & h_{\nu+1} & 1 \end{vmatrix} = 0. \tag{3.08.14}$$

3.1. The Graeffe Process. We turn now to methods for solving a single equation in a single unknown. We have seen that one can express the sum s_p of the pth powers of the roots of an algebraic equation as a rational function of the coefficients of the equation by relations (3.02.5). But we can write

$$s_p = x_1^p(1 + x_2^p/x_1^p + x_3^p/x_1^p + \cdots).$$

Hence if there should be one root that is larger than all the others, say x_1, then for a sufficiently large p all fractions within the parentheses should become negligible, and we would have approximately

$$s_p \doteq x_1^p,$$

and in particular

$$\lim_{p\to\infty} s_p^{1/p} = x_1.$$

Hence if a feasible method could be found for computing s_p for sufficiently large p, we could take the pth root of this and obtain thereby an evaluation of the largest root (in case there is such) of the equation.

The Graeffe process does this, and somewhat more. If we write the equation in the form

$$a_0x^n + a_2x^{n-2} + a_4x^{n-4} + \cdots = -a_1x^{n-1} - a_3x^{n-3} - a_5x^{n-5} - \cdots \tag{3.1.1}$$

and square both sides, we obtain

$$a_0^2x^{2n} + 2a_0a_2x^{2n-2} + (a_2^2 + 2a_0a_4)x^{2n-4} + \cdots = a_1^2x^{2n-2} + 2a_1a_3x^{2n-4} + \cdots$$

or

$$a_0^2x^{2n} + (2a_0a_2 - a_1^2)x^{2n-2} + (2a_0a_4 - 2a_1a_3 + a_2^2)x^{2n-4} + \cdots = 0.$$

Since only even powers of x occur here, this can be written

$$a_0^2y^n + (2a_0a_2 - a_1^2)y^{n-1} + (2a_0a_4 - 2a_1a_3 + a_2^2)y^{n-2} + \cdots = 0, \tag{3.1.2}$$

where

$$y = x^2.$$

Hence we obtain a new equation whose roots are the squares of the roots of the original equation. If we repeat, we obtain an equation whose roots are the fourth powers, another repetition gives one with the eighth powers, etc. After p such operations we obtain an equation whose roots are the 2^pth powers of the roots of the original:

$$a_0^{(p)}x_{(p)}^n + a_1^{(p)}x_{(p)}^{n-1} + \cdots = 0. \tag{3.1.3}$$

At any stage if we write the coefficients in sequence

$$a_0^{(p)} \quad a_1^{(p)} \quad a_2^{(p)} \quad a_3^{(p)} \quad a_4^{(p)} \quad \ldots,$$

then to get the new sequence $a_i^{(p+1)}$ we take the product of $a_0^{(p)}$ by the coefficient symmetrically placed with respect to $a_i^{(p)}$ and double, subtract the double product of $a_1^{(p)}$ by its symmetric mate, . . . , ending with $\pm a_i^{(p)2}$. Now if the roots are x_i, then

$$a_1/a_0 = -\Sigma x_i, \quad a_1^{(1)}/a_0^{(1)} = -\Sigma x_i^2, \quad \ldots, \quad a_1^{(p)}/a_0^{(p)} = -\Sigma x_i^{2^p}. \tag{3.1.4}$$

If the roots are all distinct, and x_1 has a modulus larger than that of any other root, then eventually

$$-a_1^{(p)}/a_0^{(p)} \doteq x_1^{2^p}. \tag{3.1.5}$$

Now it is also true that

$$a_2^{(p)}/a_0^{(p)} = \sum_{(i,j)} (x_i x_j)^{2^p},$$

so that by a similar argument we can take eventually

$$\begin{aligned} a_2^{(p)}/a_0^{(p)} &\doteq (x_1 x_2)^{2^p}, \\ a_2^{(p)}/a_1^{(p)} &\doteq -x_2^{2^p}, \end{aligned} \tag{3.1.6}$$

if the modulus of x_2 exceeds that of every other root except for x_1. Again

$$\begin{aligned} a_3^{(p)}/a_0^{(p)} &\doteq -(x_1 x_2 x_3)^{2^p}, \\ a_3^{(p)}/a_2^{(p)} &\doteq -x_3^{2^p}; \\ &\cdots\cdots\cdots\cdots \\ a_n^{(p)}/a_{n-1}^{(p)} &\doteq -x_n^{2^p}. \end{aligned} \tag{3.1.7}$$

If the equation has only simple real roots, all relations (3.1.7) are valid for sufficiently large p. The signs of the roots are undetermined, but these can be obtained by substitution or in other ways. But if $P(x)$ is real, and the equation has complex roots, these occur in complex conjugate pairs with equal moduli, and there may be any number of unequal roots, all having the same modulus. For example, all n roots of

$$x^n - 1 = 0$$

have unit modulus, and the method fails.

If there is one pair of complex roots whose moduli exceed the moduli of all others,

$$|x_1| = |x_2| = \rho > |x_i| \qquad (i > 2),$$

then in polar form

$$x_1 = \rho \exp i\theta,$$
$$x_2 = \rho \exp (-i\theta),$$

so that for $m = 2^p$

$$x_1^m = \rho^m \exp mi\theta,$$
$$x_2^m = \rho^m \exp (-mi\theta),$$

and

$$x_1^m + x_2^m = 2\rho^m \cos m\theta.$$

Hence for larger values of p, $\cos m\theta$ will fluctuate in value and even in sign, causing $a_1^{(p)}/a_0^{(p)}$ to do likewise. However, in $a_2^{(p)}/a_0^{(p)}$ the dominant term will be

$$x_1^m x_2^m = \rho^{2m}.$$

If we can be sure that we stop where $\cos m\theta$ is not too small, then $x_1^m + x_2^m$ will dominate the other terms in $a_1^{(p)}/a_0^{(p)}$, and we can obtain both ρ and θ, but with the quadrant of θ undecided.

This indeterminacy can be resolved if we apply the root-squaring process to the equation

$$P(y + h) = 0$$

as well as to the original equation, where h is a small fixed number (using the method of §3.04 to obtain the coefficients of y). Each root y_i of this equation is related to a root of the original equation by

$$y_i = x_i - h,$$

and if h is small enough, the moduli of y_1 and y_2 will also exceed the moduli of all the other roots. If

$$\sigma = |y_1| = |y_2|,$$

then our roots x_1 and x_2 lie in the complex plane where the circle of radius ρ about the origin intersects the circle of radius σ about the point h units to the right of the origin (or $-h$ units to the left if h is negative). This determines θ uniquely. In case there are other roots x_i with the same modulus, the corresponding roots y_i will have different moduli, and this difficulty is thereby removed.

3.11. *Lehmer's Algorithm.* The technique of investigating the roots y_i of $P(y + h) = 0$, along with the roots x_i of $P(x) = 0$, has the disadvantage of requiring two applications of the Graeffe process in addition to the special computations involved in the determination of the coefficients of y. Moreover in selecting h, one should be careful to make it small enough so that, if roots x_i and x_j are such that $|x_i| > |x_j|$, then also $|x_i - h| > |x_j - h|$.

Brodetsky and Smeal therefore make the natural proposal that h should be "infinitesimal," and Lehmer has developed an effective algorithm.

The original Graeffe process can be described in slightly different terms by saying that we start with a polynomial $P(x)$ and obtain from it a polynomial $P_1(x)$ whose zeros are the squares of those of P; from P_1 we obtain $P_2(x)$ whose zeros are the squares of those of P_1 and hence the fourth powers of those of P, ... On setting $P_0 = P$ for uniformity, one verifies that

$$P_{p+1}(x) \equiv P_p(\sqrt{x})P_p(-\sqrt{x}), \qquad p = 0, 1, 2, \ldots . \tag{3.11.1}$$

In fact

$$\begin{aligned} P_0(x) &= a_0\Pi(x - x_i), \\ P_1(x) &= a_0^2\Pi(\sqrt{x} - x_i)(-\sqrt{x} - x_i) = P_0(\sqrt{x})P_0(-\sqrt{x}) \\ &= a_0^2\Pi(-x + x_i^2), \end{aligned}$$

and the general statement follows by a simple induction.

If we write

$$\begin{aligned} Q_0(x) &= a_0\Pi(x - x_i - h), \\ Q_1(x) &= a_0^2\Pi(\sqrt{x} - x_i - h)(-\sqrt{x} - x_i - h) = Q_0(\sqrt{x})Q_0(-\sqrt{x}) \\ &= a_0^2\Pi[-x + (x_i + h)^2], \end{aligned}$$

and continue the same procedure with

$$Q_{p+1}(x) = Q_p(\sqrt{x})Q_p(-\sqrt{x}),$$

then we find inductively that $Q_p(x)$ is a polynomial whose zeros are the n quantities

$$(x_i + h)^m = x_i^m + mhx_i^{m-1} + \cdots, \qquad m = 2^p,$$

where the terms omitted contain h^2 and higher powers of h.

Lehmer's algorithm is obtained by setting

$$\begin{aligned} \phi_0(x, h) &= (x - h)^{-n}P_0(x - h) \\ &= a_0 + a_1(x - h)^{-1} + a_2(x - h)^{-2} + \cdots \end{aligned} \tag{3.11.2}$$

and defining recursively

$$\phi_{p+1}(x, h) = \phi_p(\sqrt{x}, h)\phi_p(-\sqrt{x}, h). \tag{3.11.3}$$

Then

$$Q_p(x) = (h^m - x)^n\phi_p(x, h). \tag{3.11.4}$$

Also

$$\begin{aligned} \phi_0(x, h) &= \phi_0(x, 0) - h\phi_0'(x, 0) + \cdots \\ &= a_0 + a_1x^{-1} + a_2x^{-2} + \cdots + h(b_2x^{-2} + b_3x^{-3} + \cdots) \\ &\qquad + \cdots , \end{aligned}$$

where

$$\phi_0' = \partial\phi/\partial x = -\partial\phi/\partial h,$$

and

$$b_r = (r-1)a_{r-1}, \qquad r = 2, 3, \ldots .$$

By direct calculation from (3.11.3) if

(3.11.5)
$$\begin{aligned}\phi_{p+1}(x, h) = a_0^{(p+1)} &+ a_1^{(p+1)}x^{-1} + a_2^{(p+1)}x^{-2} + \cdots \\ &+ 2^{p+1}h(b_1^{(p+1)}x^{-1} + b_2^{(p+1)}x^{-2} + \cdots) \\ &+ \cdots ,\end{aligned}$$

we obtain the recursion

(3.11.6)
$$\begin{aligned} a_r^{(p+1)} &= (-1)^r a_r^{(p)^2} + 2\sum_{\nu=0}^{r-1}(-1)^\nu a_\nu^{(p)}a_{2r-\nu}^{(p)}, \qquad r = 0, 1, \ldots , \\ b_r^{(p+1)} &= \sum_{\nu=0}^{2r-1}(-1)^\nu a_\nu^{(p)}b_{2r-\nu}^{(p)}, \qquad r = 1, 2, \ldots . \end{aligned}$$

If $a_0 = 1$, as we may suppose, then $a_0^{(p)} = 1$ for every p. From (3.11.4), Q_p and $(-x)^n\phi_p$ differ only by terms containing the factor h^{2^p}. Hence the coefficient of x^{-1} in ϕ_p is the sum of the zeros. But the zeros are of the form

$$(x_i + h)^m = x_i^m + mhx_i^{m-1} + \cdots .$$

Hence

(3.11.7)
$$\begin{aligned} a_1^{(p)} &= -\Sigma x_i^m, \\ b_1^{(p)} &= -\Sigma x_i^{m-1}. \end{aligned}$$

But if there is a root of largest modulus x_1, then for p sufficiently large it follows that

(3.11.8)
$$x_1 = a_1^{(p)}/b_1^{(p)},$$

approximately. Thus we obtain the solution directly without the need for a root extraction. Again if

$$|x_1| > |x_2| > \cdots ,$$

then

(3.11.9)
$$\begin{aligned} a_2^{(p)} &\doteq x_1^m x_2^m, \\ b_2^{(p)} &\doteq x_1^{m-1}x_2^{m-1}(x_1 + x_2), \end{aligned}$$

whence

(3.11.10)
$$x_2 \doteq 1/(b_2^{(p)}/a_2^{(p)} - b_1^{(p)}/a_1^{(p)}).$$

In like manner if

$$|x_1| > |x_2| > |x_3| > \cdots ,$$

then

(3.11.11)
$$x_3 \doteq 1/(b_3^{(p)}/a_3^{(p)} - b_2^{(p)}/a_2^{(p)}).$$

There are many special cases that can occur, and effort will be made, not to enumerate them all, but only to suggest how they can be treated. If x_1 is a root of multiplicity k and

$$|x_1| > |x_{k+1}| \geq \cdots,$$

then

$$a_1^{(p)} \doteq -kx_1^m,$$
$$b_1^{(p)} \doteq -kx_1^{m-1},$$

so that (3.11.8) still holds. Moreover,

$$a_2^{(p)} \doteq \binom{k}{2} x_1^{2m},$$
$$b_2^{(p)} \doteq 2\binom{k}{2} x_1^{2m-1},$$
$$\cdots\cdots\cdots\cdots$$
$$(-1)^k a_k^{(p)} \doteq x_1^{km},$$
$$(-1)^k b_k^{(p)} \doteq kx_1^{km-1},$$
$$(-1)^{k+1} a_{k+1}^{(p)} \doteq x_1^{km} x_{k+1}^m,$$
$$(-1)^{k+1} b_{k+1}^{(p)} \doteq x_1^{km} x_{k+1}^{m-1} + kx_1^{km-1} x_{k+1}^m.$$

Hence

$$x_{k+1} = 1/(b_{k+1}^{(p)}/a_{k+1}^{(p)} - b_k^{(p)}/a_k^{(p)}).$$

Analogous relations can be worked out for a multiple root of intermediate modulus.

The most important case of unequal roots of equal modulus occurs for a real polynomial with complex roots. Suppose, for example,

$$|x_1| = |x_2| > |x_3| > \cdots.$$

Then if $P(x)$ is real, and $x_1 \neq \pm x_2$, we can write

$$x_1 = \rho \exp(i\theta), \qquad x_2 = \rho \exp(-i\theta),$$
$$x_1^m = \rho^m \exp(mi\theta), \qquad x_2^m = \rho^m \exp(-mi\theta).$$

Since

$$\exp(\pm i\theta) = \cos\theta \pm i \sin\theta,$$

$a_1^{(p)}$ will contain the term $-2\rho^m \cos m\theta$ which will oscillate in value with increasing p and m but will dominate the other terms whenever $m\theta$ is not too far from an integral multiple of π. When this is the case, we may say

$$a_1^{(p)} \doteq -2\rho^m \cos m\theta,$$
$$b_1^{(p)} \doteq -2\rho^{m-1} \cos (m-1)\theta.$$

Also

$$a_2^{(p)} \doteq \rho^{2m},$$
$$b_2^{(p)} \doteq 2\rho^{2m-1} \cos\theta.$$

Since ρ is real and positive, this is obtainable from $a_2^{(p)}$ by a root extraction, and then θ can be found from $b_2^{(p)}$ or, in fact, from $b_1^{(p)}$ or $a_1^{(p)}$. In any event there is no ambiguity since the two possible quadrants for θ correspond to the two roots x_1 and x_2.

3.12. *Transcendental Equations.* If one sets $z = x^{-1}$, an equation equivalent to $P = 0$ is the equation

$$(3.12.1) \qquad f(z) \equiv a_0 + a_1 z + a_2 z^2 + \cdots = 0$$

in z. Each formula referring to the application of the Graeffe process in either the original form or that given by Lehmer remains valid if each x root x_i is replaced by z_i^{-1}, the reciprocal of a z root of (3.12.1). But when this is done, they are applicable also to the case when f is analytic and not necessarily a polynomial. That is to say, if among the roots of (3.12.1) which lie within the circle of convergence there is a root z_1 whose modulus is less than that of all the others, then

$$(3.12.2) \qquad \lim_{p \to \infty} a_0^{(p)}/a_1^{(p)} = -z_1^{2^p},$$

which corresponds to (3.1.5), and

$$(3.12.3) \qquad \lim_{p \to \infty} b_1^{(p)}/a_1^{(p)} = z_1,$$

which corresponds to (3.11.8). Polya's demonstration of the formulas such as (3.12.2) will now be sketched briefly.

Suppose that the series $f(z)$ converges inside a circle of radius ρ' at least and that the equation has exactly n roots, $z_1, z_2, \ldots, z_n$, of moduli less than ρ'. Choose ρ and designate the roots so that

$$|z_1| \le |z_2| \le \cdots \le |z_n| < \rho < \rho'.$$

It is no restriction to suppose that $a_0 = 1$. For any positive integer m, let

$$\omega = \exp(2\pi i/m) = \cos(2\pi/m) + i \sin(2\pi/m).$$

Thus ω is a complex mth root of unity, and one can verify that the other complex roots are $\omega^2, \ldots, \omega^{m-1}$ and that

$$1 + \omega + \omega^2 + \cdots + \omega^{m-1} = 0.$$

This done, one can verify that the product

$$(3.12.4) \qquad f(z)f(\omega z) \cdots f(\omega^{m-1}z) \equiv 1 + a_{1,m}z^m + a_{2,m}z^{2m} + \cdots$$

contains only powers of z^m. The Graeffe process takes advantage of this in the special case when m is some power of 2. The theorem states that

$$(3.12.5) \qquad \lim_{m \to \infty} z_1^m z_2^m \cdots z_n^m a_{n,m} = (-1)^n.$$

If we write

$$f = (z - z_1)(z - z_2) \cdots (z - z_n)\phi(z),$$

(3.12.6) $$f'/f = (z - z_1)^{-1} + \cdots + (z - z_n)^{-1} + b_1 + b_2 z + \cdots$$

where

$$\phi'/\phi = b_1 + b_2 z + \cdots,$$

then the series ϕ'/ϕ has a radius of convergence at least ρ'. Hence for some γ

(3.12.7) $$|b_n| \leq \gamma|\rho^{-n}|.$$

On integrating (3.12.6) from 0 to z and taking the antilogarithm,

(3.12.8) $$f(z) = (1 - z/z_1) \cdots (1 - z/z_n) \exp (b_1 z + b_2 z^2/2 + b_3 z^3/3 + \cdots),$$

(3.12.9) $$f(z)f(\omega z) \cdots f(\omega^{m-1}z) = (1 - z^m/z_1^m) \cdots (1 - z^m/z_n^m)(1 + B_{1,m}z^m + \cdots),$$

where

$$1 + B_{1,m}z^m + B_{2,m}z^{2m} + \cdots = \exp (b_m z^m + b_{2m}z^{2m}/2 + b_{3m}z^{3m}/3 + \cdots).$$

To express the coefficients $B_{p,m}$ in terms of the b_{pm}, consider the related problem

$$1 + A_1 y + A_2 y^2 + \cdots = \exp (\alpha_1 y + \alpha_2 y^2/2 + \cdots).$$

By differentiating both sides with respect to y, we get

$$A_1 + 2A_2 y + \cdots = (\alpha_1 + \alpha_2 y + \cdots)(1 + A_1 y + A_2 y^2 + \cdots).$$

Hence on multiplying and comparing coefficients, we obtain the recursion

$$\begin{aligned} A_1 &= \alpha_1, \\ 2A_2 &= \alpha_1 A_1 + \alpha_2, \\ 3A_3 &= \alpha_1 A_2 + \alpha_2 A_1 + \alpha_3, \\ &\cdots\cdots\cdots \end{aligned}$$

Hence

$$\begin{aligned} B_{1,m} &= b_m, \\ 2B_{2,m} &= b_m B_{1,m} + b_{2m}, \\ 3B_{3,m} &= b_m B_{2,m} + b_{2m} B_{1,m} + b_{3m}, \\ &\cdots\cdots\cdots \end{aligned}$$

It follows immediately from the first of these relations and (3.12.7) that

$$|B_{1,m}| \leq \gamma|\rho^{-m}|,$$

and if $\gamma \geq 1$, as we may require, one can show inductively that

$$|B_{p,m}| \leq \gamma^p|\rho^{-pm}|.$$

Now from (3.12.4) and (3.12.9)

(3.12.10) $1 + a_{1,m}z^m + a_{2,m}z^{2m} + \cdots$

$$= [1 - z^m \Sigma z_i^{-m} + \cdots + (-1)^n z^{nm} z_1^{-m} \cdots z_n^{-m}](1 + B_{1,m}z^m + \cdots).$$

Hence, by comparing coefficients of z^{nm},

$$z_1^m z_2^m \cdots z_n^m a_{n,m} = (-1)^n + (-1)^{n-1}B_{1,m} \Sigma z_i^m + (-1)^{n-2}B_{2,m} \sum_{i<j} z_i^m z_j^m + \cdots + B_{n,m} z_1^m \cdots z_n^m.$$

But

$$|B_{1,m}\Sigma z_i^m| \le n\gamma |z_n/\rho|^m,$$

$$\left|B_{2,m} \sum_{i<j} z_i^m z_j^m\right| \le \binom{n}{2} \gamma^2 |z_n/\rho|^{2m},$$

$$\cdots\cdots\cdots\cdots\cdots\cdots$$

$$|B_{n,m} z_1^m \cdots z_n^m| \le \gamma^n |z_n/\rho|^{nm}.$$

For fixed n, as m increases, the first term vanishes as $|z_n/\rho|^m$, the second as $|z_n/\rho|^{2m}$, ... Hence

(3.12.11) $$z_1^m z_2^m \cdots z_n^m a_{n,m} = (-1)^n + 0(|z_n/\rho|^m).$$

This is the required theorem.

3.2. Bernoulli's Method. The Graeffe process has the decided advantage that the exponents $m = 2^p$ themselves build up exponentially. Hence if one is fortunate enough to have the roots reasonably well separated in the original equation, he may hope that only a relatively small number of root squarings will be required. It has the further advantage that, in principle at least, it yields simultaneously all the roots of an algebraic equation and all roots within a circle of analyticity when the equation is transcendental. Hence, once the squaring has been carried sufficiently far, all solutions are obtainable by simple division or, at worst, by root extraction.

The methods now to be described do not converge nearly so fast; they give only one root, or at most a few roots, at a time; and in some cases they require some previous knowledge of the approximate locations of the roots to be determined. Nevertheless, they all have one striking advantage. Errors, once made, do not propagate but tend to die out. If there were no round-off error, they would die out completely. A gross error might cause the process to converge to some root other than the one intended, however. But the self-correcting tendency suggests that a method of this type might be useful at least for improving approximate solutions obtained by, say, Graeffe's method, but with insufficient accuracy.

Let the equation

(3.2.1) $$f(z) \equiv a_0 + a_1 z + a_2 z^2 + \cdots = 0$$

have a single root α interior to some circle about the origin throughout which $f(z)$ is analytic. Then if $g(z)$ is analytic throughout the same circle, and $g(\alpha) \neq 0$, it follows from König's theorem that $h_r/h_{r+1} \to \alpha$, where

$$g/f \equiv h_0 + h_1 z + h_2 z^2 + \cdots . \tag{3.2.2}$$

Now if

$$g(z) \equiv g_0 + g_1 z + g_2 z^2 + \cdots , \tag{3.2.3}$$

we can set

$$g_0 + g_1 z + \cdots = (a_0 + a_1 z + \cdots)(h_0 + h_1 z + \cdots)$$

and compare coefficients to obtain the recursion

$$\begin{aligned} a_0 h_0 &= g_0, \\ a_0 h_1 + a_1 h_0 &= g_1, \\ a_0 h_2 + a_1 h_1 + a_2 h_0 &= g_2, \\ \cdots\cdots\cdots&\cdots\cdots \end{aligned} \tag{3.2.4}$$

so that, if $a_0 \neq 0$, the h_r can be obtained in sequence.

In the case of an algebraic equation

$$P(x) \equiv a_0 x^n + a_1 x^{n-1} + \cdots + a_0 = 0,$$

$f(z) \equiv z^n P(z^{-1})$ is also a polynomial, and the root α is the reciprocal of some root x_i of $P = 0$. If g is taken to be some polynomial of degree less than n, then for $\nu \geq n$

$$a_0 h_\nu + a_1 h_{\nu-1} + \cdots + a_n h_{\nu-n} = 0. \tag{3.2.5}$$

However, instead of first making a nearly arbitrary selection of $g(z)$, one can just as well select $h_0, h_1, \ldots, h_{n-1}$ arbitrarily and apply only Eq. (3.2.5). One never needs to know the function g explicitly. It might happen, by chance, that the selection of $h_0, \ldots, h_{n-1}$ defines by (3.2.4) a polynomial g which vanishes at α, but if so, the sequence h_r/h_{r+1} may converge to some other root.

Bernoulli's method, properly speaking, is the method just described for an algebraic equation, though the usual derivation is somewhat different. If the roots x_i of the algebraic equation $P = 0$ are all distinct, then for any choice of $h_0, \ldots, h_{n-1}$, the n equations

$$\Sigma u_i x_i^\rho = h_\rho, \qquad \rho = 0, \ldots, n-1$$

can be solved for the u_i, since the determinant $|x_i^\rho|$ is a Vandermonde which vanishes only when two or more of the x_i are equal. If now $h_n, h_{n+1}, \ldots$ are determined by (3.2.5), then

$$\Sigma u_i x_i^\nu = h_\nu$$

for every ν. But if x_1, say, is the largest root, and $u_1 \neq 0$, then

$$h_\nu = x_1^\nu(u_1 + u_2 x_2^\nu / x_1^\nu + \cdots),$$

and for ν sufficiently large, $h_\nu \doteq u_1 x_1^\nu$, and hence $h_\nu / h_{\nu+1} \doteq x_1^{-1} = \alpha$.

To return to the general case where f may or may not be a polynomial, there may not be a circle about the origin containing only a single root. It may be, instead, that there are two conjugate complex roots α_1 and α_2, for which therefore $|\alpha_1| = |\alpha_2|$, and which, however, lie within some circle which contains no other root. If so, we can apply the extension of König's theorem, computing the h_ν as before, but forming a quadratic equation (3.08.13), for ν sufficiently large, whose roots will be α_1 and α_2 approximately. More generally, if there are n distinct roots with equal moduli, (3.08.14) can be applied. However, it may be preferable to set $z = y + u$ for some fixed u, and apply the method to the resulting equation in y.

If there is some circle about the origin which contains only the root α_1, a somewhat larger one which contains only α_1 and α_2, a still larger one containing only α_1, α_2, and α_3, . . . , then in principle one can obtain all roots α_1, α_2, α_3, . . . without a change of origin. Thus having found α_1, we can apply the extension of König's theorem, and on setting

$$(3.2.6) \qquad H_\nu^{(2)} = \begin{vmatrix} h_\nu & h_{\nu-1} \\ h_{\nu+1} & h_\nu \end{vmatrix},$$

obtain

$$H_\nu^{(2)}/H_{\nu+1}^{(2)} \to \alpha_1\alpha_2.$$

Again if

$$(3.2.7) \qquad H_\nu^{(3)} = \begin{vmatrix} h_\nu & h_{\nu-1} & h_{\nu-2} \\ h_{\nu+1} & h_\nu & h_{\nu-1} \\ h_{\nu+2} & h_{\nu+1} & h_\nu \end{vmatrix},$$

then

$$H_\nu^{(3)}/H_{\nu+1}^{(3)} \to \alpha_1\alpha_2\alpha_3.$$

Aitken has given a convenient recursion for calculating the determinants $H_\nu^{(p)}$ of successively higher order. The formula is

$$(3.2.8) \qquad H_\nu^{(p-1)}H_\nu^{(p+1)} = H_\nu^{(p)2} - H_{\nu+1}^{(p)}H_{\nu-1}^{(p)}, \qquad p = 1, 2, \ldots,$$

where for uniformity we set

$$H_\nu^{(0)} = 1, \qquad H_\nu^{(1)} = h_\nu.$$

For $p = 1$ the equation merely gives the expansion of the determinant (3.2.6). The proof is sufficiently well illustrated for $p = 2$ and is based upon a classical determinantal identity. The sixth-order determinants

$$\begin{vmatrix} h_\nu & h_{\nu-1} & 0 & h_{\nu-2} & 0 & 1 \\ h_{\nu+1} & h_\nu & 0 & h_{\nu-1} & 0 & 0 \\ h_{\nu+2} & h_{\nu+1} & 0 & h_\nu & 1 & 0 \\ h_\nu & 0 & h_{\nu-1} & h_{\nu-2} & 0 & 1 \\ h_{\nu+1} & 0 & h_\nu & h_{\nu-1} & 0 & 0 \\ h_{\nu+2} & 0 & h_{\nu+1} & h_\nu & 1 & 0 \end{vmatrix} = \begin{vmatrix} 0 & h_{\nu-1} & -h_{\nu-1} & 0 & 0 & 0 \\ 0 & h_\nu & -h_\nu & 0 & 0 & 0 \\ 0 & h_{\nu+1} & -h_{\nu+1} & 0 & 0 & 0 \\ h_\nu & 0 & h_{\nu-1} & h_{\nu-2} & 0 & 1 \\ h_{\nu+1} & 0 & h_\nu & h_{\nu-1} & 0 & 0 \\ h_{\nu+2} & 0 & h_{\nu+1} & h_\nu & 1 & 0 \end{vmatrix} = 0$$

are equal, since the second is obtainable from the first by subtracting the fourth, fifth, and sixth rows, respectively, from the first, second, and third. But the second one vanishes, since in the Laplace expansion by third-order minors every term contains a determinant with a column of zeros. The Laplace expansion of the first along the first three rows has six nonvanishing terms, but these are equal in pairs. When the three distinct terms are written out, one obtains

$$-\begin{vmatrix} h_\nu & h_{\nu-1} & h_{\nu-2} \\ h_{\nu+1} & h_\nu & h_{\nu-1} \\ h_{\nu+2} & h_{\nu+1} & h_\nu \end{vmatrix} \cdot \begin{vmatrix} h_{\nu-1} & 0 & 1 \\ h_\nu & 0 & 0 \\ h_{\nu+1} & 1 & 0 \end{vmatrix} - \begin{vmatrix} h_{\nu-1} & h_{\nu-2} & 0 \\ h_\nu & h_{\nu-1} & 0 \\ h_{\nu+1} & h_\nu & 1 \end{vmatrix} \cdot \begin{vmatrix} h_\nu & h_{\nu-1} & 1 \\ h_{\nu+1} & h_\nu & 0 \\ h_{\nu+2} & h_{\nu+1} & 0 \end{vmatrix}$$
$$+ \begin{vmatrix} h_{\nu-1} & h_{\nu-2} & 1 \\ h_\nu & h_{\nu-1} & 0 \\ h_{\nu+1} & h_\nu & 0 \end{vmatrix} \cdot \begin{vmatrix} h_\nu & h_{\nu-1} & 0 \\ h_{\nu+1} & h_\nu & 0 \\ h_{\nu+2} & h_{\nu+1} & 1 \end{vmatrix} = 0.$$

On simplifying and rearranging, we obtain (3.2.8) for $p = 2$. The general case requires the expansion of a vanishing determinant of order $2p + 2$ formed in a similar manner.

Aitken first proposed his δ^2 process, described below, as a device for accelerating the convergence of the sequence $H_\nu^{(p)}/H_{\nu+1}^{(p)}$. If the sequence

$$u_0, u_1, u_2, \ldots$$

converges geometrically to the limit u, that is to say, if for some k, $|k| < 1$, it is true that

$$u_\nu - u = k^\nu(u_0 - u),$$

then for any ν

$$\begin{vmatrix} u_{\nu-1} & u_\nu \\ u_\nu & u_{\nu+1} \end{vmatrix} / (u_{\nu-1} - 2u_\nu + u_{\nu+1}) = u.$$

The proof can be based upon the further property that, if

$$u' = u + \omega, \qquad u'_\nu = u_\nu + \omega,$$

the same identity holds when the quantities are primed. In fact, by direct substitution one finds that, when each term in the sequence is increased by ω, the entire quantity on the left is increased by ω. We can therefore take $\omega = -u$ and consider the sequence whose limit is 0. But by direct substitution, then, the determinant is seen to vanish, which proves the assertion. In §3.08 it was shown that each sequence

$u_\nu = H_\nu^{(p)}/H_{\nu+1}^{(p)}$ for fixed p converges geometrically (in the limit). Hence we may expect that the derived sequence

$$u_\nu^{(1)} = \begin{vmatrix} u_{\nu-1} & u_\nu \\ u_\nu & u_{\nu+1} \end{vmatrix} / (u_{\nu-1} - 2u_\nu + u_{\nu+1})$$

would converge more rapidly than the original one. This is Aitken's δ^2 process. A second derived sequence, $u_\nu^{(2)}$, can be formed from the $u_\nu^{(1)}$ just as the $u_\nu^{(1)}$ was formed from the u_ν. It is to be noted that in forming a term in a derived sequence one can neglect all digits on the left that are common to the three terms being used. This is because of the property that in increasing each term by ω the result is increased by ω.

We conclude this section with a brief mention of an expansion due to Whittaker. In (3.2.4) we are free to take $g_0 = a_0$, $g_1 = g_2 = \cdot \cdot \cdot = 0$. If the first $\nu + 1$ of these equations are regarded as $\nu + 1$ linear equations in the $\nu + 1$ unknowns $h_0, h_1, \ldots, h_\nu$, the solution for h_ν can be written down in determinantal form (cf. §3.32 below). Hence the ratio $h_\nu/h_{\nu+1}$ can be expressed as the ratio of two determinants. Moreover, one can write

$$\alpha = h_0/h_1 + (h_1/h_2 - h_0/h_1) + (h_2/h_3 - h_1/h_2) + \cdot \cdot \cdot ,$$

and therefore α can be written as the limit of an infinite series involving quotients of determinants. A slight transformation yields

$$\alpha = h_0/h_1 + (h_1^2 - h_0h_2)/h_1h_2 + (h_2^2 - h_1h_3)/h_2h_3 + \cdot \cdot \cdot ,$$

and the numerators in these fractions are second-order determinants whose elements are themselves determinants. One can now apply an identity of the same type as the one used to demonstrate (3.2.8) and obtain Whittaker's expansion:

$$\alpha = -\frac{a_0}{a_1} - \frac{a_0^2a_2}{a_1\begin{vmatrix} a_1 a_2 \\ a_0 a_1 \end{vmatrix}} - \frac{a_0^3\begin{vmatrix} a_2 a_3 \\ a_1 a_2 \end{vmatrix}}{\begin{vmatrix} a_1 a_2 \\ a_0 a_1 \end{vmatrix}\begin{vmatrix} a_1 a_2 a_3 \\ a_0 a_1 a_2 \\ 0 a_0 a_2 \end{vmatrix}} - \cdot \cdot \cdot . \tag{3.2.9}$$

3.3. Functional Iteration. If $\psi(x)$ has no pole which coincides with a root of

$$f(x) = 0, \tag{3.3.1}$$

and if

$$\phi(x) \equiv x - \psi(x)f(x), \tag{3.3.2}$$

then any root of (3.3.1) satisfies also

$$x = \phi(x). \tag{3.3.3}$$

In particular if $\psi(x)$ is analytic and non-null throughout some neighborhood of a root α of (3.3.1), then α is the only root of (3.3.3) in that neighborhood of α. This suggests the possibility of so choosing ψ that the sequence

$$x_{i+1} = \phi(x_i) \tag{3.3.4}$$

converges to α, provided the initial point x_0 is sufficiently close to α. In fact if the sequence (3.3.4) converges at all, it must converge to a solution of (3.3.3), since clearly the sequences x_i and $\phi(x_i)$ have the same limit. Consider first the conditions upon ϕ that will ensure the convergence of (3.3.4).

Define the ρ neighborhood $N(x_0, \rho)$ of x_0 by

$$N(x_0, \rho): |x - x_0| < \rho. \tag{3.3.5}$$

That is, the ρ neighborhood of x_0 is the set of all points x within a distance of ρ from x_0. Now if for some positive $k < 1$ and some ρ it is true that

$$|\phi(x') - \phi(x'')| \le k \qquad \text{for } x' \text{ and } x'' \text{ in } N(\alpha, \rho), \tag{3.3.6}$$

and if x_0 itself is in $N(\alpha, \rho)$, then every x_i in the sequence (3.3.4) is in $N(\alpha, \rho)$, and the sequence converges to α. For

$$x_{i+1} - \alpha = \phi(x_i) - \phi(\alpha),$$

so that

$$|x_{i+1} - \alpha| = |\phi(x_i) - \phi(\alpha)| \le k|x_i - \alpha| < |x_i - \alpha|,$$

and inductively every x_i lies in $N(\alpha, \rho)$. Also

$$|x_i - \alpha| \le k^i|x_0 - \alpha| \tag{3.3.7}$$

by an induction that can be carried out once we know that every x_i lies in $N(\alpha, \rho)$. Since $k < 1$, therefore, the distance $|x_i - \alpha|$ decreases geometrically at least.

Now suppose that for some x_0 and ρ and a positive $k < 1$ we have

$$|\phi(x') - \phi(x'')| \le k \qquad \text{for } x', x'' \text{ in } N(x_0, \rho), \tag{3.3.8}$$

the condition holding in a ρ neighborhood of x_0, while at x_0 we have

$$|x_0 - \phi(x_0)| \le (1 - k)\rho. \tag{3.3.9}$$

We do not now presuppose the existence of a solution. Instead we show that the sequence defined by (3.3.4) has a limit α which lies in $N(x_0, \rho)$ and which satisfies our equation. Hence the conditions (3.3.8) and (3.3.9) are together sufficient to assure us of the existence of a solution and that it is obtainable as the limit of our sequence.

We show first that every term in the sequence lies in $N(x_0, \rho)$. We are assured by (3.3.9) that $x_1 = \phi(x_0)$ does, since this is equivalent to saying that

$$|x_0 - x_1| \le (1 - k)\rho < \rho.$$

If also $x_2, x_3, \ldots, x_i$ all lie in $N(x_0, \rho)$, then since

$$|x_{i+1} - x_i| = |\phi(x_i) - \phi(x_{i-1})| \le k|x_i - x_{i-1}|,$$

and by induction

$$|x_{i+1} - x_i| \le k^i|x_1 - x_0| \le k^i(1 - k)\rho,$$

therefore

$$\begin{aligned} |x_{i+1} - x_0| &\le |x_{i+1} - x_i| + |x_i - x_{i-1}| + \cdots + |x_1 - x_0| \\ &\le (k^i + k^{i-1} + \cdots + 1)(1 - k)\rho = (1 - k^{i+1})\rho < \rho. \end{aligned}$$

Hence the series

$$|x_0| + |x_1 - x_0| + |x_2 - x_1| + \cdots$$

converges, and hence the series

$$x_0 + (x_1 - x_0) + (x_2 - x_1) + \cdots$$

converges absolutely. But the partial sums of the last series are the x_i. Hence the sequence (3.3.4) converges, and the limit therefore satisfies (3.3.3).

If ϕ is analytic in some neighborhood of a root α, as will be assumed throughout, and if

$$|\phi'(\alpha)| < 1, \tag{3.3.10}$$

then for any k which satisfies

$$|\phi'(\alpha)| < k < 1$$

there exists a ρ sufficiently small so that (3.3.6) will be satisfied. Hence (3.3.10) is sufficient to ensure the existence of some neighborhood of α, though possibly small, within which (3.3.4) converges. However, it appears from (3.3.7) that the convergence is more rapid for smaller k, and hence for smaller $|\phi'(\alpha)|$. Hence it would be advantageous, if possible, to make $\phi'(\alpha) = 0$. When this is the case, we shall say that the sequence (3.3.4) has second-order convergence or, more briefly, that the iteration (defined by) $\phi(x)$ is a second-order iteration. More generally, if

$$\phi'(\alpha) = \phi''(\alpha) = \cdots = \phi^{(m-1)}(\alpha) = 0, \tag{3.3.11}$$

the iteration will be said to be of order m at least, and of order m exactly if $\phi^{(m)}(\alpha) \neq 0$. While one can write in general the expansion

$$\phi(x) - \alpha = (x - \alpha)\phi'(\alpha) + (x - \alpha)^2\phi''(\alpha)/2! + \cdots, \tag{3.3.12}$$

(since ϕ is supposed analytic), when the iteration is of order m, one can write

$$(3.3.13) \qquad \phi(x) - \alpha = (x - \alpha)^m \phi^{(m)}(\alpha)/m! + \cdots .$$

In this case

$$x_{i+1} - \alpha = (x_i - \alpha)^m \phi^{(m)}(\alpha)/m! + \cdots .$$

3.31. *Some Special Cases.* Some classical methods of successive approximation are methods of functional iteration as here understood. Horner's method is not, and since it has little to recommend it in any case, it will not be described here.

3.311. *First-order iterations.* Best known of these is the *regula falsi.* This applies to real roots of real equations, algebraic or not. If $f(x')$ and $f(x'')$ have opposite signs,

$$x_2 = [x'f(x'') - x''f(x')]/[f(x'') - f(x')]$$

lies between x' and x''. The chord from the point x', $f(x')$ to x'', $f(x'')$ intersects the x axis at x_2 as one verifies easily. It is no restriction to suppose that $f(x_2)f(x'') > 0$, since otherwise we can reverse the designations of x' and x''. We now let x_2 play the role of x'' and repeat, or we let $x'' = x_1$ and regard this as the first step in the iteration. Hence we are taking

$$(3.311.1) \qquad \phi(x) \equiv [x'f(x) - xf(x')]/[f(x) - f(x')].$$

The derivative $\phi'(\alpha)$ is seen to be

$$\phi'(\alpha) = [f(x') + (\alpha - x')f'(\alpha)]/f(x').$$

In case f has continuous first and second derivatives near α, then the Taylor expansion gives

$$f(x') = f(\alpha) + (x' - \alpha)f'(\alpha) + \tfrac{1}{2}(x' - \alpha)^2 f''(x''),$$

where now x'' is some point on the interval (α, x'). Hence

$$f(x') + (\alpha - x')f'(\alpha) = \tfrac{1}{2}(x' - \alpha)^2 f''(x''),$$

and therefore

$$\phi'(\alpha) = \tfrac{1}{2}(x' - \alpha)^2 f''(x'')/f(x').$$

Hence for x' sufficiently close to α, $\phi'(\alpha)$ will be small, and there will exist an interval about α over which

$$|\phi'(x)| \le k < 1,$$

Hence once we find an initial x' close enough to α, the subsequent iteration will converge to the solution. The choice (3.311.1) of ϕ is equivalent to the choice

$$\psi \equiv (x - x')/[f(x) - f(x')]$$

in (3.3.2), as we verify directly.

Geometrically this method amounts to drawing a series of chords all passing through the point $x', f(x')$. The ith chord passes also through $x_{i-1}, f(x_{i-1})$. Instead we might take a series of parallel chords. For this

$$\psi \equiv m,$$

where m is the constant slope. Then

$$\phi(x) \equiv x - mf(x),$$

and

$$\phi'(x) \equiv 1 - mf'(x).$$

If $f'(x)$ has fixed sign throughout some neighborhood of the solution, we choose m to have the same sign and, in fact, such that throughout the interval

$$2 > mf'(x) > 0.$$

3.312. *Second-order iterations.* If we take

$$\phi(x) \equiv x - f(x)/f'(x), \tag{3.312.1}$$

which is to say

$$\psi(x) \equiv 1/f'(x), \tag{3.312.2}$$

we obtain the well-known Newton's method. The derivative is

$$\phi'(x) = 1 - [f'^2(x) - f(x)f''(x)]/f'^2(x), \tag{3.312.3}$$

whence

$$\phi'(\alpha) = 0.$$

If α is not a multiple root,

$$f'(\alpha) \neq 0,$$

and for any positive k there is a neighborhood of α throughout which

$$|\phi'(x)| \leq k.$$

The requirement often made that at the initial approximation x_0 we should have

$$f(x_0)f''(x_0) > 0$$

is not strictly necessary.

Newton's method applies to transcendental as well as to algebraic equations, and to complex roots as well as to real. However if the equation is real, then the complex roots occur in conjugate pairs, and the iteration cannot converge to a complex root unless x_0 is itself complex. But if x_0 is complex, and sufficiently close to a complex root, the iteration will converge to that root.

For algebraic equations, as each of the first two or three x_i is obtained, it is customary to diminish the roots by x_i by the process described in §3.04.

Or rather, one first diminishes by x_0; then one obtains $x_1 - x_0$ and diminishes the roots of the last equation by this amount; then obtains $x_2 - x_1$ and diminishes by this amount, etc. Since

$$f(x_i + u) = f(x_i) + uf'(x_i) + \cdots,$$

one has always that $x_{i+1} - x_i$ is the negative quotient of the constant term by the coefficient of the linear term. Hence one calculates $f(x_i)$ and $f'(x_i)$ in the process of diminishing the roots at each stage.

However, $f(x_i)$ decreases as one proceeds. When $f(x_i)/f'(x_i)$ becomes sufficiently small, one can write

$$u = -[f(x_i) + u^2 f''(x_i)/2! + \cdots]/f'(x_i),$$

which is exact. When u is small, the terms in u^2, u^3, . . . become small correction terms, and subsequent improvements in the value of u can be made quite rapidly by resubstituting corrected values.

When Newton's method is applied to the equation

$$x^2 - N = 0,$$

the result is a standard method for extracting roots in which one uses

$$\phi(x) \equiv (x + N/x)/2.$$

3.32. *Iterations of Higher Order: König's Theorem.* If one applies Newton's method to any product $q(x)f(x)$ to obtain a particular zero α of $f(x)$, one always obtains an iteration of second order at least, provided only $q(\alpha)$ is neither zero nor infinite. Hence one might expect that by proper choice of q it should be possible to obtain an iteration of third order or even higher. This is true; one can in fact obtain an iteration of any desired order, and various schemes have been devised for the purpose. Some of these will now be described. However, one must expect that an iteration of higher order is apt to require more laborious computations. The optimal compromise between simplicity of algorithm and rapidity of convergence will depend in large measure upon the nature of the available computing facilities.

Consider first König's theorem. In the expansion

$$h(z) \equiv g/f \equiv h_0 + h_1 z + h_2 z^2 + \cdots,$$

we have Taylor's expansion about the origin in which

$$h_r = h^{(r)}(0)/r!.$$

If we move the origin to some point x, we can restate the theorem in an apparently more general form, as follows:

If in some circle about x the equation

$$f(z) = 0 \tag{3.32.1}$$

has only a single root α, which is simple; if $f(z)$ and $g(z)$ are analytic throughout this circle, and $g(\alpha) \neq 0$; and if we define

$$P_r(x) \equiv r[g(x)/f(x)]^{(r-1)}/[g(x)/f(x)]^{(r)}, \tag{3.32.2}$$

then

$$\lim_{r \to \infty} P_r(x) = \alpha - x. \tag{3.32.3}$$

For $P_r(x)$ is simply the ratio of the coefficients of the Taylor expansion of $h(z)$ about the point x.

This being true, then at least for r sufficiently large it is to be expected that

$$|\alpha - x - P_r(x)| < |\alpha - x|,$$

and hence $x + P_r$ should then define a convergent iteration of some order. It turns out that the iteration is convergent for any r and in fact is of order $r + 1$.

In proof, we write the expansions

$$h(z) \equiv h_0(x) + (z - x)h_1(x) + (z - x)^2h_2(x) + \cdots$$

and

$$F(z) \equiv k_0(x) + (z - x)k_1(x) + \cdots + (z - x)^r k_r(x) + R_{r+1}(z, x).$$

Then

$$\begin{aligned} P_r(x) &= h_{r-1}(x)/h_r(x) \\ &= (\alpha - x)[F(\alpha) - R_r(\alpha, x)]/[F(\alpha) - R_{r+1}(\alpha, x)], \end{aligned}$$

and

$$\alpha - x - P_r = (\alpha - x)^{r+1}k_r(x)/[F(\alpha) - R_{r+1}(\alpha, x)].$$

Since $\alpha - x - P_r(x)$ has the factor $(\alpha - x)^{r+1}$, all derivatives of

$$\alpha - x - P_r,$$

and hence of $x + P_r(x)$, from the first to the rth will vanish at $x = \alpha$. By definition, therefore, $x + P_r(x)$ defines an iteration of order $r + 1$ at least. Note that with $g \equiv 1$, $P_1 = -f/f'$, which yields Newton's method.

For the functions $h_\nu(x)$ required in forming any $P_r(x)$, one can obtain them by differentiation, as indicated in the theorem, or by solving a recursion like that of (3.2.4). However, now the a_ν and g_ν are functions of x, coefficients in the expansions

$$\begin{aligned} f(z) &\equiv a_0(x) + (z - x)a_1(x) + \cdots, \\ g(z) &\equiv g_0(x) + (z - x)g_1(x) + \cdots. \end{aligned}$$

In the statement of Whittaker's expansion (§3.2), reference was made to the fact that the h_ν could be expressed by means of determinants.

This is equally true for the $h_\nu(x)$, and even for the $P_\nu(x)$. For this it is convenient to suppose that any desired function g has been divided into f in advance, and f now designates the quotient. The expansion is now

$$1/f(z) = h_0(x) + (z - x)h_1(x) + (z - x)^2 h_2(x) + \cdots, \tag{3.32.4}$$

so that

$$h_r(x) = [1/f(x)]^{(r)}/r!,$$

and

$$P_r(x) = h_{r-1}(x)/h_r(x).$$

Let

$$\Delta_0 = 1, \qquad \Delta_1 = a_1(x),$$

$$\Delta_r = \begin{vmatrix} a_1 & a_0 & 0 & \cdots \\ a_2 & a_1 & a_0 & \cdots \\ \cdots & \cdots & \cdots & \cdots \\ a_r & a_{r-1} & a_{r-2} & \cdots \end{vmatrix}. \tag{3.32.5}$$

Then

$$h_r = (-)^r \Delta_r a_0^{-r-1}, \tag{3.32.6}$$

and

$$P_r = -a_0 \Delta_{r-1}/\Delta_r. \tag{3.32.7}$$

This is equivalent to saying that an iteration

$$\phi_r = x + P_r \tag{3.32.8}$$

of order $r + 1$ is given by ϕ_r satisfying

$$\begin{vmatrix} \phi_r - x & a_0 & 0 & \cdots \\ -1 & a_1 & a_0 & \cdots \\ 0 & a_2 & a_1 & \cdots \\ \cdots & \cdots & \cdots & \cdots \\ 0 & a_r & a_{r-1} & \cdots \end{vmatrix} = 0. \tag{3.32.9}$$

As an example, let

$$f \equiv x^m - N.$$

Then

$$a_0 = x^m - N, \qquad a_1 = mx^{m-1}, \qquad a_2 = m(m-1)x^{m-2}/2.$$

We have for third-order iteration

$$\phi = x - a_0 a_1 \Big/ \begin{vmatrix} a_1 & a_0 \\ a_2 & a_1 \end{vmatrix},$$

which becomes after simplification

$$\phi = x[(m-1)x^m + (m+1)N]/[(m+1)x^m + (m-1)N].$$

This defines Bailey's iteration for root extraction. For square roots

$$\phi = x(x^2 + 3N)/(3x^2 + N).$$

3.33. *Iterations of Higher Order: Aitken's δ^2 Process.* Consider any sequence x_i, satisfying the iteration $\phi(x)$, with the limit α. The sequence

$$u_i = x_i - \beta \tag{3.33.1}$$

for any fixed β has the limit $\alpha - \beta$ and satisfies the iteration

$$\psi(u) \equiv \phi(u + \beta) - \beta, \tag{3.33.2}$$

since

$$x_{i+1} = u_{i+1} + \beta = \phi(x_i) = \phi(u_i + \beta).$$

In particular if $\beta = \alpha$, the sequence $\xi_i = x_i - \alpha$ represents the deviations of the approximations x_i from the limit α, and this sequence is defined by the initial deviation ξ_0, together with the iteration

$$\omega(\xi) \equiv \phi(\xi + \alpha) - \alpha. \tag{3.33.3}$$

If the iteration ϕ is of order r, then

$$\omega(\xi) \equiv \alpha_r\xi^r + \alpha_{r+1}\xi^{r+1} + \cdots . \tag{3.33.4}$$

Now let $\phi_{(1)}(x)$ and $\phi_{(2)}(x)$ represent any two iterations of the same order and not necessarily distinct. With these can be associated the iterations $\omega_{(1)}(\xi)$ and $\omega_{(2)}(\xi)$ satisfied by the deviations. The function

$$\Phi(x) \equiv \frac{\{x\phi_{(1)}[\phi_{(2)}(x)] - \phi_{(1)}(x)\phi_{(2)}(x)\}}{\{x - \phi_{(1)}(x) - \phi_{(2)}(x) + \phi_{(1)}[\phi_{(2)}(x)]\}} \tag{3.33.5}$$

also defines an iteration. This is invariant in the sense that, if one makes the substitution (3.33.1), forms the $\psi_{(1)}$ and $\psi_{(2)}$ as in (3.33.2), defines $\Psi(u)$ as in (3.33.5) with $\psi_{(1)}$, $\psi_{(2)}$, and u replacing $\phi_{(1)}$, $\phi_{(2)}$, and x, then $\Psi(u) \equiv \Phi(u + \beta) - \beta$. This can be verified by direct substitution.

Now it can be shown that, if the iterations $\phi_{(1)}$ and $\phi_{(2)}$ are of order r, then the iteration Φ is of order greater than r, provided only

$$[\phi'_{(1)}(\alpha) - 1][\phi'_{(2)}(\alpha) - 1] \neq 0. \tag{3.33.6}$$

In fact if $r = 1$, and this condition (3.33.6) is satisfied, then Φ is of order 2 at least, and if $r > 1$, then Φ is of order $2r - 1$, and condition (3.33.6) is necessarily satisfied.

Because of the invariance, it is sufficient to consider

$$\Omega(\xi) \equiv \frac{\{\xi\omega_{(1)}[\omega_{(2)}(\xi)] - \omega_{(1)}(\xi)\omega_{(2)}(\xi)\}}{\{\xi - \omega_{(1)}(\xi) - \omega_{(2)}(\xi) + \omega_{(1)}[\omega_{(2)}(\xi)]\}},$$

where the sequences of deviations satisfy

$$\omega_{(1)}(\xi) \equiv \alpha_r^{(1)}\xi^r + \cdots ,$$
$$\omega_{(2)}(\xi) \equiv \alpha_r^{(2)}\xi^r + \cdots .$$

Then

$$\omega_{(1)}[\omega_{(2)}(\xi)] \equiv \alpha_r^{(1)}(\alpha_r^{(2)}\xi^r + \cdots)^r + \cdots .$$

Hence if $r > 1$, the term of lowest degree in the numerator is the second term which is of degree $2r$, while the term of lowest degree in the denominator is the term ξ, itself of degree 1. Hence an expansion of the fraction in powers of ξ will begin with a term of degree $2r - 1$, and this exceeds r when $r > 1$.

If $r = 1$, then to form the numerator we have

$$\xi\omega_{(1)}[\omega_{(2)}(\xi)] \equiv \alpha_1^{(1)}\alpha_1^{(2)}\xi^2 + \cdots,$$
$$\omega_{(1)}(\xi)\omega_{(2)}(\xi) \equiv \alpha_1^{(1)}\alpha_1^{(2)}\xi^2 + \cdots,$$

whence on subtraction the terms in ξ^2 drop out, and the term of lowest degree in ξ is of degree 3. For the denominator we have

$$\begin{aligned}\xi - \omega_{(1)}(\xi) - \omega_{(2)}(\xi) + \omega_{(1)}[\omega_{(2)}(\xi)]& \\ &= (1 - \alpha_1^{(1)} - \alpha_1^{(2)} + \alpha_1^{(1)}\alpha_1^{(2)})\xi + \cdots \\ &= (1 - \alpha_1^{(1)})(1 - \alpha_1^{(2)})\xi + \cdots.\end{aligned}$$

But

$$\alpha_1^{(1)} = \phi'_{(1)}(\alpha), \qquad \alpha_1^{(2)} = \phi'_{(2)}(\alpha),$$

so that, if (3.33.6) is satisfied, the expansion of the denominator begins with a term in the first power of ξ. Hence the expansion of the fraction begins with a term of degree 2 at least, and the iteration is therefore of order 2 at least.

Thus given two iterations of the same order, one can always form an iteration of higher order. But it was nowhere required that $\phi_{(1)}$ and $\phi_{(2)}$ be distinct, so that from any single iteration one can form an iteration of higher order. More than this, an iteration Φ of order $r > 1$ always converges in some neighborhood about α, whereas an iteration ϕ of order 1 need not converge and will not unless $|\phi'(\alpha)| < 1$, as we have seen. Hence from any function ϕ, analytic in the neighborhood of α and satisfying $\phi(\alpha) = \alpha$, one can form an iteration which converges to α whether that defined by ϕ converges or not.

In ordinary practical application it is not desirable to form Φ explicitly. Instead, one can proceed as follows: One forms

$$x_1 = \phi(x_0), \qquad x_2 = \phi(x_1)$$

in the usual manner. However, for x_3 one takes not $\phi(x_2)$ but $\Phi(x_0)$ by

$$x_3 = (x_0x_2 - x_1^2)/(x_0 - 2x_1 + x_2).$$

In terms of the difference operator Δ, defined by

$$\Delta x_i = x_{i+1} - x_i,$$
$$\Delta^2 x_i = \Delta x_{i+1} - \Delta x_i = x_{i+2} - 2x_{i+1} + x_i.$$

this can be written

$$x_3 = x_0 - (\Delta x_0)^2/\Delta^2 x_0,$$

and more generally one can take

$$x_{3(\nu+1)} = x_{3\nu} - (\Delta x_{3\nu})^2/\Delta^2 x_{3\nu}.$$

This form brings out the fact that in practical computation any sequence of digits on the left that is identical for $x_{3\nu}$, $x_{3\nu+1}$, and $x_{3\nu+2}$ can be ignored, since it drops out in forming the differences and is restored in subtracting from $x_{3\nu}$.

Just as the iteration Φ was of order higher than ϕ, so one can form from Φ an iteration of still higher order. Thus having computed

$$x_3 = \Phi(x_0), \qquad x_6 = \Phi(x_3),$$

instead of computing $\phi(x_6)$, one could now form

$$x_7 = (x_0 x_6 - x_3^2)/(x_0 - 2x_3 + x_6).$$

In principle, iterations of arbitrarily high order can be built up by proceeding in this manner.

3.34. *Iterations of Higher Order: Schröder's Method.* The oldest method of obtaining iterations of arbitrary order seems to be that given by Schröder. Consider any simple root α of

$$f(x) = 0. \tag{3.34.1}$$

In the neighborhood of α we can set

$$y = f(x) \tag{3.34.2}$$

and let

$$x = h(y), \tag{3.34.3}$$

where

$$x \equiv h[f(x)], \qquad y \equiv f[h(y)], \tag{3.34.4}$$

identically. Hence

$$\alpha = h(0). \tag{3.34.5}$$

If we expand

$$X = h(Y)$$

in powers of $(Y - y)$, regarding x and y as fixed, we have

$$X = x + (Y - y)h_1(y) + (Y - y)^2 h_2(y) + \cdots,$$

where

$$h_r(y) = h^{(r)}(y)/r!. \tag{3.34.6}$$

Hence

$$\alpha = x - yh_1(y) + \cdots + (-y)^r h_r(y) + (-y)^{r+1} R_{r+1}(y). \tag{3.34.7}$$

Now the h_r are functions of y, but y is a function of x by (3.34.2), so that we may write

$$b_r(x) = h_r[f(x)], \tag{3.34.8}$$

with

$$b_1 = 1/f', \qquad b_r = b'_{r-1}/(rf'). \tag{3.34.9}$$

Then (3.34.7) can be written

$$\alpha \equiv x - fb_1 + \cdots + (-f)^r b_r + (-f)^{r+1} R_{r+1}(f) \tag{3.34.10}$$

identically. If we define ϕ_r by

$$\alpha \equiv \phi_r(x) + (-f)^{r+1} R_{r+1}(f), \tag{3.34.11}$$

then we see that ϕ_r provides an iteration of order $r + 1$, since its rth derivative must vanish with f. Again ϕ_1 yields Newton's method.

The quantities $b_i(x)$ required for the iteration

$$\phi_r(x) = x - fb_1 + \cdots + (-f)^r b_r \tag{3.34.12}$$

can be formed by successive differentiation as in (3.34.9), where the prime denotes differentiation with respect to x. It is also possible to write a system of recursion relations which can be used in case the successive derivatives of $f(x)$ are known or easily evaluated. If these are known, we can write the expansion of $Y = f(X)$ in powers of $(X - x)$:

$$Y - y = a_1(X - x) + a_2(X - x)^2 + a_3(X - x)^3 + \cdots,$$

where $a_i(x) = f^{(i)}(x)/i!$. If this is substituted into the expansion of $X - x$ given above, we obtain

$$\begin{aligned} X - x \equiv {} & b_1[a_1(X - x) + a_2(X - x)^2 + a_3(X - x)^3 + \cdots] \\ & + b_2[a_1^2(X - x)^2 + 2a_1a_2(X - x)^3 + \cdots] + b_3[a_1^3(X - x)^3 + \cdots] \end{aligned}$$

after replacing the $h_r(y)$ by $b_r(x)$, as in (3.34.8). Now the two sides must be equal identically so that

$$\begin{aligned} a_1b_1 &= 1, \\ a_1^2b_2 + a_2b_1 &= 0, \\ a_1^3b_3 + 2a_1a_2b_2 + a_3b_1 &= 0, \\ \cdots\cdots\cdots\cdots\cdots\cdots \end{aligned} \tag{3.34.13}$$

This is the required recursion for expressing the $b_i(x)$ in terms of the $a_i(x)$, and hence of the derivatives of f.

3.35. *Iterations of Higher Order: Polynomials.* Three distinct methods have been given for forming iterations of higher order. The δ^2 process presupposes that some iteration is known and deduces from it an iteration of higher order. The methods of Schröder and of Richmond start with the equation $f = 0$ and form directly an iteration of any prescribed order, provided only that the root α is a simple root and that f is analytic, or at least possesses derivatives of sufficiently high order, in the neighborhood of α. Clearly if $g(\alpha) \neq 0$ and is analytic in the same neighborhood, one could apply either method to the equation $F = 0$, where $F \equiv fg$ or where $F \equiv f/g$. Thus there are a great many ways of forming an iteration of any order, and doubtless many different iterations. In special cases it might be desirable to impose upon the iterations special conditions other than the order of convergence. In particular if f is a polynomial, one might ask that ϕ be a polynomial. This would be desirable in case one is using a computing machine for which division is inconvenient; or in operations with matrices, where direct inversion is to be avoided; or, as Rademacher and Schoenberg have shown, in operations with Laurent's series, where direct inversion is impossible.

We ask now whether, when f is a polynomial, one can find a function g such that, when Schröder's method is applied to the equation

$$F \equiv f/g = 0, \tag{3.35.1}$$

an iteration ϕ_r will be a polynomial. Taking first ϕ_1, we wish to choose g so that

$$(f/g)/(f/g)' = pf,$$

where p is a polynomial. This requires that

$$pf' - p(g'/g)f = 1.$$

But if $f = 0$ has only simple roots, one can always find polynomials p and q such that

$$pf' - qf = 1 \tag{3.35.2}$$

by the theorem of §3.06. Hence if g satisfies

$$g'/g = q/p, \tag{3.35.3}$$

the requirement is satisfied for ϕ_1.

Now suppose that the process yields

$$\phi_r = x - fp_1 + f^2p_2/2! - \cdots + (-f)^r p_r/r!, \qquad p_1 = p \tag{3.35.4}$$

and that all p_i, and hence all ϕ_i, are polynomials for $i \leq r$. To obtain ϕ_{r+1} we must add to this a term

$$(-f)^{r+1}p_{r+1}/(r+1)! = (-F)^{r+1}B_{r+1},$$

where the B's are obtained from F as were the b's from f in (3.34.9). Hence

$$B_r = g^r p_r/r!,$$

and

$$B_{r+1} = (g^r p_r)'/(r+1)!F'.$$

After minor manipulations and applications of (3.35.2) and (3.35.3), this gives

$$B_{r+1} = g^{r+1}(pp'_r + rqp_r)/(r+1)!$$

and therefore

$$F^{r+1}B_{r+1} = f^{r+1}(pp'_r + rqp_r)/(r+1)!.$$

Hence

$$p_{r+1} = pp'_r + rqp_r \tag{3.35.5}$$

is a polynomial, and this recursion, together with (3.35.4), defines polynomial iterations of all orders. Note that g itself is not required explicitly, but only p and q.

For illustration let

$$f \equiv 1 - x^n a, \qquad f' \equiv -nx^{n-1}a.$$

Then

$$(-x/n)(-nx^{n-1}a) + (1 - x^n a) \equiv 1.$$

Hence

$$p = -x/n, \qquad q = -1,$$
$$g'/g = n/x,$$

and a possible choice is $g = x^n$.

We can now evaluate the p_i recursively and obtain a polynomial iteration of any order for the nth root. But in this case it is easy to verify that

$$\alpha \equiv x(1-f)^{-1/n}.$$

On referring back to the argument used in §3.34 to derive Schröder's iterations in general, we may conclude that the first r terms in the expansion in powers of f must provide an iteration of order r, and this is certainly a polynomial. Direct verification shows it to be the same as is given by (3.35.4) and (3.35.5).

For the case $n = 1$, the iteration of order r is

$$\phi = x(1 + f + f^2 + \cdots + f^{r-1}), \tag{3.35.6}$$

and for $r = 2$ this becomes on expanding f

$$\phi = x(2 - ax).$$

For the case when a and x are matrices, this defines the Hotelling-Bodewig iteration for inverting a matrix.

On introducing the subject of functional iteration, it was shown that the iteration would converge to α when x_0 is any point in the complex plane within a circle about α throughout which a certain Lipschitz condition is satisfied. This condition is sufficient but by no means necessary. In particular, suppose α is real and ϕ is a polynomial. The curve

$y = \phi$ intersects the line $y = x$ at $x = y = \alpha$, and if the iteration is of order 2 or greater, then at this point the curve has a horizontal tangent. Suppose the curve intersects the line also at some point $\alpha' < \alpha$ but nowhere between α' and α. Then if $\alpha' < x_0 \leq \alpha$, the sequence converges, whereas the Lipschitz condition holds only in the vicinity of α. Likewise if the curve intersects the line at $\alpha'' > \alpha$ but nowhere between α and α'', then the sequence converges when $\alpha \leq x_0 < \alpha''$. For $x_0 \leq \alpha'$ and for $x_0 \geq \alpha''$, the sequence either diverges or converges to some other limit. Thus for the iteration $x(1 + f + f^2)$ to a^{-1}, the curve and line intersect at $x = 0$, a^{-1}, and $2a^{-1}$, and the sequence converges to a^{-1} if and only if $0 < x_0 < 2a^{-1}$. For the iteration $x(1 + f)$, the curve intersects the line only at $x = 0$ and a^{-1}, while $\phi(2a^{-1}) = 0$. This sequence converges under the same conditions, as does any iteration (3.35.6).

3.4. Systems of Equations. All methods so far described have been methods for dealing with a single equation in a single unknown. For solving a system of equations one would like to follow the Gaussian procedure in systems of linear equations, eliminating one variable from each of $n - 1$ of the equations, another from each of $n - 2$ of these, and so on until there is left a single equation in a single unknown. This of course is quite impossible ordinarily. Nevertheless, it is possible to reduce the problem to that of solving an infinite sequence of single equations. For each equation of the sequence any of the methods described above can be applied. The reduction to an infinite sequence can be accomplished by the method of steepest descent.

On the other hand, one can attempt to reduce the problem to that of solving an infinite sequence of sets of linear equations. This type of reduction is accomplished by an appropriate generalization of the method of functional iteration. These two methods will now be described. Only the case of real functions of real variables will be considered in this section.

3.41. *The Method of Steepest Descent.* The method of steepest descent applies specifically to the location of the maximum or minimum of a function of n real variables. However if we have to solve a set of n equations in n variables

$$\psi_i = 0, \tag{3.41.1}$$

the function

$$\phi = \Sigma\psi_i^2 \tag{3.41.2}$$

takes on the minimum value $\phi = 0$ at all points satisfying (3.41.1). More generally if α_{ij} are elements of a positive definite matrix, the function

$$\phi^* = \Sigma\Sigma\psi_i\alpha_{ij}\psi_j \tag{3.41.3}$$

also takes on the minimum value $\phi^* = 0$ at the same points. There are thus many ways in which the problem of solving a set of equations can be replaced by a problem in minimization. We therefore consider the problem of minimizing the function $\phi(\xi_1, \xi_2, \ldots, \xi_n)$, or briefly $\phi(x)$, where x is the vector whose components are the ξ_i. The partial derivatives

$$\phi_{\xi_i} = \partial\phi/\partial\xi_i$$

are components of a vector which we shall designate as ϕ_x and take to be a column vector. This is known as the gradient of ϕ, often denoted by grad ϕ or $\nabla\phi$, and its direction at any point x is normal to that surface

$$\phi = \text{const} \tag{3.41.4}$$

which passes through the point x.

In the neighborhood of the point x which minimizes ϕ, the surfaces (3.41.4) are closed if, as we suppose, ϕ is continuous, since if we take the constant sufficiently close to the minimum value, then along any ray through the point x the function ϕ can only increase, and at some point along the ray the function will first take on the value of the assigned constant.

Now suppose that x_0 represents some initial estimate to the required x, close enough to it so that the surface

$$\phi(x) = \phi(x_0) \tag{3.41.5}$$

is closed, and that the point x lies in the region enclosed by the surface. Then choose an arbitrary vector u, subject only to the requirement that at x_0

$$u^\mathsf{T}\phi_x(x_0) \neq 0. \tag{3.41.6}$$

This means that the direction u is not tangent to the surface (3.41.5) at x_0. It therefore cuts through the surface and so intersects surfaces at which ϕ has smaller (as well as larger) values than at x_0. Determine

$$x_1 = x_0 - \lambda u \tag{3.41.7}$$

as that point on the line through x_0 in the direction of u at which ϕ takes on its least value. Thus we minimize the function

$$\phi_1(\lambda) = \phi(x_0 - \lambda u) \tag{3.41.8}$$

of the single variable λ. If the line in question should happen to pass through the required point x, then we shall have solved our problem, but in general this will not happen. However, x_1 will be on a surface

$$\phi(x) = \phi(x_1) \tag{3.41.9}$$

at which $\phi(x)$ has a value smaller than it has at x_0. The equation for locating x_1 is obtained by equating to zero the derivative of $\phi_1(\lambda)$:

$$u^{\mathsf{T}}\phi_x(x_0 - \lambda u) = 0. \tag{3.41.10}$$

This is not satisfied by $\lambda = 0$ because of condition (3.41.6). Equation (3.41.10) states that at the point x_1 the line through x_0 in the direction u is tangent to the surface (3.41.9), and this is geometrically evident. At x_1, choose a new direction u, not tangent to the surface, and proceed sequentially.

In this way is obtained a monotonically decreasing sequence $\phi(x_i)$ that is bounded below by the minimum value $\phi(x)$. The sequence therefore has a limit at some point x_∞. If $\phi_x(x_\infty) = 0$, then x_∞ minimizes ϕ, and necessarily $x_\infty = x$ if x_0 is close enough to x so that ϕ has a minimum at one point only within the closed surface (3.41.5). But if $\phi_x(x_\infty) \neq 0$, then we can find a direction u not tangent to the surface through x_∞ and so proceed farther. But then no limit would have been reached, and hence the limit $x_\infty = x$.

In the method of steepest descent, strictly so-called, one chooses always $u = \phi_x$. Since ϕ_x is the direction of most rapid variation of ϕ, this choice may be expected to lead to convergence after the fewest steps. However, each step is fairly cumbersome. As with linear systems, it is much simpler to take for each u one of the reference vectors e_i, either in rotation as in the Seidel process, or selected by some other criterion. It is in keeping with the method of relaxation to examine the components of ϕ_x and select the largest. If this is the ith, then one takes $u = e_i$, and Eq. (3.41.10) reduces to

$$\phi_{\xi_i}(x_0 - \lambda e_i) = 0, \tag{3.41.11}$$

which amounts to solving the ith equation for the ith variable as a function of current estimates of the other variables.

Equation (3.41.11) in λ will ordinarily be nonlinear, and one of the methods of successive approximation described above for an equation in one variable will generally have to be applied. Newton's method, for example, gives

$$\lambda' = \phi_{\xi_i}(x_0)/\phi_{\xi_i\xi_i}(x_0) \tag{3.41.12}$$

as the first approximation to λ. This will not necessarily reduce ϕ_{ξ_i} to zero, but it should reduce the magnitude of the vector ϕ_x. If so, it is sufficient to take $x_1' = x_0 - \lambda' e_i$ and look for the largest component of ϕ_x at this point.

One should note carefully that the equations $\phi_{\xi_i} = 0$ being solved explicitly by this method are not in general the same as Eqs. (3.41.1), but they are satisfied by any solution x of (3.41.1). The new equations may be more complicated than the original. If Eqs. (3.41.1) themselves arise from a minimizing (or a maximizing) problem, then they can be used as they are, without forming the function ϕ by (3.41.2).

3.42. *Functional Iteration.* Newton's method generalizes directly to systems of equations. Consider first the general functional iteration in n variables. Let $g(x)$ stand for the vector whose elements are $\gamma_i(\xi_1, \xi_2, \ldots, \xi_n)$. Thus g is a function of the vector x. Suppose for some constant vector a it is true that

$$g(a) = a, \tag{3.42.1}$$

and consider the sequence defined by some x_0 and

$$x_{i+1} = g(x_i). \tag{3.42.2}$$

Under what circumstances will this sequence of vectors converge to the vector a?

A sufficient condition for this can be given by means of a Lipschitz condition. If, as in Chap. 2, we let $b(v)$ represent the magnitude of the numerically greatest element in the vector v, then the sequence (3.42.2) converges to the vector a, provided that for some $k < 1$ and for some ρ it is true that

$$b[g(x') - g(x'')] < kb(x' - x'') \tag{3.42.3}$$

for every x' and x'' satisfying

$$b(x' - a) < \rho, \qquad b(x'' - a) < \rho,$$

if also $b(x_0 - a) < \rho$. The proof is made exactly as in the one-dimensional case if one uses the maximum absolute value whenever the absolute value was used before.

Again, if for some $k < 1$ the same condition (3.42.3) holds whenever x' and x'' satisfy

$$b(x' - x_0) < \rho, \qquad b(x'' - x_0) < \rho,$$

and if also

$$b[g(x_0) - x_0] < (1 - k)\rho,$$

then we can conclude that the sequence (3.42.2) has a limit, which we may call a, that

$$b(a - x_0) < \rho,$$

and that a satisfies (3.42.1). Hence under these circumstances we are assured that a solution exists.

Now consider the system of n equations in n variables

$$\phi_i(x) = 0, \tag{3.42.4}$$

where x represents the vector of elements ξ_j. If each function ϕ_i is analytic in the neighborhood of some point x_0, then

(3.42.5) $$\phi_i(x) = \phi_i(x_0) + \Sigma(\xi_j - \xi_{0,j})\partial\phi_i(x_0)/\partial\xi_j + \tfrac{1}{2}\Sigma\Sigma(\xi_j - \xi_{0,j})(\xi_k - \xi_{0,k})\partial^2\phi_i(x_0)/\partial\xi_j\,\partial\xi_k + \cdots .$$

If Eqs. (3.42.4) have a solution x sufficiently close to x_0, that is, for which $b(x - x_0)$ is sufficiently small, we might expect that by solving the equations

$$0 = \phi_i(x_0) + \Sigma(\xi_{1,j} - \xi_{0,j})\partial\phi_i(x_0)/\partial\xi_j$$

for the quantities $\xi_{1,j}$ we would obtain an approximation to the ξ_j that is better than the $\xi_{0,j}$. If f is the vector whose elements are the ϕ_i, and if we introduce the matrix

(3.42.6) $$f_x(x_0) = [\partial\phi_i(x_0)/\partial\xi_j],$$

which is the Jacobian matrix of the functions ϕ_i evaluated at x_0, these equations have the form

(3.42.7) $$f_x(x_0)x_1 = f_x(x_0)x_0 - f(x_0),$$

and if $f_x(x_0)$ is nonsingular, then they have the solution

(3.42.8) $$x_1 = x_0 - f_x^{-1}(x_0)f(x_0),$$

which is the direct generalization of the iteration given by Newton's method.

The iteration does converge, and it is not necessary that derivatives of all orders of the ϕ_i should exist. However if n is at all large, the repeated evaluation of the inverse matrix f_x^{-1} or the repeated solution of linear systems of the type (3.42.7) is certainly undesirable. Consequently a somewhat more general theorem will be proved.

Suppose all functions ϕ_i have continuous first partial derivatives in the region of n space being considered, and suppose moreover that this region is convex. If ϕ is any one of these functions, and ϕ_x is the row vector of its first partial derivatives, then for any x' and x'' in the region

$$d\phi[x' + \theta(x'' - x')]/d\theta = \phi_x[x' + \theta(x'' - x')](x'' - x').$$

Hence

$$\int_0^1 \phi_x[x' + \theta(x'' - x')](x'' - x')d\theta = \phi(x'') - \phi(x').$$

Written for all the functions, this identity becomes

(3.42.9) $$f(x'') = f(x') + \int_0^1 f_x[x' + \theta(x'' - x')](x'' - x')d\theta.$$

For brevity write

(3.42.10) $$F(x', x'') = \int_0^1 f_x[x' + \theta(x'' - x')]d\theta,$$

Then

$$F(x', x') = f_x(x').$$

If $F(x_0, x_0)$ is nonsingular, then $F(x', x'')$ will remain nonsingular for x' and x'' in a sufficiently small neighborhood of x_0. Define the vector

$$g(x) \equiv x - f_x^{-1}(x_0)f(x). \tag{3.42.11}$$

Then

$$g(x_0) - x_0 = -f_x^{-1}(x_0)f(x_0),$$

and

$$\begin{aligned} g(x'') - g(x') &= (x'' - x') - f_x^{-1}(x_0)[f(x'') - f(x')] \\ &= (x'' - x') - f_x^{-1}(x_0)F(x', x'')(x'' - x') \\ &= [I - F^{-1}(x_0, x_0)F(x', x'')](x'' - x'). \end{aligned}$$

For $x' = x'' = x_0$, the matrix within the brackets vanishes. For ρ not too large it will be true that for some $k < 1$

$$b[I - F^{-1}(x_0, x_0)F(x', x'')] < k/n,$$

when $b(x' - x_0) < \rho$ and $b(x'' - x_0) < \rho$. Then

$$b[g(x'') - g(x')] < kb(x'' - x').$$

Also

$$b[g(x_0) - x_0] \le nb[f_x^{-1}(x_0)]b[f(x_0)].$$

Hence the iteration will converge if x_0 is close enough to the solution x so that

$$nb[f_x^{-1}(x_0)]b[f(x_0)] < (1 - k)\rho.$$

This shows that, if the functions $\phi_i(x)$ have continuous first partial derivatives in the neighborhood of a solution x, and if the Jacobian f_x is nonsingular in the neighborhood of this solution, then the iteration (3.42.11) converges whenever x_0 is chosen sufficiently close to x. It is true a fortiori that the Newtonian iteration defined by

$$g(x) \equiv x - f_x^{-1}(x)f(x) \tag{3.42.12}$$

also converges.

Unfortunately, the practical question as to the convergence of the computations modeled on this method is left undecided in general, and round-off errors may make any degree of accuracy impossible. Even in the case $n = 1$, the error in the calculation of x_{i+1} from x_i depends upon the errors present in the computation of f and of f_x, and the accuracy with which these can be calculated depends entirely upon the nature of the functions. If the errors in the calculation of f and f_x can be estimated, then one can estimate the error in the quotient f/f_x or in thes olution $f_x^{-1}f$. This is the error in the correction $x_{i+1} - x_i$, and when it becomes large by comparison with the computed correction, the further applications of the iteration are of no value.

3.5. Complex Roots and Methods of Factorization. For a single variable, Newton's method and its generalizations apply equally to the determination of real and of complex roots. Also by appropriate application Graeffe's method and Bernoulli's method will yield complex roots. However, they are somewhat inconvenient, and a number of special methods have been devised for finding the real and imaginary parts of complex solutions.

Any analytic function $f(z)$ of a complex variable $z = x + iy$ can always be written in the form

$$f(x + iy) \equiv u(x, y) + iv(x, y), \tag{3.5.1}$$

where u and v are real functions of the real variables x and y. In particular if f is a polynomial, one can write Taylor's series

$$f(x + iy) \equiv f(x) + iyf'(x) - y^2f''(x)/2! - iy^3f'''(x)/3! + \cdots, \tag{3.5.2}$$

whence

$$\begin{aligned} u(x, y) &\equiv f(x) - y^2f''(x)/2! + y^4f^{iv}(x)/4! - \cdots, \\ v(x, y) &\equiv y[f'(x) - y^2f'''(x)/3! + \cdots], \end{aligned}$$

so that u and v/y are functions of x and y^2.

Any solution z of

$$f(x) = 0 \tag{3.5.3}$$

must be in the form $z = x + iy$, where x and y are real and satisfy

$$u(x, y) = v(x, y) = 0. \tag{3.5.4}$$

These are real equations in the real variables x and y, and can be treated by either of the methods described in §3.4.

In the case of a polynomial f, it is possible to eliminate y, obtaining a single equation in x to be solved for real roots only. By substitution one can obtain the equations in the associated y's. For brevity write

$$\begin{aligned} u &\equiv a_0 + a_2y^2 + \cdots + a_{2m}y^{2m} = 0, \\ y^{-1}v &\equiv a_1 + a_3y^2 + \cdots + a_{2m+1}y^{2m} = 0. \end{aligned}$$

Multiply the first equation by a_1, the second by a_0, subtract, and divide through by y^2. The resulting equation is of degree $m - 1$ in y^2. If $a_{2m+1} \neq 0$, multiply the first equation by a_{2m+1}, the second by a_{2m}, and subtract. This gives a second equation of degree $m - 1$ in y^2, and these two can be treated as were the original two. Eventually there results an equation of degree 0 in y^2. If $a_{2m+1} = 0$, continue with the equation resulting from the first elimination along with $y^{-1}v = 0$.

For applying Newton's method or one of its generalizations to the original equation, one can also separate ϕ into its real and imaginary parts, writing

$$\phi(x + iy) = \psi(x, y) + i\omega(x, y). \tag{3.5.5}$$

By a slight modification of the sequence $z_{i+1} = \phi(z_i)$, one can write

$$x_{i+1} = \psi(x_i, y_i), \qquad y_{i+1} = \omega(x_{i+1}, y_i).$$

When $f(z)$ is a real polynomial, all complex roots occur in conjugate pairs, and the roots of a conjugate pair satisfy a real quadratic equation. Hence a real polynomial $f(z)$ can be completely factored into real quadratic and linear factors. If the coefficients of a real quadratic factor of $f(z)$ are known, it is then a simple matter to find its zeros, whether they may be real or complex.

Let

$$d(z) \equiv z^2 + az + b. \tag{3.5.6}$$

If z_j satisfies

$$z_j^2 = -az_j - b,$$

i.e., if z_j is a zero of $d(z)$, then z_j satisfies also

$$\begin{aligned} z_j^3 &= -a(-az_j - b) - bz_j \\ &= (a^2 - b)z_j + ab, \end{aligned}$$

and in general any positive integral power of z_j is expressible as a linear function of z_j whose coefficients are polynomials in a and b. Hence also $f(z_j)$ is so expressible:

$$f(z_j) = r_1(a, b)z_j + r_0(a, b).$$

If $d(z)$ is not a perfect square, it has a zero $z_k \neq z_j$ and

$$f(z_k) = r_1(a, b)z_k + r_0(a, b).$$

If z_j and z_k are zeros also of $f(z)$ (whether complex or not), then

$$f(z_j) = f(z_k) = 0,$$

and for z_j and z_k distinct this implies that

$$r_1(a, b) = r_0(a, b) = 0. \tag{3.5.7}$$

Here are two equations in the two unknowns a and b which must be satisfied by the coefficients of a quadratic factor $d(z)$ of $f(z)$.

The polynomials r_0 and r_1 in a and b could be determined equally well in a slightly different manner. Dropping the subscript on the z, if z is any zero of $d(z)$, we can say that

$$z^n = -az^{n-1} - bz^{n-2}.$$

Hence if $f(z)$ is of degree n, this substitution reduces it to a polynomial of degree $n - 1$ in the zeros of $d(z)$, and the coefficients of the polynomial are themselves polynomials in a and b. Likewise

$$z^{n-1} = -az^{n-2} - bz^{n-3},$$

and by continuing sequentially in this way, $f(z)$ is again reduced to the form

$$f(z) = r_1(a, b)z + r_0(a, b), \tag{3.5.8}$$

valid for any z satisfying $d(z) = 0$. But now one can see easily that this is precisely the process for forming the remainder after division of $f(z)$ by $d(z)$. Hence for all z it is true that

$$f(z) \equiv (z^2 + az + b)Q_1(z) + zr_1(a, b) + r_0(a, b), \tag{3.5.9}$$

where $Q_1(z)$ is the quotient. Hence if a and b satisfy (3.5.7), the division is exact. Hence the conditions (3.5.7) are sufficient, as well as necessary, for $d(z)$ to divide $f(z)$.

For the solution of Eqs. (3.5.7) Hitchcock gives an iteration which is, in fact, an application of Newton's method. If one now divides $Q_1(z)$ by $d(z)$, then $f(z)$ can be written

$$f(z) \equiv (z^2 + az + b)^2Q(z) + (z^2 + az + b)q(z; a, b) + r(z; a, b), \tag{3.5.10}$$

where

$$\begin{aligned} q(z; a, b) &\equiv zq_1(a, b) + q_0(a, b), \\ r(z; a, b) &\equiv zr_1(a, b) + r_0(a, b), \end{aligned} \tag{3.5.11}$$

and $q(z; a, b)$ is the remainder after the second division.

Let z represent either zero of $d(z)$. Then from differentiating (3.5.10) with respect to a and to b, since f is itself independent of a and b, it follows that

$$\begin{aligned} 0 &= zq + \partial r/\partial a, \\ 0 &= q + \partial r/\partial b. \end{aligned}$$

In detail these equations are

$$\begin{aligned} z^2q_1 + zq_0 + z\partial r_1/\partial a + \partial r_0/\partial a &= 0, \\ zq_1 + q_0 + z\partial r_1/\partial b + \partial r_0/\partial b &= 0. \end{aligned}$$

These equations must hold for any zero z of $d(z)$. Hence the first can be written

$$z(\partial r_1/\partial a + q_0 - aq_1) + (\partial r_0/\partial a - bq_1) = 0,$$

and the second

$$z(\partial r_1/\partial b + q_1) + (\partial r_0/\partial b + q_0) = 0.$$

If $d(z)$ is not a perfect square, then the coefficients of z and the terms free of z must vanish separately. Hence

$$\begin{aligned} \partial r_1/\partial a &= aq_1 - q_0, & \partial r_0/\partial a &= bq_1, \\ \partial r_1/\partial b &= -q_1, & \partial r_0/\partial b &= -q_0. \end{aligned} \tag{3.5.12}$$

These are the partial derivatives required for the application of Newton's method to the solution of (3.5.7), and they are obtained by two divisions of $f(z)$ by $(z^2 + az + b)$.

Hence if a_α and b_α represent approximations to the coefficients a and b of an exact division $d(z)$, then improved approximations $a_{\alpha+1}$, $b_{\alpha+1}$ can be obtained by solving

$$\begin{aligned} (a_\alpha q_1 - q_0)(a_{\alpha+1} - a_\alpha) - q_1(b_{\alpha+1} - b_\alpha) &= -r_1, \\ b_\alpha q_1(a_{\alpha+1} - a_\alpha) - q_0(b_{\alpha+1} - b_\alpha) &= -r_0, \end{aligned} \tag{3.5.13}$$

where q_0, q_1, r_0, r_1 are the quantities obtained after division of $f(z)$ twice by $z^2 + a_\alpha z + b_\alpha$. When the process is carried out in this way, the general forms of the polynomials r_0 and r_1 in a and b are not obtained. Instead their numerical values and those of their partial derivatives are obtained by the divisions which use the current numerical approximations a_α and b_α.

This method can be generalized to yield a factor of arbitrary degree. If one writes down formally a factorization of $f(z)$ into factors with unknown coefficients, then by expressing that $f(z)$ is to equal identically the product of these factors one obtains a set of equations relating the unknown coefficients. Let the unknown coefficients be represented as $a_1, a_2, \ldots, a_N$ taken in any order, and let the conditions be written

$$\psi_1 = \psi_2 = \cdots = \psi_N = 0,$$

where each ψ is a polynomial in the a's. If with each ψ_i one can associate an a_j in such a way that $\psi_i = 0$ is easily solved for a_j as a function of the other a's, one can use this fact to define formally an iterative scheme for evaluating the coefficients in the factorization, and many different such schemes have been proposed. Generally, however, the question of convergence is left open.

3.6. Bibliographic Notes. The author is indebted to Professor Schwerdtfeger for numerous references on iterative methods for both linear and nonlinear equations, as well as for a copy of some lecture notes. And at this point reference may be made to Blaskett and Schwerdtfeger (1945) on the Schröder iterations.

König (1884) published the theorem that is now classic in the theory of functions. Hadamard (1892) elaborated this and related notions, with reference to the location and characterization of singular points of analytic functions, and made application to the evaluation of zeros. A more recent discussion is that of Golomb (1943). Aitken (1926, 1931, 1936–1937b) discusses extensively the use of Bernoulli's method and the δ^2

process. Whittaker and Robinson (1940, and other editions) discuss Bernoulli's method and Whittaker's method.

The general convergence theorem for functional iteration is given by Hildebrandt and Graves (1927; see also Graves, 1946). Numerous special methods for solving equations of all types, like Picard's method for differential equations, are methods of functional iteration. Hamilton (1946) gives a number of special convergence theorems and (1950) gives an analytic derivation of the algorithm derived geometrically by Richmond (1944) and previously obtained by Critchfield and Beck (1935). Schröder (1870), Runge (1885), and others had given similar algorithms for algebraic equations.

If one defines the functions ϕ_j by $x_i = \phi(x_{i-1}) = \phi_j(x_{i-j})$, these functions are of a class called "permutable" or "commutative," on which there is an extensive literature. If x is a real or complex number, or a point in a general space, and satisfies $x = \phi(x)$, then x is said to be a "fixed point" of the transformation ϕ, and comes into consideration in the topological literature.

Collatz (1950), Wenzl (1952), and others have described iterations converging to a power of a root. On questions relating to errors and rates of convergence, see Ostrowski (1936, 1937, 1938) and Bodewig (1949).

The iteration for quadratic factors of a polynomial was given by Hitchcock (1938, 1939, 1944); generalizations are given by Lin (1941, 1943), Luke and Ufford (1951), Friedman (1949), where the factors are of arbitrary degree. There are many ways by which to define an iteration, but the treatment of convergence becomes more difficult when both factors are higher than quadratic.

As it applies to algebraic equations, Graeffe's method is frequently treated both in the textbooks and in the periodicals, and several references are given in the bibliography. The differential technique was given by Brodetsky and Smeal (1924) and was applied to transcendental equations by Lehmer (1945). For an elegant treatment of convergence in the general case see Polya (1915), and for a treatment of error see Ostrowski (1940).

Most of the basic principles in the theory of equations cited here are to be found in standard textbooks, but for properties of the Vandermonde determinants and identities relating the several types of symmetric functions see Muir's "Theory."

CHAPTER 4

THE PROPER VALUES AND VECTORS OF A MATRIX

4. The Proper Values and Vectors of a Matrix

The characteristic function of a matrix A was defined in Chap. 2 as the polynomial

$$(4.0.1) \qquad \phi(\lambda) \equiv |A - \lambda I| \equiv (-1)^n(\lambda^n + \alpha_1\lambda^{n-1} + \cdots + \alpha_n)$$

obtained by expanding the determinant of the matrix $A - \lambda I$. The characteristic equation is the equation $\phi = 0$, and its roots are the proper values of the matrix. For any proper value λ, the matrix $A - \lambda I$ is singular, whence the equation

$$(4.0.2) \qquad Ax = \lambda x$$

has at least one nontrivial solution x, and any solution is a proper vector associated with the proper value λ. Of fundamental importance to the study of proper values and vectors is the Cayley-Hamilton theorem, which states that

$$(4.0.3) \qquad \phi(A) = 0$$

identically. In words, any matrix satisfies its own characteristic equation. In special cases a matrix A may satisfy an equation of lower degree, say

$$(4.0.4) \qquad \psi(\lambda) = 0.$$

Equation (4.0.4) of the lowest degree satisfied by A is called the minimal equation, and the polynomial $\psi(\lambda)$, the minimal function. The minimal function $\psi(\lambda)$ divides $\phi(\lambda)$, and on the other hand every proper value is a root of the minimal equation.

To show that $\psi(\lambda)$ divides $\phi(\lambda)$, let

$$\phi(\lambda) = q(\lambda)\psi(\lambda) + r(\lambda),$$

where $q(\lambda)$ and $r(\lambda)$ are polynomials, and $r(\lambda)$ is of lower degree than $\psi(\lambda)$. Since $\psi(A) = 0$, and $\phi(A) = 0$, it follows that $r(A) = 0$, and this can be true only if $r(\lambda) = 0$. The same argument can be used to show that $\psi(\lambda)$ divides any polynomial $\omega(\lambda)$ for which $\omega(A) = 0$.

Before showing that all proper values satisfy (4.0.4), we show that, if $h(\lambda)$ is the highest common divisor of the elements of adj $(A - \lambda I)$, then

(4.0.5) $$\phi(\lambda) = h(\lambda)\psi(\lambda).$$

We know that

(4.0.6) $$(A - \lambda I) \text{ adj } (A - \lambda I) = \phi(\lambda)I.$$

If $P(\lambda)$ is defined by

$$\text{adj } (A - \lambda I) = h(\lambda)P(\lambda),$$

then $P(\lambda)$ is a matrix whose elements are polynomials in λ, and these polynomials in λ have no nonconstant common divisor. Now (4.0.6) becomes

$$h(\lambda)(A - \lambda I)P(\lambda) = \phi(\lambda)I.$$

But $A - \lambda I$ and $P(\lambda)$ are matrices whose elements are polynomials in λ. Hence

(4.0.7) $$(A - \lambda I)P(\lambda) = m(\lambda)I,$$

where $m(\lambda) = \phi(\lambda)/h(\lambda)$ is a polynomial.

When A replaces λ in (4.0.7), the left member vanishes. Hence $m(A) = 0$, and therefore $\psi(\lambda)$ divides $m(\lambda)$. It remains to prove that $m(\lambda)$ divides $\psi(\lambda)$. We can find a polynomial matrix $Q(\lambda)$ and a constant matrix Q_0 such that

$$\psi(\lambda)I \equiv (A - \lambda I)Q(\lambda) + Q_0$$

identically by expanding and equating coefficients of like powers of λ. But since $\psi(A) = 0$, it follows, on replacing λ by A, that $Q_0 = 0$. Hence

$$\psi(\lambda)I \equiv (A - \lambda I)Q(\lambda).$$

Since $\psi(\lambda)$ divides $m(\lambda)$, we can write

$$m(\lambda) = k(\lambda)\psi(\lambda),$$

whence by (4.0.7)

$$m(\lambda)I = k(\lambda)\psi(\lambda)I = k(\lambda)(A - \lambda I)Q(\lambda).$$

By comparing this with (4.0.7) we conclude that

$$k(\lambda)Q(\lambda) = P(\lambda).$$

Hence every element of $P(\lambda)$ is divisible by the polynomial $k(\lambda)$. Hence $k(\lambda)$ is a constant, and therefore $m(\lambda)$ and $\psi(\lambda)$ differ at most by a constant multiplier. If we now take the determinants of the two sides of (4.0.7), we have

$$\phi(\lambda)|P(\lambda)| = [m(\lambda)]^n,$$

whence every linear factor of $\phi(\lambda)$ must be also a factor of $\psi(\lambda)$.

If x is any vector, then since $\psi(A) = 0$, it is certainly true that

$$\psi(A)x = 0.$$

For a given vector $x \neq 0$ let $h(\lambda)$ be a polynomial for which it is also true that $h(A)x = 0$. Then if $d(\lambda)$ is the highest common divisor of $h(\lambda)$ and $\psi(\lambda)$, it is true that $d(A)x = 0$. In fact, one can find polynomials $p(\lambda)$ and $q(\lambda)$ such that

$$p(\lambda)\psi(\lambda) + q(\lambda)h(\lambda) = d(\lambda),$$

whence

$$p(A)\psi(A) + q(A)h(A) = d(A),$$

and the conclusion follows on multiplying this identity into x. There is therefore a polynomial of lowest degree $h(\lambda)$ for which $h(A)x = 0$. This must divide $\psi(\lambda)$, since otherwise the highest common divisor $d(\lambda)$ is of still lower degree than $h(\lambda)$, contrary to the hypothesis.

If λ_i is any proper value, then

$$\psi_i(\lambda) = \psi(\lambda)/(\lambda - \lambda_i)$$

is a polynomial, since, as was shown above, every proper value satisfies $\psi = 0$. Since ψ is the polynomial of lowest degree for which $\psi(A) = 0$, it follows that $\psi_i(A) \neq 0$, whence for some x it is true that $\psi_i(A)x \neq 0$. But

$$\psi(A) = (A - \lambda_i I)\psi_i(A),$$
$$(A - \lambda_i I)\psi_i(A)x = 0,$$

and therefore $\psi_i(A)x$ is a proper vector corresponding to the proper value λ_i. That is to say, any non-null linear combination of the columns of $\psi_i(A)$ is a proper vector.

Let λ_i be a ν_i-fold root of $\psi = 0$, and now let

(4.0.8) $$\psi_i(\lambda) = \psi(\lambda)/(\lambda - \lambda_i)^{\nu_i}.$$

Then $(\lambda - \lambda_i)^{\nu_i}$ and $\psi_i(\lambda)$ are relatively prime, whence for some polynomials $p(\lambda)$ and $q(\lambda)$

$$p(\lambda)(\lambda - \lambda_i)^{\nu_i} + q(\lambda)\psi_i(\lambda) \equiv 1,$$

identically. Hence

$$p(A)(A - \lambda_i I)^{\nu_i} + q(A)\psi_i(A) \equiv I.$$

Let $y \neq 0$ be a proper vector corresponding to λ_i. Then

(4.0.9) $$\psi_i(A)q(A)y = y.$$

Hence y is a linear combination of the columns of $\psi_i(A)$. If z is any linear combination of these columns, then the last nonvanishing vector in the sequence

$$
\begin{aligned}
z_0 &= z, \\
z_1 &= (A - \lambda_i I)z_0, \\
z_2 &= (A - \lambda_i I)z_1, \\
&\cdots\cdots\cdots
\end{aligned}
\tag{4.0.10}
$$

is a proper vector associated with λ_i, and all vectors of the sequence are principal vectors.

Among the schemes for finding the proper values of a matrix, some lead directly to the characteristic function ϕ, to the minimal function ψ, or to some divisor $\omega(\lambda)$ of the minimal function. When this function is equated to zero, the resulting equation is then to be solved by any convenient method. The scheme for finding the polynomial ϕ, ψ, or ω, as the case may be, may or may not have associated with it a scheme for finding the proper vectors. If the scheme provides only some ω, and not necessarily ψ or ϕ, it may be necessary to reapply the scheme in order to obtain the remaining proper values.

Other schemes are iterative in character, depending upon the repeated multiplication of a matrix by a vector. A scheme of this type ordinarily leads to a sequence of vectors having a proper vector as its limit and to a sequence of scalars whose limit is the associated proper value. Before describing these methods in detail, we shall introduce a few further preliminaries.

4.01. *Bounds for the Proper Values of a Matrix.* Since a nonsymmetric matrix may have complex proper values, and hence complex proper vectors, it is necessary to give further consideration to complex matrices. The natural generalization of a symmetric real matrix is a Hermitian complex matrix. The matrix A is Hermitian in case it is equal to its own conjugate transpose, *i.e.*, to the matrix obtained when every element is replaced by its complex conjugate, and the resulting matrix is then transposed. Let a bar represent the conjugate (as is customary), and an asterisk represent the conjugate transpose. Then the matrix A is Hermitian in case

$$
A^* = \bar{A}^{\mathsf{T}} = A. \tag{4.01.1}
$$

If A is Hermitian, and x is any vector, real or complex, then x^*Ax is a real number. For if we take its complex conjugate, we have $\bar{x}^*\bar{A}\bar{x}$; but this is a scalar and is equal to its own transpose $\bar{x}^{\mathsf{T}}\bar{A}^{\mathsf{T}}\bar{x}^{*\mathsf{T}} = x^*A^*x^{**}$. However $x^{**} = x$, and the theorem is proved. Hence we can define a positive definite Hermitian matrix as a Hermitian matrix for which $x^*Ax > 0$ whenever $x \neq 0$, and a non-negative semidefinite matrix as one for which $x^*Ax \geq 0$ for every x. Only a singular matrix can be semidefinite without being definite. Clearly a Hermitian matrix all of whose elements are real is a symmetric matrix.

Analogous to a real orthogonal matrix, *i.e.*, a matrix V such that $V^{\mathsf{T}}V = I = VV^{\mathsf{T}}$, is a unitary matrix U, which is one such that

$$U^*U = I = UU^*. \tag{4.01.2}$$

A unitary matrix with real elements is orthogonal.

The proper values of a Hermitian matrix are all real, since if $Ax = \lambda x$, then $x^*Ax = \lambda x^*x$, and both x^*Ax and x^*x are real numbers. Also, if complex vectors x and y are said to be orthogonal when $x^*y = y^*x = 0$, then proper vectors associated with distinct proper values of a Hermitian matrix are orthogonal. For if

$$Ax = \lambda x, \qquad Ay = \mu y,$$

then

$$y^*Ax = \lambda y^*x, \qquad x^*Ay = \mu x^*y.$$

But

$$y^*Ax = x^*Ay, \qquad y^*x = x^*y,$$

whence if $\lambda \neq \mu$, this implies that $x^*y = 0$.

If A is Hermitian, there exists a unitary matrix U such that

$$U^*AU = \Lambda, \tag{4.01.3}$$

where Λ is a diagonal matrix whose elements are the proper values of A, and where the columns of U are the proper vectors of A. This corresponds to the case of the real symmetric matrix, and the argument can be made by paraphrasing that given in the real case.

If A is any matrix, Hermitian or not, any scalar of the form x^*Ax/x^*x for $x \neq 0$ is said to lie in the field of values of A. Any proper value of A lies in its field of values. For if $Ax = \lambda x$, then $x^*Ax = \lambda x^*x$.

If A is any matrix, then A^*A is Hermitian and semidefinite; it is also positive definite if A is nonsingular, for then $Ax \neq 0$ whenever $x \neq 0$, and hence $x^*A^*Ax = (Ax)^*(Ax) > 0$. If the proper values of A^*A are $\rho_1^2 \geq \rho_2^2 \geq \cdots \geq \rho_n^2 \geq 0$, and λ is any proper value of A, then

$$\rho_1^2 \geq \bar{\lambda}\lambda \geq \rho_n^2. \tag{4.01.4}$$

For if $Ax = \lambda x$, then $x^*A^* = \bar{\lambda}x^*$, and hence

$$x^*A^*Ax = \bar{\lambda}\lambda x^*x.$$

Hence $\bar{\lambda}\lambda$ is in the field of values of A^*A. But if α is a
the field of values of A^*A, then for some a with $a^*a = 1$
$a^*A^*Aa = \alpha$. If

$$U^*A^*AU = P^2,$$

where P^2 is the diagonal matrix whose elements are the ρ_i^2, and if $a = Ub$, then

$$\alpha = b^*U^*A^*AUb = b^*P^2b = \Sigma\bar{\beta}_i\beta_i\rho_i^2,$$

all products $\bar{\beta}_i\beta_i$ are real and positive, and hence α is a weighted mean of the ρ_i^2. Hence α cannot exceed the greatest nor be exceeded by the least of the ρ_i^2:

$$\rho_1^2 \geq \alpha \geq \rho_n^2.$$

Since $\bar{\lambda}\lambda$ is such an α, the relation (4.01.4) now follows.

If λ is a proper value of A with multiplicity ν, then $\lambda + \mu$ is a proper value of $A + \mu I$ with multiplicity ν. For this reason, the following classical theorem can provide information as to the limits of the proper values of a matrix: If for every i

$$2|\alpha_{ii}| > \sum_j |\alpha_{ij}|, \tag{4.01.5}$$

then the matrix A is nonsingular.

If the matrix were singular, then the equation $Ax = 0$ would have a nontrivial solution. Among the elements of x, let ξ_i be an element of greatest modulus, $|\xi_i| \geq |\xi_j|$ for all j. Then

$$\left|\sum_{j\neq i} \alpha_{ij}\xi_j\right| \leq \sum_{j\neq i} |\alpha_{ij}||\xi_j| \leq |\xi_i| \sum_{j\neq i} |\alpha_{ij}| < |\xi_i||\alpha_{ii}|.$$

But

$$\sum \alpha_{ij}\xi_j = 0, \qquad \alpha_{ii}\xi_i = -\sum_{j\neq i} \alpha_{ij}\xi_j,$$

whence

$$|\alpha_{ii}||\xi_i| = \left|\sum_{j\neq i} \alpha_{ij}\xi_j\right|.$$

Hence we have a contradiction, and the theorem is proved.

In some cases the theorem remains valid even when some, but not all, of the inequalities in (4.01.5) become equalities. Suppose, for example, that

$$|\alpha_{11}| > \sum_{j=2}^{n} |\alpha_{1j}|, \qquad 2|\alpha_{ii}| \geq \sum_j |\alpha_{ij}|, \qquad i > 1. \tag{4.01.6}$$

Then if ξ_i has the same significance as before, there is at least one ξ_k for which $|\xi_k| < |\xi_i|$. For

$$|\alpha_{11}||\xi_1| \leq \sum_{j=2}^{n} |\alpha_{1j}||\xi_j|,$$

whence on applying the hypothesis it is clear that, for some j, $|\xi_j| > |\xi_1|$, and the ξ's are therefore not all equal in modulus. Now

$$|\alpha_{ii}||\xi_i| \leq \sum_{j \neq i} |\alpha_{ij}||\xi_j|,$$

but this is inconsistent with (4.01.6) unless $\alpha_{ik} = 0$ for every k such that $|\xi_k| < |\xi_i|$. If this is so, and if there are ν values of j such that $|\xi_j| = |\xi_i|$, then there are $n - \nu$ values of k for which $\alpha_{ik} = 0$. But also by the same argument $\alpha_{jk} = 0$ for each such k and every j for which $|\xi_j| = |\xi_i|$. By performing a suitable permutation on the rows of A and the same permutation on the columns, we can assume that

$$\alpha_{jk} = 0, \qquad \begin{array}{l} j = n - \nu + 1, n - \nu + 2, \ldots, n, \\ k = 1, 2, \ldots, \nu. \end{array}$$

The matrix is then in the form

$$A = \begin{pmatrix} P & Q \\ 0 & R \end{pmatrix}, \tag{4.01.7}$$

where P and R are square matrices. Hence if the matrix is not one which can be given the form (4.01.7) by any permutation of rows, accompanied by the same permutation of columns, then the conditions

$$2|\alpha_{ii}| \geq \sum_j |\alpha_{ij}| \tag{4.01.8}$$

with a proper inequality for at least one value of i are sufficient to ensure the nonsingularity of the matrix.

Obviously the above argument can be applied to A^{T}.

Now let

$$P_i = \sum_{j \neq i} |\alpha_{ij}|, \qquad Q_i = \sum_{j \neq i} |\alpha_{ji}|. \tag{4.01.9}$$

If we apply the above results to the matrix $A - \lambda I$, it is clear that, if λ is a proper value, $A - \lambda I$ becoming singular, then either

$$|\lambda - \alpha_{ii}| = P_i \tag{4.01.10}$$

for every i, or else for some i it is true that $|\lambda - \alpha_{ii}| < P_i$. In either event it is true that the proper value lies within or on the boundary of at least one of the n circles in the complex plane defined by (4.01.10). On applying the same argument to A^{T}, we conclude that every proper value lies within or on the boundary of at least one of the n circles defined by

$$|\lambda - \alpha_{ii}| = Q_i. \tag{4.01.11}$$

4.1. Iterative Methods. These methods provide for the direct manipulation of the matrix itself or of some matrix simply related to it, without necessitating the explicit development of the characteristic or other polynomial. We begin with the relatively simple case of a Hermitian matrix.

4.11. *The Matrix A Is Hermitian.* If the proper vectors u_i of a Hermitian matrix were all known, these could be normalized to unit length $u_i^* u_i = 1$, and they would form the columns of the unitary matrix U such that

$$U^*AU = \Lambda, \qquad U^*U = I, \tag{4.11.1}$$

where Λ is the diagonal matrix of proper values. We may assume the u_i to be so ordered that

$$\lambda_1 \geq \lambda_2 \geq \cdots \geq \lambda_n. \tag{4.11.2}$$

If x is any vector and y satisfies

$$x = Uy, \qquad y = U^*x,$$

then

$$x^*x = y^*U^*Uy = y^*y.$$

Hence the field of values of A and that of Λ are identical, and if α is in this field of values, then

$$\lambda_1 \geq \alpha \geq \lambda_n.$$

If $p(\lambda)$ is any polynomial in λ, the matrix $p(A)$ has the same proper vectors as A itself, and its proper values are $p(\lambda_i)$. In particular, the matrix A^2 is necessarily non-negative, definite or semidefinite. Moreover, for μ sufficiently large, $A + \mu I$ is positive definite. It is therefore no essential restriction to assume, when convenient, that A is positive definite or is at least semidefinite.

In §2.06 the trace tr (A), which is the sum of the diagonal elements, was shown to be equal to the coefficient of λ^{n-1} in the characteristic equation, except possibly for the sign, and to be equal to the sum of the proper values,

$$\operatorname{tr}(A) = \Sigma\lambda_i.$$

More generally, if $p(\lambda)$ is any polynomial, then from (4.11.1) it follows that

$$\operatorname{tr}[p(A)] = \Sigma p(\lambda_i). \tag{4.11.3}$$

Moreover, if ν is any integer, and A is nonsingular,

$$\operatorname{tr}(A^{-\nu}) = \Sigma\lambda_i^{-\nu}. \tag{4.11.4}$$

The norm $N(A)$ of a real matrix A, symmetric or not, was defined to be the square root of the sum of the squares of the elements. Equivalently, this is the square root of the trace of $A^{\mathsf{T}}A$. For a complex matrix, Hermitian or not, it can be defined by

$$N^2(A) = \text{tr } (A^*A). \tag{4.11.5}$$

As so defined, tr (A^*A) is a positive real number, and for $N(A)$ the positive root is to be taken. Hence if A is Hermitian,

$$N^2(A) = \text{tr } (A^2) = \text{tr } (\Lambda^2) = \Sigma\lambda_i^2.$$

Now $N^2(A)$ is the sum of the squares of the moduli of all elements of A. Obviously this is equal to the sum of the squares of the diagonal elements of A if and only if A is a diagonal matrix. Hence

$$\Sigma\alpha_{ii}^2 \leq N^2(A),$$

and the equality holds only when A is diagonal. Now the unitary transform V^*AV of A, where V is any unitary matrix, has the same norm as does A, whereas in general the sum of the squares of the diagonal elements is not the same. Hence among all unitary transforms of a Hermitian matrix A, the transform (4.11.1) maximizes the sum of the squares of the diagonal elements or minimizes the sum of the squares of the moduli of the off-diagonal elements.

4.111. *The largest and smallest proper values.* If A is non-negative semidefinite, then for $\nu > 0$

$$\begin{aligned} \lambda_i^\nu \leq \text{tr } (A^\nu) = \Sigma\lambda_i^\nu \leq n\lambda_1^\nu, \\ \lambda_1 \leq [\text{tr } (A^\nu)]^{1/\nu} \leq \lambda_1 n^{1/\nu}. \end{aligned} \tag{4.111.1}$$

But as ν increases, $n^{1/\nu} \to 1$. Hence

$$[n^{-1} \text{ tr } (A^\nu)]^{1/\nu} \leq \lambda_1 \leq [\text{tr } (A^\nu)]^{1/\nu}, \tag{4.111.2}$$

and

$$\lim_{\nu\to\infty} [n^{-1} \text{ tr } (A^\nu)]^{1/\nu} = \lim_{\nu\to\infty} [\text{tr } (A^\nu)]^{1/\nu} = \lambda_1. \tag{4.111.3}$$

The two sequences obtained from the successive powers A^ν of A approach λ_1 from above and from below. The most effective application of this algorithm is made by successively squaring the matrix A and forming

$$A, A^2, A^4, A^8, \ldots,$$

with ν taking on the values 2^p. The degree of convergence is measured by the ratio $n^{1/\nu}:1$ or by $n^{2^{-p}}$.

If $x = Uy$ is any vector, then

$$A^\nu x = U\Lambda^\nu U^*x = U\Lambda^\nu y = \Sigma u_i\lambda_i^\nu\eta_i. \tag{4.111.4}$$

If $\lambda_1 > \lambda_i \geq 0$ for $i = 2, 3, \ldots, n$, or if $\lambda_1 > \lambda_2$ and $\lambda_1 > |\lambda_n|$, then since

$$A^\nu x = \lambda_1^\nu[u_1\eta_1 + (\lambda_2/\lambda_1)^\nu u_2\eta_2 + \cdots + (\lambda_n/\lambda_1)^\nu u_n\eta_n],$$

as ν increases all terms but the first within the brackets approach zero, and in the limit,

$$(4.111.5) \qquad A^\nu x \rightarrow \lambda_1^\nu \eta_1 u_1,$$

provided only $\eta_1 \neq 0$. That is to say, as ν increases, the vector $A^\nu x$ approaches a vector in the direction of the first proper vector u_1. It is necessary only to normalize to obtain u_1 itself.

If $\eta_1 = 0 \neq \eta_2$, the same argument shows that $A^\nu x$ approaches a vector in the direction of u_2, provided λ_2 exceeds $\lambda_3, \ldots, \lambda_n$ numerically.

To square a matrix of order n requires n^3 multiplications, whereas to multiply a matrix by a vector requires only n^2. If

$$(4.111.6) \qquad x_\nu = A^\nu x = Ax_{\nu-1},$$

then for large ν it appears from (4.111.5) that $x_{\nu+1} \doteq \lambda_1 x_\nu$. Hence, although if $\nu = 2^p$, ν increases rapidly with p, it may be more advantageous to form the sequence x_ν directly as in (4.111.6) than to square the matrix several times and then multiply by x. Moreover, a blunder made in computing x_ν will be corrected in the course of subsequent multiplications by A. The two methods are related essentially as are Graeffe's and Bernoulli's methods for solving ordinary equations. It might be pointed out further that, if by a rare chance it should happen that $\eta_1 = 0$, nevertheless round-off will introduce a component along u_1 in the x_ν, and this component will build up eventually, though perhaps slowly.

Let

$$(4.111.7) \qquad \alpha_p = x_\nu^* x_{p-\nu} = y_\nu^* y_{p-\nu}.$$

Then α_p is independent of ν, and in particular

$$\alpha_p = x^*A^p x = y^*\Lambda^p y = \Sigma\lambda_i^p\eta_i\bar{\eta}_i.$$

Hence, if $\eta_1 \neq 0$,

$$\alpha_{p+1} = \Sigma\lambda_i^{p+1}\eta_i\bar{\eta}_i \leq \lambda_1\Sigma\lambda_i^p\eta_i\bar{\eta}_i = \lambda_1\alpha_p,$$

$$(4.111.8) \qquad \alpha_{p+1}/\alpha_p \leq \lambda_1,$$

and

$$(4.111.9) \qquad \alpha_{p+1}/\alpha_p \rightarrow \lambda_1.$$

Thus α_{p+1}/α_p approaches λ_1 from below.

Since in the limit $x_{\nu+1} = \lambda_1 x_\nu$, the ratio of any element of $x_{\nu+1}$ to the corresponding element of x_ν provides also an estimate of λ_1, when ν is sufficiently large. The agreement among these n ratios is an indication of the nearness to the limit.

As ν increases, the elements of x_ν and of A^ν become large if $\lambda_1 > 1$, or small if $\lambda_1 < 1$. Hence in actual computation it is convenient to modify the sequence of x_ν or the sequence of A^ν by introducing factors κ_ν to hold the quantities within range. Thus one could compute the series

$$x'_{\nu+1} = \kappa_\nu A x'_\nu,$$

where each κ_ν may be selected according to convenience. For successively squaring the matrix, Bargmann, Montgomery, and von Neumann propose the sequence

$$B_0 = \rho A/\text{tr}\,(A),$$
$$B_{p+1} = \rho B_p^2/\text{tr}\,(B_p),$$

where ρ is a scalar slightly less than unity. For $\rho = 1$, all quantities would be not greater than unity in principle, though this might fail as a result of round-off. By a suitable choice of ρ, one can be sure of staying within the range from -1 to $+1$ even with round-off.

If $\mu > \lambda_1$, then $A' = \mu I - A$ is Hermitian, positive definite, and has the greatest proper value $\lambda'_1 = \mu - \lambda_n$. Hence no special discussion is needed for finding the smallest proper value.

4.112. *Accelerating convergence.* Of the schemes for accelerating convergence, the simplest is Aitken's δ^2 process, described in Chap. 3. This can be applied to the sequence α_{p+1}/α_p for finding λ_1, to the sequence of ratios of corresponding elements of $x_{\nu+1}$ and x_ν, and to the sequence x_ν for the proper vector.

In the sequence x_ν, the rapidity of the approach to the limit depends upon the smallness of all ratios λ_i/λ_1 for $i > 1$. Any matrix A' which reduces the greatest of these ratios will provide more rapid convergence. One of the simplest choices might be A^2 or A^4. This would mean taking one or two matrix products, requiring n^3 multiplications each, and following this with a sequence of products of a matrix by a vector, requiring n^2 multiplications each. Having obtained, say, the vector $x_{2\nu}$, one could then obtain $Ax_{2\nu} = x_{2\nu+1}$ to find λ_1 without a root extraction.

Another possibility is to replace A by

$$A' = A + \mu I,$$

and hence the proper values λ_i by

$$\lambda'_i = \lambda_i + \mu,$$

where μ is judiciously selected.

Assuming, as always, the ordering (4.11.2), and for present purposes that the strict inequality $\lambda_1 > \lambda_2$ holds, the best choice of μ will be that for which the greatest of the ratios $|\lambda_i'/\lambda_1'|$, $i > 1$, is least. But the greatest of these is either $|\lambda_2'/\lambda_1'|$ or $|\lambda_n'/\lambda_1'|$ or both. The optimal μ is therefore that for which these ratios are equal, since a different selection of μ would increase one of these ratios even though it decreases the other. Hence the optimal μ is

$$\mu = -(\lambda_2 + \lambda_n)/2.$$

To make the strictly optimal choice, one must know λ_2 and λ_n exactly, but enough information may be at hand to permit a good choice.

The iteration of the linear polynomial $A + \mu I$ and of the very special quadratic polynomial A^2 in place of A is sometimes advantageous. The question arises then whether A^2, or even an $A^2 + \mu I$, for some μ is necessarily the best quadratic polynomial, and more generally what is the best polynomial of any given degree. It turns out that the best polynomial is given by the Chebyshev polynomial of the prescribed degree (cf. §5.12).

If a λ' and a λ'' are known such that $\lambda_1 > \lambda' \geq \lambda_2 \geq \cdots \geq \lambda_n \geq \lambda''$, then it is no restriction to suppose that $\lambda' = -\lambda'' = 1$. For if this is not the case, one can replace A by $A' = (\lambda' - \lambda'')^{-1}[2A - (\lambda' + \lambda'')I]$. Hence assume this to have been done, and assume further that a δ is known such that

(4.112.1) $$\lambda_1 \geq \delta > 1 \geq \lambda_2 \geq \cdots \geq \lambda_n \geq -1.$$

Then let

(4.112.2) $$\begin{aligned} T_m(\lambda) &= \cos\,[m \text{ arc } \cos \lambda], \\ S_m(\lambda) &= T_m(\lambda)/T_m(\delta). \end{aligned}$$

Then $S_m(\delta) = 1$, and $S_m(\lambda_1) \geq 1$. On the other hand, for any λ satisfying $1 \geq \lambda \geq -1$ and in particular for $\lambda = \lambda_i$, $i > 1$, it is true that $|S_m(\lambda)| < 1$. Indeed, the argument developed in §5.12 can be modified to show that, of all polynomials $q(\lambda)$ of degree m satisfying $q(\delta) = 1$, $S_m(\lambda)$ is that polynomial whose maximal absolute value on the interval from -1 to $+1$ is least. Hence among all polynomials of degree m that might be used for the iteration, $S_m(A)$ is the best choice that can be made on the basis of the information contained in the hypothesis. In other words, in the sequence

$$x_0' = x, \qquad x_{\nu+1}' = S_m(A)x_\nu',$$

the components along $u_2, \ldots, u_n$ damp out as rapidly as one can make them by applying a polynomial matrix in A of degree m. As a final check, and to nullify the effect of any accumulated round-off, one may take any x_ν' as x_0 and apply the matrix A itself one or more times.

The notion of using a Chebyshev polynomial may be carried a step further. The vector x_ν' is the result of ν applications of the polynomial

$S_m(A)$ of degree m and hence is the result of applying a polynomial $[S_m(A)]^\nu$ of degree $m\nu$. Clearly the direct application of the polynomial $S_{m\nu}$ would give better results.

Hence, let us return to the original sequence

$$x_0, x_1, x_2, \ldots, x_\nu.$$

This is the sequence

$$x, Ax, A^2x, \ldots, A^\nu x.$$

Instead of accepting x_ν itself, for some ν, as giving the best approximation for the direction of u_1, one could ask for a linear combination of all $\nu + 1$ vectors that will give as good an approximation as possible. With the same hypothesis (4.112.1), the best available linear combination is

$$x' = \sigma_0 x_0 + \sigma_1 x_1 + \cdots + \sigma_\nu x_\nu,$$

where

$$S_\nu(\lambda) = \sigma_0 + \sigma_1\lambda + \cdots + \sigma_\nu\lambda^\nu.$$

Again all elements of Ax' should be in the approximate ratio λ_1 to corresponding elements of x', which fact serves both as a check and as a means of obtaining λ_1 directly without the extraction of a root.

4.113. *Intermediate proper values.* Already attention has been drawn to the resemblance between the iterative methods for finding proper values and the methods of Graeffe and of Bernoulli for solving equations. Indeed, Aitken bases his discussion of the method of Bernoulli for algebraic equations upon the iteration of a matrix for which the given equation is the characteristic equation. In principle both methods, Graeffe's and Bernoulli's, yield all the roots of an algebraic equation.

In §4.21 a particular direct method will be described for finding the coefficients of the characteristic equation. That method does not itself yield the roots of this equation. However, if the method is applied to A^ν for any ν, then one obtains the equation whose roots are the λ_i^ν. Hence if ν is sufficiently large, the coefficients of the equation are approximately equal to λ_1^ν, $\lambda_1^\nu\lambda_2^\nu$, $\lambda_1^\nu\lambda_2^\nu\lambda_3^\nu$, . . . , if the λ's are numbered in order of decreasing magnitude. If the matrix is Hermitian, then all roots are real. For the treatment of multiple roots, reference is to be made to the discussion of Graeffe's method in Chap. 3. Unfortunately this method does not yield the associated proper vectors.

For obtaining proper vectors as well as proper values, suppose $\lambda_i > \lambda_{i+1}$. Take any μ such that $0 < \lambda_i - \mu < \mu - \lambda_{i+1}$. Then the numerically smallest proper value of $A - \mu I$ is $\lambda_i - \mu$. Also $(\lambda_i - \mu)^2$ is the smallest proper value of the positive definite matrix $(A - \mu I)^2$. Hence the problem of evaluating the proper value λ_i and its associated proper vector is reduced to that of evaluating the smallest proper value of the positive definite matrix $(A - \mu I)^2$, and this in turn can be reduced to that of finding the largest proper value of a related matrix, as described in §4.112. If $\lambda_i - \mu > \mu - \lambda_{i+1} > 0$, then $(\lambda_{i+1} - \mu)^2$ is the smallest proper value of $(A - \mu I)^2$.

If μ is any number on the interval from λ_1 to λ_n but sufficiently far from either λ_1 or λ_n, then $(A - \mu I)^2$ will have $(\lambda_i - \mu)^2$ as its smallest proper value for some $i = 2, \ldots, n - 1$. Hence any proper value and vector can be obtained independently of all others and so that no errors present in one will affect the others.

A more common procedure for obtaining the intermediate proper values is to obtain them in sequence, making each depend upon the larger values already found and the vectors associated with them. The vectors u_i are mutually orthogonal and of unit length, which means in the complex case that

$$u_i^* u_j = \delta_{ij}.$$

The case of a real symmetric matrix A with real proper vectors u_i is a special case. If the trial vector x has no component in the direction of u_1, and if $\lambda_2 > |\lambda_i|$ for $i = 3, 4, \ldots, n$, then x_ν approaches u_2 in direction, and α_{p+1}/α_p, as well as the ratio of any element of $x_{\nu+1}$ to the corresponding element of x_ν, approaches λ_2. But if u_1 is known, and x' is any vector, then

$$x = x' - u_1 u_1^* x' \tag{4.113.1}$$

is orthogonal to u_1. In fact, this is merely the vector which remains when the orthogonal projection of x' upon the unit vector u_1 is subtracted from x'. Hence when u_1 has been found, a new sequence x_ν can be determined beginning with a vector x orthogonal to u_1. All vectors x_ν will be orthogonal to u_1, except when round-off introduces spurious components along u_1, but these can be removed from time to time by applying (4.113.1). Hence λ_2 and u_2 can be obtained from this sequence of vectors, just as were λ_1 and u_1 before.

In general, for evaluating any λ_i and u_i, where λ_i exceeds in absolute value every proper value not already found, one begins with a vector x from which has been subtracted the orthogonal projection upon the subspace determined by those proper vectors already found.

Another method for finding the intermediate proper values and their associated vectors is to replace the matrix A by another having $\lambda_2, \lambda_3, \ldots, \lambda_n$, but not λ_1, as proper values. The matrix

$$A_1 = A - \lambda_1 u_1 u_1^*$$

is Hermitian, and if $i \neq 1$, then

$$A_1 u_i = A u_i - \lambda_1 u_1 u_1^* u_i = A u_i = \lambda_i u_i,$$

whereas

$$A_1u_1 = \lambda_1u_1 - \lambda_1u_1 = 0.$$

Hence A_1 has the proper values 0, associated with the proper vector u_1, and λ_i, $i > 1$, associated with the proper vector u_i. Thus A_1 satisfies the requirements. It is useful to note that

$$A_1^2 = A^2 - \lambda_1^2u_1u_1^*,$$

and hence inductively

$$A_1^\nu = A^\nu - \lambda_1^\nu u_1u_1^*. \tag{4.113.2}$$

Thus if the powers of A have been formed in the process of arriving at λ_1 and u_1, to form the same powers of A_1 it is necessary only to subtract a scalar multiple of the singular matrix $u_1u_1^*$.

When λ_2 and u_2 are found, one can form

$$A_2 = A_1 - \lambda_2u_2u_2^*$$

and proceed to find λ_3 and u_3.

In this method the matrices A_1, A_2, . . . are all of order n, with 0 as a multiple proper value replacing each of the proper values already found. It is possible, however, to reduce the order of the matrices. Thus when λ_1 and u_1 are found, let u_2' be any unit vector orthogonal to u_1; u_3' any unit vector orthogonal to both u_1 and u_2'; . . . ; and finally u_n' one of the two unit vectors orthogonal to $u_1, u_2', \ldots, u_{n-1}'$. Then the matrix

$$U_1 = (u_1, u_2', \ldots, u_n')$$

is unitary, and one verifies that

$$U_1^*AU_1 = \begin{pmatrix} \lambda_1 & 0 \\ 0 & A_1 \end{pmatrix}, \tag{4.113.3}$$

where now A_1 is of order $n - 1$ and Hermitian. The matrix $U_1^*AU_1$ has the same proper values as A; hence the proper values of A_1 are λ_2, λ_3, . . . , λ_n. Let v be any proper vector of A_1:

$$A_1v = \lambda v.$$

Then

$$U_1^*AU_1\begin{pmatrix} 0 \\ v \end{pmatrix} = \begin{pmatrix} \lambda_1 & 0 \\ 0 & A_1 \end{pmatrix}\begin{pmatrix} 0 \\ v \end{pmatrix} = \begin{pmatrix} 0 \\ \lambda v \end{pmatrix} = \lambda\begin{pmatrix} 0 \\ v \end{pmatrix}.$$

Hence

$$AU_1\begin{pmatrix} 0 \\ v \end{pmatrix} = \lambda U_1\begin{pmatrix} 0 \\ v \end{pmatrix},$$

and $U_1\begin{pmatrix} 0 \\ v \end{pmatrix}$ is a proper vector of A. Hence if the largest proper value of A_1 and its associated proper vector are found, the corresponding proper vector of A can be found directly. The next step is to replace A_1 by a matrix of the form

$$\begin{pmatrix} \lambda_2 & 0 \\ 0 & A_2 \end{pmatrix},$$

where A_2 is of order $n - 2$.

A simple construction of the matrix U_1 is given by Feller and Forsythe. Write u_1 in the form

$$u_1 = \begin{pmatrix} \omega \\ w \end{pmatrix}, \tag{4.113.4}$$

where w is a vector of $n - 1$ elements and ω a scalar. Then it is possible to choose μ so that

$$U_1 = \begin{pmatrix} \omega & -w^* \\ w & I - \mu w w^* \end{pmatrix} \tag{4.113.5}$$

is unitary. For this we must have

$$I = U_1 U_1^* = \begin{pmatrix} \omega & -w^* \\ w & I - \mu w w^* \end{pmatrix} \begin{pmatrix} \bar{\omega} & w^* \\ -w & I - \bar{\mu} w w^* \end{pmatrix},$$

and hence by direct calculation

$$\mu = (1 - \bar{\omega})/(1 - \bar{\omega}\omega). \tag{4.113.6}$$

Hence when λ_1 and u_1 are found, the transformation (4.113.3), where the unitary matrix U_1 is defined by (4.113.5), (4.113.4), and (4.113.6), replaces A by a Hermitian matrix A_1 of lower order, whose proper values are the same as those of A which are not yet known, and whose proper vectors v_i yield those of A by the simple relation

$$u_i = U_1 \begin{pmatrix} 0 \\ v_i \end{pmatrix}. \tag{4.113.7}$$

4.114. *Equal and nearly equal roots.* Suppose $\lambda_1 = \lambda_2 > \lambda_3$. Then associated with λ_1 is a two-dimensional set of proper vectors, and any two mutually orthogonal unit vectors in this set can be taken as u_1 and u_2. Given any starting vector x, let u_1 be the direction of its projection on this set. Then x can be written in the form

$$x = \eta_1 u_1 + \eta_3 u_3 + \cdots + \eta_n u_n,$$

while

$$x_\nu = A^\nu x = \eta_1 \lambda_1^\nu u_1 + \eta_3 \lambda_3^\nu u_3 + \cdots + \eta_n \lambda_n^\nu u_n,$$

and for ν sufficiently large

$$x_\nu \doteq \eta_1 \lambda_1^\nu u_1.$$

If z is any other vector, then

$$z = \omega_1 u_1 + \omega_2 u_2 + \cdots + \omega_n u_n,$$

where in general $\omega_2 \neq 0$. Hence for large ν

$$z_\nu = A^\nu z \doteq (\omega_1 u_1 + \omega_2 u_2)\lambda_1^\nu.$$

Both vectors, z and x, will be effective in yielding λ_1, but in general distinct vectors will lead to distinct proper vectors associated with the same proper value. However, a third vector w will yield a w_ν for large ν such that x_ν, z_ν, and w_ν are linearly dependent.

On the other hand, if λ_1 is a triple root, three starting vectors will lead to the same λ_1 but to three independent proper vectors, while four starting vectors would approach a set of four linearly dependent vectors.

If λ_1 and λ_2 are nearly, but not exactly, equal, then with each will be associated a unique proper vector. Consider two starting vectors x and z. For ν sufficiently large, we shall have approximately

$$\begin{aligned} x_\nu &\doteq \eta_1\lambda_1^\nu u_1 + \eta_2\lambda_2^\nu u_2, \\ z_\nu &\doteq \omega_1\lambda_1^\nu u_1 + \omega_2\lambda_2^\nu u_2. \end{aligned}$$

Let

$$\alpha_\nu = x_0^* x_\nu, \qquad \beta_\nu = z_0^* z_\nu.$$

Then

$$\begin{aligned} \alpha_\nu &\doteq \bar\eta_1\eta_1\lambda_1^\nu + \bar\eta_2\eta_2\lambda_2^\nu, \\ \beta_\nu &\doteq \bar\omega_1\omega_1\lambda_1^\nu + \bar\omega_2\omega_2\lambda_2^\nu. \end{aligned} \tag{4.114.1}$$

Then the two matrices

$$\begin{pmatrix} \alpha_\nu & \lambda_1^\nu & \lambda_2^\nu \\ \alpha_{\nu+1} & \lambda_1^{\nu+1} & \lambda_2^{\nu+1} \\ \alpha_{\nu+2} & \lambda_1^{\nu+2} & \lambda_2^{\nu+2} \end{pmatrix}, \quad \begin{pmatrix} \beta_\nu & \lambda_1^\nu & \lambda_2^\nu \\ \beta_{\nu+1} & \lambda_1^{\nu+1} & \lambda_2^{\nu+1} \\ \beta_{\nu+2} & \lambda_1^{\nu+2} & \lambda_2^{\nu+2} \end{pmatrix}$$

are both singular, and hence λ_1 and λ_2 must satisfy the equation

$$\begin{vmatrix} 1 & \alpha_\nu & \beta_\nu \\ \lambda & \alpha_{\nu+1} & \beta_{\nu+1} \\ \lambda^2 & \alpha_{\nu+2} & \beta_{\nu+2} \end{vmatrix} = 0. \tag{4.114.2}$$

The method can be extended to the case when three or more of the proper values are nearly equal.

When the powers A^ν are formed explicitly, there are already at hand the vectors x_ν of n distinct sequences since the ith column of A^ν is $A^\nu e_i$. The ith diagonal element of A^ν is the α_ν for the starting vector e_i.

If the matrices A^ν of the sequence approach rank 1, then all columns approach the proper vector associated with the single largest proper value. If they approach rank 2, then the two largest proper values are equal or nearly so. Pick out any two diagonal elements and consider their values in consecutive matrices of the series. Let these be α_ν and β_ν in A^ν, $\alpha_{\nu+1}$ and $\beta_{\nu+1}$ in $A^{\nu+1}$. From (4.114.1), the matrix

$$\begin{pmatrix} \alpha_\nu & \beta_\nu \\ \alpha_{\nu+1} & \beta_{\nu+1} \end{pmatrix} = \begin{pmatrix} \lambda_1^\nu & \lambda_2^\nu \\ \lambda_1^{\nu+1} & \lambda_2^{\nu+1} \end{pmatrix} \begin{pmatrix} \bar\eta_1\eta_1 & \bar\omega_1\omega_1 \\ \bar\eta_2\eta_2 & \bar\omega_2\omega_2 \end{pmatrix}.$$

Hence if the determinant of this matrix vanishes for large ν, then $\lambda_1 = \lambda_2$. If it does not vanish, the roots are nearly equal, by comparison with the other proper values, but not exactly equal, and they can be found by solving Eqs. (4.114.2).

4.115. *Rotational reduction to diagonal form.* A second-order Hermitian matrix has the form

$$B = \begin{pmatrix} \beta_1 & \beta e^{i\phi} \\ \beta e^{-i\phi} & \beta_2 \end{pmatrix}, \qquad \beta \geq 0, \tag{4.115.1}$$

and its characteristic equation is

$$\mu^2 - (\beta_1 + \beta_2)\mu + \beta_1\beta_2 - \beta^2 = 0. \tag{4.115.2}$$

The discriminant of this quadratic is

$$(\beta_1 - \beta_2)^2 + 4\beta^2,$$

and since all the β's are real, this can vanish only in case $\beta_1 - \beta_2 = \beta = 0$. Hence the only second-order Hermitian matrices with coincident proper values are scalar matrices, which are necessarily diagonal.

In complex 2 space, a unitary vector can be written in the form

$$v = \begin{pmatrix} e^{i\omega_1} \cos \theta \\ e^{i\omega_2} \sin \theta \end{pmatrix}.$$

Assume $\beta > 0$, and let μ_1 and $\mu_2 < \mu_1$ be the two roots of (4.115.2). Then if v is a proper vector associated with μ_1,

$$(\beta_1 - \mu_1)e^{i\omega_1} \cos \theta + \beta e^{i(\phi + \omega_2)} \sin \theta = 0.$$

This can be satisfied by taking

$$\omega_1 = \omega + \phi/2, \qquad \omega_2 = \omega - \phi/2,$$

with ω arbitrary, and

$$\tan \theta = (\mu_1 - \beta_1)/\beta = \beta/(\mu_1 - \beta_2). \tag{4.115.3}$$

The second expression is obtained by using the second row of B, and the two expressions are equivalent since (4.115.2) can be written

$$(\mu - \beta_1)(\mu - \beta_2) = \beta^2.$$

Because of this relation, either root μ_i must exceed both β_1 and β_2 or be exceeded by both. But since the sum of the roots is

$$\mu_1 + \mu_2 = \beta_1 + \beta_2,$$

it follows that the larger root μ_1 must exceed both, and the smaller root μ_2 must be exceeded by both. Hence

$$\tan \theta > 0,$$

and θ can be taken to lie in the first quadrant. Hence θ is uniquely determined. By applying a standard trigonometric identity, one obtains

$$\tan 2\theta = 2\beta/(\beta_1 - \beta_2), \tag{4.115.4}$$

an expression which does not involve μ.

The proper vector v_1 associated with μ_1 can now be written

$$v_1 = \begin{pmatrix} e^{i\phi/2} \cos \theta \\ e^{-i\phi/2} \sin \theta \end{pmatrix}, \tag{4.115.5}$$

where ϕ is defined by (4.115.1) and θ by (4.115.4) with the additional convention that $0 < \theta < \pi$.

If one uses (4.115.4) and then applies trigonometric identities to find the functions of θ, then from (4.115.3) and the expression for the sum of the roots it follows that

$$\mu_1 = \beta_1 + \beta \tan \theta, \qquad \mu_2 = \beta_2 - \beta \tan \theta. \tag{4.115.6}$$

Since v_2 is orthogonal to v_1, it is easy to obtain

$$v_2 = \begin{pmatrix} -e^{i\phi/2} \sin \theta \\ e^{i\phi/2} \cos \theta \end{pmatrix}, \tag{4.115.7}$$

so that the unitary matrix V which diagonalizes B to M is

$$V = \begin{pmatrix} e^{i\phi/2} \cos \theta & -e^{i\phi/2} \sin \theta \\ e^{-i\phi/2} \sin \theta & e^{-i\phi/2} \cos \theta \end{pmatrix}, \tag{4.115.8}$$

where

$$V^*BV = M. \tag{4.115.9}$$

One verifies directly that the sums of squares of the diagonal elements of B and M are related by

$$\mu_1^2 + \mu_2^2 = \beta_1^2 + \beta_2^2 + 2\beta^2.$$

Now suppose B represents any principal minor of A. For simplicity let this be the minor taken from the first two rows and columns of A. The matrix

$$U_1 = \begin{pmatrix} V & 0 \\ 0 & I \end{pmatrix}$$

is easily seen to be unitary. If we write A in the form

$$A = \begin{pmatrix} B & B_1^* \\ B_1 & B_{11} \end{pmatrix},$$

then

$$A_1 = U_1^* A U_1 = \begin{pmatrix} M & V^*B_1^* \\ B_1 V & B_{11} \end{pmatrix}.$$

Hence in the transform A_1 of A by the unitary matrix U_1 the sum of the squares of the diagonal elements is increased by $2\beta^2$. The same would hold if B were any principal minor with non-null off-diagonal elements α_{ij} and α_{ji}, except that the matrix U must be formed by placing the elements of V in the ith and jth rows and columns.

The sum of the squares of the diagonal elements of A is maximal among all the unitary transforms of A for the diagonal matrix Λ. Hence an infinite sequence of transforms A_ν of A, each produced from the preceding by a plane rotation as just described, and chosen to nullify a pair of off-diagonal elements of greatest magnitude, will approach Λ as a limit, and the infinite product of unitary matrices U_ν will approach the matrix U of proper vectors. The ordering of the proper values down the main diagonal of Λ is taken care of by the convention that μ_1 must be the larger of the two proper values of each B, provided the rows and columns in B are ordered as they are in A.

4.12. *The Matrix A Is Arbitrary.* If the proper values of the non-symmetric real matrix or the non-Hermitian complex matrix A are all distinct, then there exists a nonsingular matrix W such that

(4.12.1) $$W^{-1}AW = \Lambda,$$

and Λ is again a diagonal matrix of the proper values of A. In this case, the matrix W is not orthogonal or unitary. Nevertheless, it is still true that

(4.12.2) $$A^\nu = W\Lambda^\nu W^{-1},$$

where ν is any integer if A is nonsingular or any non-negative integer when A is singular. Moreover,

(4.12.3) $$p(A) = Wp(\Lambda)W^{-1},$$

where $p(\lambda)$ is any polynomial. Hence let $x = Wy$ be any vector. Then

$$A^\nu x = W\Lambda^\nu y = \Sigma w_i\lambda_i^\nu \eta_i.$$

If there is a single proper value of largest modulus, let this be λ_1. Then just as in the Hermitian case for large ν

(4.12.4) $$x_\nu = A^\nu x \doteq \lambda_1^\nu \eta_1 w_1,$$

provided only $\eta_1 \neq 0$.

For a non-Hermitian matrix, there are two sets of proper vectors, a set of row vectors and a set of column vectors. Corresponding to any proper value λ_i there is a column vector w_i and a row vector w^i such that

(4.12.5) $$Aw_i = \lambda_i w_i, \qquad w^i A = \lambda_i w^i.$$

The w^i are in fact the rows of W^{-1}, as the w_i are the columns of W, or may be taken so when properly normalized. Let $u = vW^{-1}$ be any row vector. Then

$$uA^\nu = v\Lambda^\nu W^{-1} = \Sigma\lambda_i^\nu \phi_i w^i,$$

if the ϕ_i are the elements of v. Hence, again, for sufficiently large ν

$$u_\nu = uA^\nu \doteq \lambda_1^\nu \phi_1 w^1, \tag{4.12.6}$$

provided $\phi_1 \neq 0$. It is not necessary that w_1 and w^1 be of unit length, but for these to be a column of W and a row of W^{-1}, respectively, it is necessary that

$$w^1 w_1 = 1.$$

For sufficiently large ν, consecutive vectors in the sequence x_ν and also consecutive vectors in the sequence u_ν are approximately linearly dependent. Hence the ratio of any two corresponding components can be used to provide an approximation to λ_1, and the agreement among these ratios provides evidence as to the nearness to the limit. The δ^2 process can be applied to accelerate the convergence to all limits, λ_1, w_1, and w^1.

Suppose now that λ_1 and λ_2 are equal in modulus, but exceed in modulus all other proper values. Then in the limit

$$\begin{aligned} x_\nu &\doteq \lambda_1^\nu \eta_1 w_1 + \lambda_2^\nu \eta_2 w_2, \\ u_\nu &\doteq \lambda_1^\nu\, \phi_1 w^1 + \lambda_2^\nu \phi_2 w^2. \end{aligned} \tag{4.12.7}$$

Consider the sequence of scalars

$$\alpha_\nu = ux_\nu = u_\nu x. \tag{4.12.8}$$

For large ν this becomes

$$\alpha_\nu \doteq \lambda_1^\nu \phi_1 \eta_1 + \lambda_2^\nu \phi_2 \eta_2.$$

Then in the limit the matrix

$$\begin{pmatrix} \alpha_\nu & \lambda_1^\nu & \lambda_2^\nu \\ \alpha_{\nu+1} & \lambda_1^{\nu+1} & \lambda_2^{\nu+1} \\ \alpha_{\nu+2} & \lambda_1^{\nu+2} & \lambda_2^{\nu+2} \end{pmatrix}$$

is singular. If either u or x is replaced by a different vector, another sequence of scalars β_ν can be defined, and by a familiar argument λ_1 and λ_2 must satisfy the quadratic equation

$$\begin{vmatrix} 1 & \alpha_\nu & \beta_\nu \\ \lambda & \alpha_{\nu+1} & \beta_{\nu+1} \\ \lambda^2 & \alpha_{\nu+2} & \beta_{\nu+2} \end{vmatrix} = 0. \tag{4.12.9}$$

The α's and the β's can indeed be individual components in the x's or in the u's. The extension to a larger number of proper values of equal modulus is direct, and applies, moreover, also to the case of moduli that are nearly equal.

In case $\lambda_1 = \lambda_2$, the coefficients in the quadratic (4.12.9) all vanish. This case will be considered later.

Suppose the proper value λ_1 of greatest modulus and its associated vectors w_1 and w^1 have been found. Suppose λ_2 exceeds all other proper values in modulus. Its proper vector w_2 is orthogonal to w^1, and w^2 is orthogonal to w_1. Hence one can proceed as above but with starting vectors x, orthogonal to w^1, and u, orthogonal to w_1:

$$w^1x = uw_1 = 0.$$

Or, one can replace the matrix A by

$$A_1 = A - \lambda_1 w_1 w^1,$$

whose proper values are $0, \lambda_2, \lambda_3, \ldots, \lambda_n$.

Turning now to the case of multiple values, it may or may not be possible to diagonalize the matrix when such occur. If $\lambda_1 = \lambda_2 \neq \lambda_i$ for $i > 2$, and if two independent proper vectors exist associated with λ_1, then in general a starting vector x will have some component which is a linear combination of these vectors. The iteration can proceed and will yield λ_1 and some linear combination of w_1 and w_2. Likewise a starting vector u will yield λ_1 and some linear combination of w^1 and w^2. A different starting vector x will yield a different linear combination of w_1 and w_2, and a different starting vector u a different linear combination of w^1 and w^2. When powers A^ν of the matrix are computed explicitly, then one has the effect of iterating simultaneously upon n distinct column vectors, the columns of A, and upon n distinct row vectors, the rows of A. In the limit if corresponding columns (or rows) in consecutive powers A^ν are linearly independent, then the largest proper values are equal, or nearly equal, in modulus. But if these are linearly dependent, whereas the matrix A^ν has rank > 1, then the largest proper value is a multiple root.

However, in the case of a multiple proper value, it can happen that the number of linearly independent proper vectors associated with this proper value is less than the multiplicity. If so, then the matrix cannot be diagonalized. The general form was discussed in §2.06, and is shown in (2.06.20), (2.06.21), and (2.06.22).

It is clear that among the iterates $A^\nu x$ of any vector x only a finite number can be linearly independent. There will be some $m \leq n$ such that for $\nu \geq m$, $A^\nu x$ is expressible as a linear combination of $x, Ax, \ldots, A^{m-1}x$. Hence all iterates lie in some subspace of dimension $m \leq n$. In this subspace there is at most one (in general there will be exactly one) proper vector associated with each distinct proper value. The fact can be deduced from the discussion in §2.06, but will be brought out more clearly in §4.23. Furthermore, associated with each λ_i there is a principal vector of highest grade in the subspace, say $u_i^{(n_i)}$ of grade n_i, the vectors

$$\begin{aligned} u_i^{(n_i-1)} &= (A - \lambda_i I)u_i^{(n_i)}, \\ u_i^{(n_i-2)} &= (A - \lambda_i I)^2 u_i^{(n_i)} = (A - \lambda_i I)u_i^{(n_i-1)}, \\ &\cdots\cdots\cdots\cdots\cdots\cdots \end{aligned}$$

are of progressively lower grade, and lie also in the subspace, and in particular $u_i^{(1)}$ is the unique proper vector in the subspace.

Consider the effect of the iteration upon these principal vectors. If we write

$$U_i = (u_i^{(1)} u_i^{(2)} \cdots u_i^{(n_i)}),$$

then one verifies that

$$AU_i = U_i \Lambda_i,$$

where

$$\Lambda_i = \begin{pmatrix} \lambda_i & 1 & 0 & \cdots \\ 0 & \lambda_i & 1 & \cdots \\ \cdots & \cdots & \cdots & \cdots \end{pmatrix} = \lambda_i I + I_1,$$

the matrix Λ_i being of order n_i. Hence

$$A^2 U_i = AU_i \Lambda_i = U_i \Lambda_i^2 = U_i(\lambda_i I + I_1)^2,$$

and in general

$$A^\nu U_i = U_i(\lambda_i I + I_1)^\nu.$$

The auxiliary unit matrix I_1 vanishes in the n_ith and higher powers.

Associated with each distinct λ_i will be a particular matrix U_i of principal vectors, and x can be expressed as a sum

$$x = \Sigma U_i x^{(i)},$$

where $x^{(i)}$ is a vector of as many elements as there are columns in U_i. Hence

$$A^\nu x = \Sigma U_i(\lambda_i I + I_1)^\nu x^{(i)}.$$

If there is a proper value λ_1 of modulus exceeding all other $|\lambda_i|$, then in the limit

$$x_\nu = A^\nu x \doteq U_1(\lambda_1 I + I_1)^\nu x^{(1)}.$$

Since

$$\begin{aligned} (\lambda_1 I + I_1)^{n_1} - n_1\lambda_1(\lambda_1 I + I_1)^{n_1-1} + \binom{n_1}{2}\lambda_1^2(\lambda_1 I + I_1)^{n_1-2} + \cdots \\ + \lambda_1^{n_1} I = [(\lambda_1 I + I_1) - \lambda_1 I]^{n_1} = I_1^{n_1} = 0, \end{aligned}$$

it follows that in the limit any $1 + n_1$ consecutive vectors in the sequence $A^\nu x$ will be linearly dependent, and in fact the coefficients expressing the dependence relation are the coefficients of the powers of λ in the expansion of $(\lambda - \lambda_1)^{n_1}$. This fact provides a means for computing λ_1.

Given λ_1, it is possible to form the combinations

$$\begin{aligned}\Delta_{\lambda_1} x_\nu &= x_{\nu+1} - \lambda_1 x_\nu, \\ \Delta^2_{\lambda_1} x_\nu &= x_{\nu+2} - 2\lambda_1 x_{\nu+1} + \lambda_1^2 x_\nu, \\ &\cdots\cdots\cdots\cdots\cdots\end{aligned}$$

We find

$$\begin{aligned}\Delta_{\lambda_1} x_\nu &= U_1[\lambda_1 I + I_1 - \lambda_1 I](\lambda_1 I + I_1)^\nu x \\ &= U_1 I_1(\lambda_1 I + I_1)^\nu x,\end{aligned}$$

and more generally

$$\Delta^r_{\lambda_1} x_\nu = U_1 I_1^r(\lambda_1 I + I_1)^\nu x.$$

But since $I_1^{n_1} = 0$, therefore $\Delta^{n_1}_{\lambda_1} x_\nu = 0$. Moreover, $I_1^{n_1-1}$ differs from the null matrix only in the last element of the first row. Hence in the product $U_1 I_1^{n_1-1}$ only one column is non-null, and this is the proper vector $u^{(1)}$. Hence $\Delta^{n_1-1}_{\lambda_1} x_\nu$ is equal to $u^{(1)}$ except for a nonessential scalar multiplier. Thus even in this case, it is possible to obtain from the iteration both the largest proper value and an associated proper vector.

4.2. Direct Methods. By a direct method will be meant a method for obtaining explicitly the characteristic function, the minimal function, or some divisor, possibly coupled with a method for obtaining any proper vector in a finite number of steps, given the associated proper value. The method to be used in evaluating the zeros of the function is left open.

Naturally one such method would be direct expansion of the determinant $|A - \lambda I|$ to obtain the characteristic function. This done, and the equation solved, one could proceed to solve the several sets of homogeneous equations $(A - \lambda I)x = 0$, where λ takes on each of the proper values. Such a naïve method might be satisfactory for simple matrices of order 2 or 3, but for larger matrices the labor would quickly become astronomical.

In discussing iterative methods, it was convenient to consider separately Hermitian and non-Hermitian matrices. This was primarily because of the fact that for Hermitian matrices the proper values are known to be real, though a further point in favor of the Hermitian matrix is the fact that it can always be diagonalized. For the application of direct methods, however, the occurrence of complex proper values introduces no difficulty in principle, though naturally they complicate the task of solving the equation once it is obtained. Intrinsic difficulties arise in the use of direct methods only with the occurrence of multiple proper values. Fortunately, however, one begins in the same way regardless of whether multiplicities are present or not. If present, the fact reveals itself as one proceeds. If a given direct method applies at all to non-Hermitian matrices, then its application to any diagonalizable matrix can be discussed about as easily as can the application to the special case of a Hermitian matrix. Consequently, each method or class of methods will be described for whatever cases it may cover, Hermitian or not, before passing on to another.

One ingenious and simple method that has been used to find the characteristic equation may be mentioned at the outset. This is to evaluate the determinant $\phi(\lambda) = |A - \lambda I|$ for each of $n + 1$ selected values of λ, and then write the interpolation polynomial of degree n which they determine. This might be advantageous when the matrix is small, and values of λ could be found for which the evaluation is especially simple. However, it provides no assistance for the computation of the proper vectors.

4.21. *Symmetric Functions of the Roots.* If we write

$$f(\lambda) = (-)^n\phi(\lambda) = \lambda^n - \gamma_1\lambda^{n-1} + \gamma_2\lambda^{n-2} + \cdots + (-)^n\gamma_n, \tag{4.21.1}$$

then the coefficients γ_h are the "elementary" symmetric polynomials in the proper values. That is to say, γ_h is the sum of the products h at a time of the λ_i. In particular

$$\gamma_1 = \Sigma\lambda_i = \text{tr}\,(A). \tag{4.21.2}$$

Newton's identities (3.02.5) express the sums of the powers s_i as polynomials in these elementary polynomials, where the γ_i here take the places of the σ_i in (3.02.5). But

$$s_1 = \gamma_1 = \text{tr}\,(A),$$
$$s_2 = \Sigma\lambda_i^2 = \text{tr}\,(A^2),$$

and in general

$$s_h = \Sigma\lambda_i^h = \text{tr}\,(A^h). \tag{4.21.3}$$

Hence by taking powers of A up to and including the nth, one can compute the sums of powers of the λ_i, and thence by applying Newton's identities find the γ_h. Hence to find the s_h by this method requires $(n - 1)$ matrix products; each matrix product requires n^3 multiplications; hence altogether to find the coefficients in (4.21.1) requires approximately n^4 multiplications.

To improve the algorithm and obtain further information, consider

$$C(\lambda) = \text{adj}\,(\lambda I - A) = C_0\lambda^{n-1} - C_1\lambda^{n-2} + C_2\lambda^{n-3} - \cdots \pm C_{n-1}. \tag{4.21.4}$$

Then

$$C(\lambda)(\lambda I - A) = (-1)^n\phi(\lambda)I. \tag{4.21.5}$$

On expanding and comparing coefficients of λ on the two sides of this equation, one finds

$$
\begin{aligned}
C_0 &= I, \\
C_1 + C_0A &= \gamma_1 I, \\
C_2 + C_1A &= \gamma_2 I, \\
&\cdots\cdots\cdots \\
C_{n-1} + C_{n-2}A &= \gamma_{n-1} I, \\
C_{n-1}A &= \gamma_n I.
\end{aligned} \tag{4.21.6}
$$

Now γ_1 is given by (4.21.2). Hence C_1 can be found from the second of (4.21.6). On multiplying this by A and taking the trace of both sides, one finds in view of (4.21.3)

$$\text{tr}\,(C_1A) + s_2 = \gamma_1 s_1.$$

Comparison with the second of (3.02.5) shows that

$$2\gamma_2 = \text{tr}\,(C_1A).$$

Hence γ_2 can be found and therefore, by the third equation, C_2. In general

$$C_h = \gamma_h I - \gamma_{h-1}A + \gamma_{h-2}A^2 - \cdots \pm A^h. \tag{4.21.7}$$

On multiplying by A, taking the trace, and comparing with (3.02.5), one finds that

$$h\gamma_h = \text{tr}\,(C_{h-1}A). \tag{4.21.8}$$

Hence the coefficients γ_h and the matrices C can be obtained in the sequence $C_0 = I, \gamma_1, C_1, \gamma_2, C_2, \ldots, \gamma_n$. The final equation in (4.21.6) serves as a check. Note that, since each C_h is a polynomial in A, it is commutative with A:

$$AC_h = C_hA.$$

As a byproduct of this computation, one obtains the determinant

$$|A| = \gamma_n,$$

the adjoint

$$\text{adj}\,A = C_{n-1},$$

and the inverse

$$A^{-1} = C_{n-1}/\gamma_n.$$

If $C(\lambda_i) \neq 0$, then any non-null column of $C(\lambda_i)$ is a proper column vector, and any non-null row of $C(\lambda_i)$ is a proper row vector associated with λ_i, since by (4.21.5) and the commutativity of the C_h and A we have

$$C(\lambda_i)(\lambda_i I - A) = (\lambda_i I - A)C(\lambda_i) = f(\lambda_i)I = 0.$$

But if λ_i is a simple root, then necessarily $C(\lambda_i) \neq 0$, and there is at least one non-null column and at least one non-null row. For suppose $C(\lambda_i) = 0$. By differentiating (4.21.5) and setting $\lambda = \lambda_i$, one obtains

$$(\lambda_i I - A)C'(\lambda_i) = (-1)^n\phi'(\lambda_i)I,$$

and on taking determinants of both sides,

$$(-1)^n\phi(\lambda_i)|C'(\lambda_i)| = [(-1)^n\phi'(\lambda_i)]^n.$$

But $\phi(\lambda_i) = 0$, whereas $\phi'(\lambda_i) \neq 0$, and we have therefore reached a contradiction. Hence the method yields at least those proper row and column vectors that are associated with simple proper values.

In forming the matrices $C_2, C_3, \ldots, C_{n-1}$, each requires n^3 multiplications, making $n^3(n-2)$ in all for forming the characteristic function. Given a λ_i, to form $C(\lambda_i)$ one can form

$$C_0\lambda_i,\ (C_0\lambda_i + C_1)\lambda_i,\ (C_0\lambda_i^2 + C_1\lambda_i + C_2)\lambda_i,\ \ldots,$$

which requires $n^2(n-2)$ multiplications for each λ_i, and hence again $n^3(n-2)$ multiplications when the λ_i are all distinct. Altogether this is $2n^3(n-2)$ multiplications. However, if one forms only a single row and a single column, and these are non-null, this can be reduced to a total of $n^2(n^2-4)$.

Suppose next that λ_i is a double root, but not a triple root. It may still happen that $C(\lambda_i) \neq 0$. If so, then again any non-null column is a proper column vector, and any non-null row a proper row vector. Since by differentiation of (4.21.5)

$$C(\lambda) + (\lambda I - A)C'(\lambda) = (-1)^n\phi'(\lambda)I, \tag{4.21.9}$$

and since λ_i is a double root, therefore

$$(\lambda_i I - A)C'(\lambda_i) = -C(\lambda_i) \neq 0,$$

and certainly therefore $C'(\lambda_i) \neq 0$. However,

$$(\lambda I - A)C(\lambda) + (\lambda I - A)^2C'(\lambda) = (-1)^n\phi'(\lambda)(\lambda I - A),$$

and therefore

$$(\lambda_i I - A)^2C'(\lambda_i) = 0. \tag{4.21.10}$$

Hence there is at least one non-null column x and at least one non-null row u of $C'(\lambda_i)$, and

$$u(\lambda_i I - A)^2 = 0, \qquad (\lambda_i I - A)^2x = 0,$$

whereas

$$u(\lambda_i I - A) \neq 0, \qquad (\lambda_i I - A)x \neq 0.$$

Thus u and x are principal vectors of grade 2 associated with λ_i.

Suppose, on the other hand, that $C(\lambda_i) = 0$. Then by (4.21.9) it follows that

$$(\lambda_i I - A)C'(\lambda_i) = 0, \tag{4.21.11}$$

since λ_i is a double root. Then any non-null column of $C'(\lambda_i)$, if there is such, is a proper column vector, and any non-null row a proper row vector. We can show that $C'(\lambda_i)$ has rank 2.

We know from Chap. 2 that a double root has associated at most two proper vectors. Hence $(\lambda_i I - A)$ has rank $n - 2$ at least. But $0 = C(\lambda_i) = \text{adj}\,(\lambda_i I - A)$ so that $(\lambda_i I - A)$ has rank $n - 2$ at most. Hence in this case $(\lambda_i I - A)$ has rank $n - 2$ exactly. Hence by (4.21.11) $C'(\lambda_i)$ has rank 2 at most. Let B be a constant matrix of maximal rank such that

$$C'(\lambda_i)B = 0.$$

Since $C'(\lambda_i)$ has rank 2 at most, B has rank $n - 2$ at least. If we differentiate (4.21.9), set $\lambda = \lambda_i$, and multiply by B, we obtain

$$(\lambda_i I - A)C''(\lambda_i)B = (-1)^n\phi''(\lambda_i)B.$$

The rank of the right member of this equation is the same as the rank of B, since $\phi''(\lambda_i) \neq 0$, and this cannot exceed the rank of any matrix factor on the left. But $(\lambda_i I - A)$ has rank $n - 2$, whence the rank of B cannot exceed $n - 2$. Hence B has rank $n - 2$ exactly, and therefore $C'(\lambda_i)$ has rank 2 exactly.

Thus when λ_i is a double root (and not a triple root), either $C(\lambda_i) \neq 0$, in which case there exists a non-null column of $C(\lambda_i)$ and a non-null column of $C'(\lambda_i)$, the first being a proper, and the second a principal, column vector associated with λ_i; or else $C(\lambda_i) = 0$, in which case $C'(\lambda_i)$ has rank 2 and any two linearly independent columns are proper vectors. Corresponding statements can be made for the rows.

The argument can be extended to the case of a root of arbitrary multiplicity.

4.22. *Methods of Enlargement.* Suppose A_{n-1} is a principal minor of A, say that taken from the first $n - 1$ rows and columns of A, and let

$$A = A_n = \begin{pmatrix} A_{n-1} & a_{n-1} \\ a'_{n-1} & \alpha_{n-1} \end{pmatrix}. \tag{4.22.1}$$

Then

$$\lambda I_n - A_n = \begin{pmatrix} \lambda I_{n-1} - A_{n-1} & -a_{n-1} \\ -a'_{n-1} & \lambda - \alpha_{n-1} \end{pmatrix}. \tag{4.22.2}$$

Let

$$\phi_n(\lambda) = |\lambda I_n - A_n|, \tag{4.22.3}$$

and

$$B_n(\lambda) = \text{adj}\,(\lambda I_n - A_n) = \begin{pmatrix} F_{n-1}(\lambda) & f_{n-1}(\lambda) \\ f'_{n-1}(\lambda) & \phi_{n-1}(\lambda) \end{pmatrix}. \tag{4.22.4}$$

Here the prime does not denote differentiation, but is merely a distinguishing mark. Note in particular that

$$\phi_{n-1}(\lambda) = |\lambda I_{n-1} - A_{n-1}|.$$

Now $\phi_{n-1}(\lambda)$ is of degree $n - 1$ in λ. However, each element of $f_{n-1}(\lambda)$ and each element of $f'_{n-1}(\lambda)$ is of degree $n - 2$ at most. Hence if we note that

$$(\lambda I_n - A_n)\begin{pmatrix} f_{n-1}(\lambda) \\ \phi_{n-1}(\lambda) \end{pmatrix} = \phi_n(\lambda)e_n,$$

or in more detail

$$\begin{aligned} (\lambda I_{n-1} - A_{n-1})f_{n-1}(\lambda) - a_{n-1}\phi_{n-1}(\lambda) &= 0, \\ -a'_{n-1}f_{n-1}(\lambda) + (\lambda - \alpha_{n-1})\phi_{n-1}(\lambda) &= \phi_n(\lambda), \end{aligned} \tag{4.22.5}$$

it follows that, when $\phi_{n-1}(\lambda)$ is known, all coefficient vectors of $f_{n-1}(\lambda)$ can be obtained by comparing coefficients, and hence $\phi_n(\lambda)$ can be formed from the last equation. In fact

$$\lambda f_{n-1}(\lambda) = A_{n-1}f_{n-1}(\lambda) + a_{n-1}\phi_{n-1}(\lambda),$$

so that beginning with the coefficient of λ^{n-2} in $f_{n-1}(\lambda)$, which is simply a_{n-1}, the vector coefficients can be obtained in sequence.

If one first forms $\phi_1(\lambda) = \lambda - \alpha_{11}$, one can then by this scheme form $f_1(\lambda)$ and hence $\phi_2(\lambda)$; then $f_2(\lambda)$, and hence $\phi_3(\lambda)$, . . . , eventually obtaining $\phi_n(\lambda)$.

Given $\phi_{n-1}(\lambda)$, the product $A_{n-1}f_{n-1}(\lambda)$ requires $(n - 1)^3$ multiplications; $a_{n-1}\phi_{n-1}(\lambda)$ requires $(n - 1)^2$; $a'_{n-1}f_{n-1}(\lambda)$ requires $(n - 1)^2$; and $\alpha_{n-1}\phi_{n-1}(\lambda)$ requires $n - 1$. Altogether this is $n^2(n - 1)$. When they are summed over all values from 2 to n, we have a total of

$$n(n^2 - 1)(3n + 2)/12$$

multiplications, or approximately $n^4/4$. The advantage over the other method lies in the fact that not the entire adjoint but only one column of it has been computed.

However, this also accounts for a major disadvantage, when proper vectors as well as proper values are needed. If for any λ_i the vector $f_{n-1}(\lambda_i)$ and the scalar $\phi_{n-1}(\lambda_i)$ do not both vanish, then a proper vector associated with λ_i is

$$\begin{pmatrix} f_{n-1}(\lambda_i) \\ \phi_{n-1}(\lambda_i) \end{pmatrix}.$$

Undoubtedly this covers the majority of cases that arise in actual practice. But this column alone is insufficient for obtaining all the proper and principal vectors when λ_i is a multiple root, and moreover no general method has been provided for the theoretically possible case of a simple root λ_i for which this column (but not the entire adjoint) vanishes.

The proper row vectors can be obtained in the "usual" case by using equations corresponding to (4.22.5) to compute $f'_{n-1}(\lambda)$, and hence $f'_{n-1}(\lambda_i)$.

The "escalator method" also proceeds to matrices of progressively higher order, but it requires the actual solution of the characteristic equation at each stage. Consider only the case of a symmetric matrix, for which moreover all proper values are distinct. Thus suppose for the symmetric matrix A all proper values and all proper vectors are known:

$$AU = U\Lambda. \tag{4.22.6}$$

If A is bordered by a column vector and its transpose to form a symmetric matrix of next higher order, we wish to solve the system

$$\begin{pmatrix} A & a \\ a^{\mathsf{T}} & \alpha \end{pmatrix}\begin{pmatrix} y \\ \eta \end{pmatrix} = \lambda \begin{pmatrix} y \\ \eta \end{pmatrix}. \tag{4.22.7}$$

Hence

$$\begin{aligned} Ay + a\eta &= \lambda y, \\ a^{\mathsf{T}}y + \alpha\eta &= \lambda\eta. \end{aligned} \tag{4.22.8}$$

Let

$$y = Uw. \tag{4.22.9}$$

Then

$$\begin{aligned} AUw + a\eta &= \lambda Uw, \\ U\Lambda w + a\eta &= \lambda Uw, \\ \Lambda w + U^{\mathsf{T}}a\eta &= \lambda w, \\ w = (\lambda I - \Lambda)^{-1}&U^{\mathsf{T}}a\eta. \end{aligned} \tag{4.22.10}$$

Since η can be taken as an arbitrary scale factor, (4.22.9) and (4.22.10) determine the proper vector once the proper value λ is known. However, from the second equation (4.22.8) it follows that

$$a^{\mathsf{T}}Uw = (\lambda - \alpha)\eta,$$

and, on substituting (4.22.10),

$$a^{\mathsf{T}}U(\lambda I - \Lambda)^{-1}U^{\mathsf{T}}a = \lambda - \alpha. \tag{4.22.11}$$

This equation in scalar form can be written

$$\Sigma(a^{\mathsf{T}}u_i)^2/(\lambda - \lambda_i) = \lambda - \alpha, \tag{4.22.12}$$

and its roots are the proper values of the bordered matrix. If the λ_i are arranged on the λ axis in the order $\lambda_1 > \lambda_2 > \cdot \cdot \cdot > \lambda_n$, then as λ varies from $-\infty$ to λ_n, the left member decreases from 0 to $-\infty$; as λ varies from λ_n to λ_{n-1}, the left member decreases from $+\infty$ to $-\infty$; . . . ; as λ varies from λ_1 to $+\infty$, the left member decreases from $+\infty$ to 0. The right member increases linearly throughout. Hence (4.22.12) has exactly one root $\lambda < \lambda_n$; exactly one root λ between each pair of consecutive λ_i; and exactly one root $\lambda > \lambda_1$. With the roots thus isolated, Newton's method is readily applied for evaluating them.

4.23. *Finite Iterations.* If b_0 is an arbitrary non-null vector, then in the sequence

$$b_0, b_1 = Ab_0, \qquad b_2 = Ab_1, \ . \ . \ . \ ,$$

at most n of the vectors are linearly independent. Suppose the first $m \leq n$ of the vectors are linearly independent, but the first $m + 1$ are linearly dependent. Then b_m is expressible as a linear combination of the other m, and hence for some scalars β_i it is true that

$$b_m - \beta_1 b_{m-1} + \beta_2 b_{m-2} - \cdots \pm \beta_m b_0 = 0. \tag{4.23.1}$$

Hence

$$(A^m - \beta_1 A^{m-1} + \beta_2 A^{m-2} - \cdots \pm \beta_m I)b_0 = 0, \tag{4.23.2}$$

which is to say that

$$p(A)b_0 = 0, \tag{4.23.3}$$

where

$$p(\lambda) = \lambda^m - \beta_1 \lambda^{m-1} + \cdots \pm \beta_m. \tag{4.23.4}$$

If

$$d(\lambda) = \lambda^\nu + \delta_1 \lambda^{\nu-1} + \cdots + \delta_\nu$$

is the highest common divisor of $p(\lambda)$ and $\psi(\lambda)$, then $d(A)b_0 = 0$. That is to say

$$(A^\nu + \delta_1 A^{\nu-1} + \cdots + \delta_\nu I)b_0 = 0,$$

or

$$b_\nu + \delta_1 b_{\nu-1} + \cdots + \delta_\nu b_0 = 0.$$

But the vectors $b_0, b_1, \ . \ . \ . \ , b_{m-1}$ are linearly independent, whence $\nu = m$ and $d = p$. Hence $p(\lambda)$ divides the minimal function $\psi(\lambda)$ and hence also the characteristic function $\phi(\lambda)$. In particular if $m = n$, then $p(\lambda) = (-1)^n \phi(\lambda)$. When this is true, therefore, one can form the characteristic equation by first performing the n iterations Ab_i and then by solving a system of n linear equations. When this is not so, one can at least obtain a divisor of the characteristic function by performing a smaller number of iterations and by solving a system of lower order. However, in addition one must test at each step the independence of the vectors already found, as long as the number is below $n + 1$.

A great improvement is provided by Lanczos's "method of minimized iterations." Let b_0 and c_0 be arbitrary nonorthogonal non-null vectors. In case A is symmetric, take $b_0 = c_0$; if A is Hermitian, take $b_0 = c_0$ and in the following discussion replace the transpose by the conjugate transpose. In other cases b_0 and c_0 may or may not be the same. Form b_1 as a linear combination of b_0 and Ab_0, orthogonal to c_0; and c_1 as a linear combination of c_0 and $A^{\mathsf{T}}c_0$, orthogonal to b_0. Thus

$$b_1 = Ab_0 - \alpha_0 b_0,$$

where

$$0 = c_0^{\mathsf{T}}b_1 = c_0^{\mathsf{T}}Ab_0 - \alpha_0 c_0^{\mathsf{T}}b_0;$$

and

$$c_1^{\mathsf{T}} = c_0^{\mathsf{T}}A - \delta_0 c_0^{\mathsf{T}},$$

where

$$0 = c_1^{\mathsf{T}}b_0 = c_0^{\mathsf{T}}Ab_0 - \delta_0 c_0^{\mathsf{T}}b_0.$$

But then

$$\alpha_0 = \delta_0 = c_0^{\mathsf{T}}Ab_0/c_0^{\mathsf{T}}b_0.$$

Next, choose b_2 as a linear combination of b_0, b_1, and Ab_1, orthogonal to both c_0 and c_1; and c_2 as a linear combination of c_0, c_1, and $A^{\mathsf{T}}c_1$, orthogonal to both b_0 and b_1. Then

$$b_2 = Ab_1 - \alpha_1 b_1 - \beta_0 b_0$$

where

$$0 = c_0^{\mathsf{T}}b_2 = c_0^{\mathsf{T}}Ab_1 - \beta_0 c_0^{\mathsf{T}}b_0,$$
$$0 = c_1^{\mathsf{T}}b_2 = c_1^{\mathsf{T}}Ab_1 - \alpha_1 c_1^{\mathsf{T}}b_1.$$

Hence

$$\alpha_1 = c_1^{\mathsf{T}}Ab_1/c_1^{\mathsf{T}}b_1, \qquad \beta_0 = c_0^{\mathsf{T}}Ab_1/c_0^{\mathsf{T}}b_0.$$

But from the relations already derived,

$$c_0^{\mathsf{T}}Ab_1 = c_1^{\mathsf{T}}b_1 = c_1^{\mathsf{T}}Ab_0.$$

Hence

$$\beta_0 = c_1^{\mathsf{T}}b_1/c_0^{\mathsf{T}}b_0,$$

and if

$$c_2^{\mathsf{T}} = c_1^{\mathsf{T}}A - \alpha_1 c_1^{\mathsf{T}} - \beta_0 c_0^{\mathsf{T}},$$

then it follows that

$$0 = c_2^{\mathsf{T}}b_0 = c_2^{\mathsf{T}}b_1.$$

The step breaks down in case $c_1^{\mathsf{T}}b_1 = 0$, since this is the denominator in α_1. But this means that

$$0 = c_0^{\mathsf{T}}Ab_1 = c_0^{\mathsf{T}}A(A - \alpha_0 I)b_0,$$

and when α_0 is replaced by its value, this is, apart from the factor $(c_0^{\mathsf{T}}b_0)^{-1}$, equal to the determinant of the matrix product

$$\begin{pmatrix} c_0^{\mathsf{T}} \\ c_0^{\mathsf{T}}A \end{pmatrix} (b_0 \quad Ab_0).$$

Hence this can vanish only if the pair b_0, Ab_0 or else the pair c_0, $A^{\mathsf{T}}c_0$ is linearly dependent.* Suppose for the present that this is not the case.

*This conclusion, and the more general ones, do not follow. It can be shown, however, that if neither pair (more generally neither set) is linearly dependent, then after a slight perturbation of either starting vector, b_0 or c_0, the determinant will not vanish. One could say, in pseudotechnical language, that failure has zero probability.

We now show that α_2 and β_1 can be chosen so that

$$b_3 = Ab_2 - \alpha_2 b_2 - \beta_1 b_1$$

is orthogonal to all three vectors c_0, c_1, and c_2 and so that

$$c_3 = A^{\mathsf{T}}c_2 - \alpha_2 c_2 - \beta_1 c_1$$

is orthogonal to b_0, b_1, and b_2, provided only that the vectors b_0, Ab_0, A^2b_0, and also the vectors c_0, $A^{\mathsf{T}}c_0$, $A^{\mathsf{T}2}c_0$, are linearly independent triples. First we note that

$$\begin{aligned} c_0^{\mathsf{T}}Ab_2 &= (c_1^{\mathsf{T}} + \alpha_0 c_0^{\mathsf{T}})b_2 = 0, \\ c_2^{\mathsf{T}}Ab_0 &= c_2^{\mathsf{T}}(b_1 + \alpha_0 b_0) = 0, \end{aligned}$$

so that c_3 is orthogonal to b_0, and b_3 to c_0, independently of α_2 and β_1. Next

$$c_1^{\mathsf{T}}Ab_2 = (c_2^{\mathsf{T}} + \alpha_1 c_1^{\mathsf{T}} + \beta_0 c_0^{\mathsf{T}})b_2 = c_2^{\mathsf{T}}b_2 = c_2^{\mathsf{T}}Ab_1.$$

Hence

$$c_1^{\mathsf{T}}b_3 = c_2^{\mathsf{T}}b_2 - \beta_1 c_1^{\mathsf{T}}b_1,$$

whence c_1 and b_3 are orthogonal if

$$\beta_1 = c_2^{\mathsf{T}}b_2/c_1^{\mathsf{T}}b_1.$$

But in this event

$$c_3^{\mathsf{T}}b_1 = c_2^{\mathsf{T}}b_2 - \beta_1 c_1^{\mathsf{T}}b_1 = 0.$$

Finally,

$$\begin{aligned} c_2^{\mathsf{T}}b_3 &= c_2^{\mathsf{T}}Ab_2 - \alpha_2 c_2^{\mathsf{T}}b_2, \\ c_3^{\mathsf{T}}b_2 &= c_2^{\mathsf{T}}Ab_2 - \alpha_2 c_2^{\mathsf{T}}b_2, \end{aligned}$$

and both vanish provided

$$\alpha_2 = c_2^{\mathsf{T}}Ab_2/c_2^{\mathsf{T}}b_2.$$

This is always possible if c_2 and b_2 are not orthogonal. But the matrix product

$$\begin{pmatrix} c_0^{\mathsf{T}} \\ c_0^{\mathsf{T}}A \\ c_0^{\mathsf{T}}A^2 \end{pmatrix} (b_0 \quad Ab_0 \quad A^2b_0) = \begin{pmatrix} c_0^{\mathsf{T}}b_0 & c_0^{\mathsf{T}}Ab_0 & c_0^{\mathsf{T}}A^2b_0 \\ c_0^{\mathsf{T}}Ab_0 & c_0^{\mathsf{T}}A^2b_0 & c_0^{\mathsf{T}}A^3b_0 \\ c_0^{\mathsf{T}}A^2b_0 & c_0^{\mathsf{T}}A^3b_0 & c_0^{\mathsf{T}}A^4b_0 \end{pmatrix}$$

has a determinant which by successive reductions of rows and columns yields

$$\begin{vmatrix} c_0^{\mathsf{T}}b_0 & c_0^{\mathsf{T}}Ab_0 & c_0^{\mathsf{T}}A^2b_0 \\ c_0^{\mathsf{T}}Ab_0 & c_0^{\mathsf{T}}A^2b_0 & c_0^{\mathsf{T}}A^3b_0 \\ c_0^{\mathsf{T}}A^2b_0 & c_0^{\mathsf{T}}A^3b_0 & c_0^{\mathsf{T}}A^4b_0 \end{vmatrix} = \begin{vmatrix} c_0^{\mathsf{T}}b_0 & c_1^{\mathsf{T}}b_0 & c_2^{\mathsf{T}}b_0 \\ c_0^{\mathsf{T}}b_1 & c_1^{\mathsf{T}}b_1 & c_2^{\mathsf{T}}b_1 \\ c_0^{\mathsf{T}}b_2 & c_1^{\mathsf{T}}b_2 & c_2^{\mathsf{T}}b_2 \end{vmatrix} = (c_0^{\mathsf{T}}b_0)(c_1^{\mathsf{T}}b_1)(c_2^{\mathsf{T}}b_2).$$

Since we have assumed $(c_0^{\mathsf{T}}b_0)(c_1^{\mathsf{T}}b_1) \neq 0$, it follows that $c_2^{\mathsf{T}}b_2 = 0$ if and only if either the triple c_0, $A^{\mathsf{T}}c_0$, $A^{\mathsf{T}2}c_0$ or else the triple b_0, Ab_0, A^2b_0 is linearly dependent.

We can proceed inductively, forming

$$(4.23.5) \qquad \begin{aligned} b_{i+1} &= (A - \alpha_i)b_i - \beta_{i-1}b_{i-1}, \\ c_{i+1} &= (A^\mathsf{T} - \alpha_i)c_i - \beta_{i-1}c_{i-1}, \end{aligned}$$

where

$$(4.23.6) \qquad \alpha_i = c_i^\mathsf{T}Ab_i/c_i^\mathsf{T}b_i, \qquad \beta_{i-1} = c_i^\mathsf{T}b_i/c_{i-1}^\mathsf{T}b_{i-1},$$

until possibly at some stage c_i and b_i are orthogonal. When c_i and b_i are not orthogonal, then b_{i+1} is orthogonal to every vector $c_0, c_1, \ldots, c_i$, and c_{i+1} is orthogonal to every vector $b_0, b_1, \ldots, b_i$.

Necessarily, there is a smallest $m \leq n$ for which $c_m^\mathsf{T}b_m = 0$, since if the vectors $c_0, c_1, \ldots, c_{n-1}$ are all linearly independent, then only $b_n = 0$ is orthogonal to them all, and this vector is orthogonal to c_n, whatever c_n may be. Suppose the relation holds for some $m < n$. Then either the set $b_0, Ab_0, \ldots, A^mb_0$ or else the set $c_0, A^\mathsf{T}c_0, \ldots, (A^\mathsf{T})^mc_0$ is linearly dependent. For definiteness suppose it to be the former. The set from which A^mb_0 is omitted is linearly independent, for if it were not, the m selected would not be the smallest possible. Hence A^mb_0 is expressible as a linear combination of the m vectors $b_0, \ldots, A^{m-1}b_0$. Hence Ab_{m-1} is some linear combination of the vectors $b_0, b_1, \ldots, b_{m-1}$:

$$Ab_{m-1} = \mu_{m-1}b_{m-1} + \mu_{m-2}b_{m-2} + \cdots + \mu_0b_0.$$

Hence

$$\begin{aligned} b_m &= Ab_{m-1} - \alpha_{m-1}b_{m-1} - \beta_{m-2}b_{m-2} \\ &= (\mu_{m-1} - \alpha_{m-1})b_{m-1} + (\mu_{m-2} - \beta_{m-2})b_{m-2} + \mu_{m-3}b_{m-3} + \cdots + \mu_0b_0. \end{aligned}$$

But then

$$\begin{aligned} 0 &= c_0^\mathsf{T}b_m = \mu_0c_0^\mathsf{T}b_0, \\ 0 &= c_1^\mathsf{T}b_m = \mu_1c_1^\mathsf{T}b_1, \\ &\cdots\cdots\cdots\cdots \\ 0 &= c_{m-1}^\mathsf{T}b_m = (\mu_{m-1} - \alpha_{m-1})c_{m-1}^\mathsf{T}b_{m-1}. \end{aligned}$$

By hypothesis all the vector products on the right are non-null, and therefore

$$0 = \mu_0 = \mu_1 = \cdots = \mu_{m-1} - \alpha_{m-1},$$

whence $b_m = 0$ and

$$(4.23.7) \qquad 0 = (A - \alpha_{m-1}I)b_{m-1} - \beta_{m-2}b_{m-2}.$$

Thus if the vectors $b_0, Ab_0, \ldots, A^mb_0$ are linearly dependent, then (4.23.7) holds; correspondingly if the iterates of c_0 by A^T are linearly dependent, then it follows that

$$(4.23.8) \qquad 0 = (A^\mathsf{T} - \alpha_{m-1}I)c_{m-1} - \beta_{m-2}c_{m-2}.$$

Hence the recursion (4.23.5) and (4.23.6) can be continued until for some $i = m$ either (4.23.7) or else (4.23.8) holds.

Consider now the sequence of polynomials

$$\begin{aligned} p_0(\lambda) &= 1, \\ p_1(\lambda) &= (\lambda - \alpha_0)p_0(\lambda), \\ p_2(\lambda) &= (\lambda - \alpha_1)p_1(\lambda) - \beta_0 p_0(\lambda), \\ &\cdots\cdots\cdots\cdots\cdots \\ p_{i+1}(\lambda) &= (\lambda - \alpha_i)p_i(\lambda) - \beta_{i-1}p_{i-1}(\lambda), \\ &\cdots\cdots\cdots\cdots\cdots \end{aligned} \tag{4.23.9}$$

where the α's and β's are defined by (4.23.6). One verifies inductively that

$$p_i(A)b_0 = b_i, \qquad p_i(A^{\mathsf{T}})c_0 = c_i. \tag{4.23.10}$$

Hence either $p_m(A)b_0 = 0$, or else $p_m(A^{\mathsf{T}})c_0 = 0$. In either case $p_m(\lambda)$ is a divisor of the minimal function $\psi(\lambda)$, its coefficients are provided without the necessity for solving a system of equations, and moreover the test for dependence of the successive sets of iterates of b_0 and of c_0 is performed automatically in the course of the computation.

Suppose now that the proper values are all distinct. Then if W is the matrix of proper vectors,

$$W^{-1}AW = \Lambda, \tag{4.23.11}$$

and Λ is diagonal. Also

$$W^{-1}p_i(A)W = p_i(\Lambda). \tag{4.23.12}$$

If only one of b_m and c_m vanishes while the other does not, one can by a different choice of b_0 (if $b_m = 0$) or of c_0 (if $c_m = 0$) obtain a longer sequence. Hence suppose that $b_m = c_m = 0$. Moreover $p_m(\lambda)$ has only simple zeros:

$$p(\lambda) = p_m(\lambda) = (\lambda - \lambda_1)(\lambda - \lambda_2) \cdots (\lambda - \lambda_m). \tag{4.23.13}$$

Then

$$q_i(\lambda) = p(\lambda)/(\lambda - \lambda_i) \tag{4.23.14}$$

is a polynomial, and there is no nonconstant factor common to all the q_i. Hence by a theorem in §2.06 there exist polynomials $f_i(\lambda)$ such that

$$\Sigma f_i(\lambda)q_i(\lambda) = 1.$$

Hence

$$\Sigma f_i(A)q_i(A) = I = \Sigma f_i(A^{\mathsf{T}})q_i(A^{\mathsf{T}}), \tag{4.23.15}$$

and therefore

$$\Sigma f_i(A)q_i(A)b_0 = b_0, \qquad \Sigma f_i(A^{\mathsf{T}})q_i(A^{\mathsf{T}})c_0 = c_0. \tag{4.23.16}$$

Also, since polynomials in A (or those in A^{T}) are commutative with one another,

$$\Sigma f_i(A)q_i(A)b_j = b_j, \qquad \Sigma f_i(A^{\mathsf{T}})q_i(A^{\mathsf{T}})c_j = c_j.$$

But

$$(A - \lambda_i I)f_i(A)q_i(A)b_j = f_i(A)p(A)b_j = 0.$$

Hence $f_i(A)q_i(A)b_j$ is a proper vector of A associated with the proper value λ_i, so that each b_j is expressed as a linear combination of proper vectors.

Again, if the first of (4.23.16) is solved for any proper vector

$$f_i(A)q_i(A)b_0,$$

it is expressed as a linear combination of vectors $b_0, Ab_0, A^2b_0, \ldots$, and these in turn are expressible as linear combinations of the b_j. Hence each proper vector appearing in the first of (4.23.16) is expressible as a linear combination of the b_j. Likewise each proper vector appearing in the second of (4.23.16) is expressible as a linear combination of the c_j. Let u_i represent the proper vectors of A, v_i the proper vectors of A^{T}, which appear in (4.23.16). Then

$$b_0 = \Sigma u_i, \qquad c_0 = \Sigma v_i. \tag{4.23.17}$$

Since the u_i and v_i are proper vectors, therefore

$$\begin{aligned} b_j &= p_j(A)b_0 = \sum_i p_j(A)u_i = \sum_i p_j(\lambda_i)u_i, \\ c_j &= p_j(A^{\mathsf{T}})c_0 = \sum_i p_j(A^{\mathsf{T}})v_i = \sum_i p_j(\lambda_i)v_i. \end{aligned} \tag{4.23.18}$$

Hence if we let

$$P = \begin{pmatrix} p_0(\lambda_1) & \cdots & p_{m-1}(\lambda_1) \\ \cdots & \cdots & \cdots \\ p_0(\lambda_m) & \cdots & p_{m-1}(\lambda_m) \end{pmatrix}, \tag{4.23.19}$$

$$\begin{aligned} B &= (b_0 \quad \cdots \quad b_{m-1}), \qquad & C &= (c_0 \quad \cdots \quad c_{m-1}), \\ U &= (u_1 \quad \cdots \quad u_m), \qquad & V &= (v_1 \quad \cdots \quad v_m), \end{aligned} \tag{4.23.20}$$

then (4.23.18) can be written

$$B = UP, \qquad C = VP. \tag{4.23.21}$$

But

$$V^{\mathsf{T}}U = D,$$

where D is a diagonal matrix. Hence

$$V^{\mathsf{T}}B = DP, \qquad C^{\mathsf{T}}U = P^{\mathsf{T}}D.$$

But we already know that the u's can be expressed in terms of the b's, and the v's in terms of the c's:

$$U = BH, \qquad V^{\mathsf{T}} = K^{\mathsf{T}}C^{\mathsf{T}}.$$

Moreover,

$$C^{\mathsf{T}}B = \Delta$$

is also a diagonal matrix. Hence

$$C^{\mathsf{T}}U = \Delta H, \qquad V^{\mathsf{T}}B = K^{\mathsf{T}}\Delta,$$

$$H = \Delta^{-1}C^{\mathsf{T}}U = \Delta^{-1}P^{\mathsf{T}}D, \qquad K^{\mathsf{T}} = V^{\mathsf{T}}B\Delta^{-1} = DP\Delta^{-1},$$

(4.23.22) $$U = B\Delta^{-1}P^{\mathsf{T}}D, \qquad V^{\mathsf{T}} = DP\Delta^{-1}C^{\mathsf{T}}.$$

The diagonal matrix D is determined only by scale factors in the vectors u_i and v_i and these can be chosen as convenient. Thus the m proper vectors of A which appear in (4.23.17) can be expressed as linear combinations of the b_i, and the m proper vectors of A^{T} (or proper row vectors of A) can be expressed as linear combinations of the c_i (or of the c_i^{T}).

When $m < n$, in general this will be because $\phi(\lambda)$ has multiple zeros. Suppose still that $p_m(\lambda)$ has only simple zeros and take a b_0' orthogonal to all the c_j. Then Ab_0' is also orthogonal to all the c_j, since

$$c_j^{\mathsf{T}}Ab_0' = (A^{\mathsf{T}}c_j)^{\mathsf{T}}b_0',$$

and $A^{\mathsf{T}}c_j$ is itself a linear combination of the c's. Likewise if c_0' is orthogonal to all the b's of the original sequence, so also will $A^{\mathsf{T}}c_0'$ be orthogonal to all the b's of this sequence. Hence one can develop sequences b_i' and c_i', and these vectors will be independent of those hitherto found. They will yield new proper vectors associated with the multiple roots. Moreover, the new sequences will in general have fewer than m members each. If they have fewer than $n - m$ members, a third pair of sequences must be started with b_0'' orthogonal to all previous c's, and c_0'' orthogonal to all previous b's. Since each such sequence will contain at least one member, the initial vector of the sequence, the process will eventually terminate and yield all proper vectors.

If, instead of imposing the orthogonality requirement upon b_0' and c_0', one required only that $b_0' \neq b_0$ and $c_0' \neq c_0$, then in general new sequences of m terms each will result, proper vectors associated with simple roots will be found over again, but a new proper vector for A and one for A^{T} will be found associated with each multiple root. This course might be preferred in order to avoid the computation of the orthogonal starting vectors.

When the zeros of $p_m(\lambda)$ are not all distinct, it is still possible to obtain the proper vectors in much the same way. A resolution such as (4.23.16) can be employed to show that b_0 and c_0 are expressible as sums of principal vectors. Here $q_i(\lambda)$ is the quotient of $p(\lambda)$ by the highest power of $(\lambda - \lambda_i)$ it contains as a factor. Each principal vector which appears in

(4.23.16) is expressible as a linear combination of the b_i (or of the c_i). Also, if x_i is the principal vector of A associated with λ_i, then $(A - \lambda_i I)x_i$, $(A - \lambda_i I)^2 x_i$, . . . can each be expressed as a linear combination of the b_i. But every vector of this sequence is a principal vector or vanishes, and one is a proper vector. Hence associated with each zero of $p_m(\lambda)$ there is a proper vector of A which is a linear combination of the b's and a proper vector of A^T which is a linear combination of the c's. Let u_i be the proper vector of A associated with λ_i. Then

$$u_i = \sum_j \omega_{ij} b_j.$$

Hence

$$c_j^\mathsf{T} u_i = \omega_{ij} c_j^\mathsf{T} b_j.$$

But

$$c_j = p_j(A^\mathsf{T}) c_0,$$

whence

$$c_j^\mathsf{T} u_i = c_0^\mathsf{T} p_j(A) u_i = c_0^\mathsf{T} u_i p_j(\lambda_i), \tag{4.23.23}$$

since u_i is a proper vector. Hence

$$\omega_{ij} = c_0^\mathsf{T} u_i p_j(\lambda_i) / c_j^\mathsf{T} b_j, \tag{4.23.24}$$

and therefore

$$u_i = c_0^\mathsf{T} u_i \sum_j p_j(\lambda_i) b_j / c_j^\mathsf{T} b_j.$$

But $c_0^\mathsf{T} u_i$ is a scale factor which can be chosen arbitrarily. Thus if in (4.23.19) the matrix P is taken to be rectangular, one row for each λ_i, and the matrices U and V contain only the proper vectors, then (4.23.22) holds also in this more general case. Again if $m < n$, one can select vectors b_0' orthogonal to every c_j, and c_0' orthogonal to every b_j, and form new sequences to provide any proper vectors not already found.

4.24. *The Triple-diagonal Form for a Symmetric Matrix.* In the method outlined in the last section, consider again for the moment the case $m = n$. Let

$$B = (b_0 \quad b_1 \quad \cdots \quad b_{n-1}), \qquad C = (c_0 \quad c_1 \quad \cdots \quad c_{n-1}).$$

Then from the defining relations we have $C^\mathsf{T} B$ as a diagonal matrix D whose diagonal elements are

$$\delta_0 = c_0^\mathsf{T} b_0 \neq 0, \; \ldots, \; \delta_{n-1} = c_{n-1}^\mathsf{T} b_{n-1} \neq 0. \tag{4.24.1}$$

Hence

$$C^\mathsf{T} B = D,$$
$$B^{-1} = D^{-1} C^\mathsf{T}, \qquad C^{-1} = D^{-1} B^\mathsf{T}.$$

Also we found that

$$\begin{aligned} c_i^{\mathsf{T}}Ab_i &= \alpha_i c_i^{\mathsf{T}}b_i = \alpha_i\delta_i, \\ c_i^{\mathsf{T}}Ab_{i-1} &= c_{i-1}^{\mathsf{T}}Ab_i = c_i^{\mathsf{T}}b_i = \delta_i, \\ c_i^{\mathsf{T}}Ab_j &= 0, \qquad |i-j| > 1. \end{aligned}$$

Hence the product $C^{\mathsf{T}}AB$ is a matrix for which the only non-null elements lie along, just above or just below, the main diagonal:

$$C^{\mathsf{T}}AB = \begin{pmatrix} \alpha_0\delta_0 & \delta_1 & 0 & 0 & \cdots \\ \delta_1 & \alpha_1\delta_1 & \delta_2 & 0 & \cdots \\ 0 & \delta_2 & \alpha_2\delta_2 & \delta_3 & \cdots \\ \cdots & \cdots & \cdots & \cdots & \cdots \end{pmatrix}. \tag{4.24.2}$$

Hence

$$B^{-1}AB = D^{-1}C^{\mathsf{T}}AB = \begin{pmatrix} \alpha_0 & \beta_0 & 0 & \cdots \\ 1 & \alpha_1 & \beta_1 & \cdots \\ 0 & 1 & \alpha_2 & \cdots \\ \cdots & \cdots & \cdots & \cdots \end{pmatrix}, \tag{4.24.3}$$

since by (4.23.6) we know that $\beta_{i-1} = \delta_i/\delta_{i-1}$. One should expect that after reducing the matrix to this form a considerable step has been taken toward complete diagonalization.

In case A is symmetric, if $c_0 = b_0$, then $C = B$, and every $\delta_i > 0$. Hence $D^{1/2}$ is a real diagonal matrix. If we set

$$U = BD^{-1/2},$$

then U is an orthogonal matrix, and

$$S = U^{\mathsf{T}}AU = \begin{pmatrix} \alpha_0 & \beta_0^{1/2} & 0 & \cdots \\ \beta_0^{1/2} & \alpha_1 & \beta_1^{1/2} & \cdots \\ \cdots & \cdots & \cdots & \cdots \end{pmatrix}. \tag{4.24.4}$$

For the case of a symmetric matrix, we now consider another method of obtaining the triple-diagonal form (4.24.4).

Before doing so, however, consider the characteristic function. Let

$$\begin{aligned} p_0(\lambda) &= 1, \\ p_1(\lambda) &= \lambda - \alpha_0, \\ p_2(\lambda) &= \begin{vmatrix} \lambda - \alpha_0 & -\beta_0^{1/2} \\ -\beta_0^{1/2} & \lambda - \alpha_1 \end{vmatrix} = (\lambda - \alpha_1)p_1(\lambda) - \beta_0 p_0(\lambda), \\ p_3(\lambda) &= \begin{vmatrix} \lambda - \alpha_0 & -\beta_0^{1/2} & 0 \\ -\beta_0^{1/2} & \lambda - \alpha_1 & -\beta_1^{1/2} \\ 0 & -\beta_1^{1/2} & \lambda - \alpha_2 \end{vmatrix} = (\lambda - \alpha_2)p_2(\lambda) - \beta_1 p_1(\lambda). \\ &\cdots\cdots\cdots\cdots\cdots\cdots \end{aligned} \tag{4.24.5}$$

Thus the polynomials $p_i(\lambda)$ are the expansions of the determinants of the first principal minors of the matrix $\lambda I - S$. Note that $p_{i+1}(\lambda)$ and $p_i(\lambda)$

cannot have a common factor, for if they did, this factor would be contained also in $p_{i-1}(\lambda)$, hence also in $p_{i-2}(\lambda)$, . . . , hence also in $p_0(\lambda)$, which is absurd. Also at any ρ for which $p_i(\rho) = 0$, $p_{i+1}(\rho)$ and $p_{i-1}(\rho)$ have opposite signs, since each $\beta_i > 0$.

We can show that between any consecutive zeros of $p_i(\lambda)$ there is a zero of $p_{i+1}(\lambda)$, and further that $p_{i+1}(\lambda)$ has a zero to the left, and one to the right, of all those of $p_i(\lambda)$. Hence between any two consecutive zeros of $p_{i+1}(\lambda)$ there is exactly one zero of $p_i(\lambda)$. Since $p_1(\alpha_0) = 0$, and since $p_2(\alpha_0) < 0$, while $p_2(\pm\infty) = +\infty$, the statement is certainly true for $i = 1$. Suppose it demonstrated for $p_3, p_4, \ldots, p_{i+1}$ and consider p_{i+2}. Let ρ_1 and ρ_2 be consecutive zeros of $p_{i+1}(\lambda)$. Then

$$p_{i+2}(\rho_1) = -\beta_i p_i(\rho_1), \qquad p_{i+2}(\rho_2) = -\beta_i p_i(\rho_2).$$

The hypothesis implies that $p_2, p_3, \ldots, p_{i+1}$ can have only simple zeros and that p_i has one and only one zero between the consecutive zeros ρ_1 and ρ_2 of p_{i+1}; hence $p_i(\rho_1)$ and $p_i(\rho_2)$ have opposite signs, and hence $p_{i+2}(\rho_1)$ and $p_{i+2}(\rho_2)$ have opposite signs. Therefore p_{i+2} has an odd number of zeros between ρ_1 and ρ_2.

Next suppose ρ is the greatest zero of p_{i+1}. Then $p_{i+2}(\rho) = -\beta_i p_i(\rho)$. But p_i has no zero exceeding ρ, and $p_i(\infty) = +\infty$. Hence $p_i(\rho) > 0$, and therefore $p_{i+2}(\rho) < 0$. Hence $p_{i+2}(\rho)$ has an odd number of zeros exceeding ρ. But p_{i+2} is of degree $i + 2$. The hypothesis of the induction implies that p_{i+1} has $i + 1$ real and distinct zeros; these divide the real λ axis into i segments and two rays extending to $+\infty$ and to $-\infty$, respectively. We have shown that each segment and one of the rays each has on it an odd number of zeros of p_{i+2}. Hence each can contain only one, and the remaining zero lies on the other ray.

Thus the polynomials $p_i(\lambda)$ have all the properties required of a Sturm sequence, as in §3.05, though they are not formed in the same way. Hence by counting the number of variations in sign exhibited by the sequence $p_i(\lambda)$ at each of two values of λ and by taking the difference, one has the exact number of proper values of the matrix A contained on the interval between these values.

Now return to the symmetric matrix A and consider, *ab initio*, the problem of reducing it to a triple-diagonal form. Suppose A is 3×3. Then if $\alpha_{12} \neq 0$, one can find an orthogonal matrix of the form

$$U = \begin{pmatrix} 1 & 0 & 0 \\ 0 & c & -s \\ 0 & s & c \end{pmatrix}, \tag{4.24.6}$$

where c and s are the sine and cosine of some angle, such that in the transformed matrix

$$A' = U^{\mathsf{T}} A U$$

the elements $\alpha'_{13} = \alpha'_{31} = 0$. In fact, one finds

$$\alpha'_{13} = \alpha'_{31} = c\alpha_{13} - s\alpha_{12}, \tag{4.24.7}$$

and hence one has only to choose

$$\kappa = \alpha_{13}/\alpha_{12}, \qquad c = (1 + \kappa^2)^{-1/2}, \qquad s = c\kappa, \tag{4.24.8}$$

which can always be done if $\alpha_{12} \neq 0$. It turns out then that

$$\alpha'_{12} = \alpha'_{21} = c\alpha_{12} + s\alpha_{13} = \alpha_{12}/c,$$

so that as a result of the transformation

$$|\alpha'_{12}| > |\alpha_{12}| > 0.$$

For an arbitrary symmetric matrix A, let it be partitioned

$$A = \begin{pmatrix} A_{11} & A_{12} \\ A_{21} & A_{22} \end{pmatrix}, \tag{4.24.9}$$

where A_{11} is of order 3. Suppose $\alpha_{12} \neq 0$. Let

$$V = \begin{pmatrix} U & 0 \\ 0 & I \end{pmatrix}, \tag{4.24.10}$$

where U is the orthogonal matrix of order 3 defined by (4.24.6) and (4.24.8). Then V is orthogonal, and

$$V^T A V = \begin{pmatrix} U^T A_{11} U & U^T A_{12} \\ A_{21} U & A_{22} \end{pmatrix}. \tag{4.24.11}$$

Hence the matrix V transforms the symmetric matrix A, arbitrary except that $\alpha_{12} \neq 0$, into a matrix A' in which $\alpha'_{13} = 0$. Moreover, the transformation leaves unaffected all elements in the first row of A_{12}, and in the first column of A_{21}, as well as the entire submatrix A_{22}.

Now if in A' the element $\alpha'_{14} \neq 0$, one can interchange the third and fourth columns and the third and fourth rows, apply a similar transformation, interchange again, if desired, and obtain a matrix A'' in which

$$\alpha''_{13} = \alpha''_{31} = \alpha''_{14} = \alpha''_{41} = 0.$$

By continuing in this fashion, all elements but the first two in the first row and all elements but the first two in the first column can be caused to vanish.

Having achieved this, one can now operate with the submatrix of order $n - 1$ obtained after leaving out the first row and the first column of the matrix. Eventually, therefore, one obtains the required triple-diagonal form. When the characteristic equation is solved, the proper vectors associated with each proper value are obtained by a direct solution of the homogeneous equations.

If it should happen that any $\beta_i = 0$ on the off diagonals, then it is sufficient to consider separately the characteristic equation and proper vectors of the submatrix above and to the left of the vanishing β_i and of that below and to the right.

4.3. Bibliographic Notes. A good development of the topics outlined in §4.0 can be found in MacDuffee (1943). Of the many papers on bounds see especially Taussky (1949), Brauer (1946, 1947, 1948), Parker (1951), Ostrowski (1952), and Price (1951).

Papers by Aitken (1931, 1936–1937*b*) are classic. A more recent general discussion of iterative methods is by Semendiaev (1950). Bargmann, Montgomery, and von Neumann (1946) discuss the use of the trace of powers of A and consider in particular the prevention of "overflow" in spite of round-off. The use of polynomials to accelerate convergence was proposed by Flanders and Shortley (1950). On the use of transformations and orthogonalization for successive proper values see Hotelling (1943) and Feller and Forsythe (1951). Kohn (1949) describes without proof an iteration which converges to an arbitrary (but random) proper value.

Aitken uses the Δ operator for obtaining successive proper values. On equal and nearly equal roots see Rosser, Lanczos, Hestenes, and Karush (1951). Rotational diagonalization, for which equality or near equality is no difficulty, was used by Kelley (1935), but is in fact much older. The method was discussed by Goldstine in August, 1951, at a symposium at the Institute for Numerical Analysis and is to be published with detailed error analysis in a paper by Goldstine, Murray, and von Neumann.

The method of §4.21 is due to Frame (1949, and an unpublished paper). It was also published by Fettis (1950), but without detail or consideration of multiple roots. The recursion defined by (4.22.5) was given by Bryan (1950). On the escalator method, see Morris and Head (1942), and for a more general (brief) treatment, Vinograde (1951). Lanczos (1951*a* and 1951*b*) gave the method of minimized iteration. The triple-diagonal form with a valuable treatment of error is discussed by Givens (1951 and a forthcoming memorandum).

That the polynomials $p_j(x)$ form a Sturm sequence is a classical result (see Browne, 1930).

A method for the simultaneous improvement of approximation to all proper values is given by Jahn (1948) and Collar (1948).

CHAPTER 5

INTERPOLATION

5. Interpolation

This book falls naturally into two parts: one part dealing with the solution of equations and systems of equations, the other part dealing with the approximate representation of functions. We come now to the second part.

It may be that a function or its integral or its derivative is not easily evaluated; or that one knows nothing but a limited number of its functional values, and perhaps these only approximately. In either case one may require an approximate representation in some form that is readily evaluated or integrated or otherwise manipulated. If one knows only certain functional values, which, however, are presumed exact, the approximate representation may be required to assume the same functional values corresponding to the given values of the argument. The problem is then one of interpolation. If the given functional values cannot be taken as exact, then a somewhat simpler representation will be accepted, and one which is not required to take the same, but only approximately the same, functional values. This is smoothing or curve fitting. Even a function that is easy to evaluate may not be easy to integrate. For approximate quadrature, therefore, the usual method is to obtain an approximate representation in terms of functions that are easily integrated.

In general, the method is to select from some class of simple functions $\phi(x)$ a limited number, $\phi_0(x), \phi_1(x), \ldots, \phi_n(x)$, and attempt to approximate the required function $f(x)$ by a linear combination of these functions. Thus we wish to find constants γ_i such that the function

$$\Phi(x) \equiv \gamma_0\phi_0(x) + \gamma_1\phi_1(x) + \cdots + \gamma_n\phi_n(x) \tag{5.0.1}$$

is in some sense a reasonable approximation to $f(x)$. If we agree to use $n + 1$ functions ϕ, then we have $n + 1$ constants γ at our disposal, and we can impose $n + 1$ conditions for their determination. In interpolation the conditions imposed are that $f(x)$ and $\Phi(x)$ shall be equal at each of $n + 1$ distinct values x_i of the abscissas. Thus we require that

$$f(x_i) = \Phi(x_i) \qquad (i = 0, 1, \ldots, n). \tag{5.0.2}$$

The points x_i on the axis of abscissas will be called the fundamental points. A particularly simple choice of functions $\phi_i(x)$ for most computational purposes is

$$\phi_i = x^i,$$

so that $\Phi(x)$ is a polynomial of degree n.

Another possibility is to require that (5.0.2) shall hold for $i = 0, \ldots, r$, while for $j = r + 1, \ldots, n$ we require that

$$f'(x_j) = \Phi'(x_j). \tag{5.0.3}$$

In this event all the x_j must be distinct, but they may coincide with some of the x_i. We could equally well require that higher derivatives of f and Φ shall be equal for certain values of x, provided that altogether we have exactly $n + 1$ independent and consistent conditions imposed on the γ's. Still other types of conditions may be used, and some will be discussed later.

To return to (5.0.2), if we write

$$y_i = f(x_i) \tag{5.0.4}$$

for brevity, then Eqs. (5.0.2) can be written

$$y_j = \Sigma\gamma_i\phi_i(x_j). \tag{5.0.5}$$

If the determinant

$$\Delta = |\phi_i(x_j)| \neq 0, \tag{5.0.6}$$

then these equations have a unique solution which can be written down on applying Cramer's rule. It is clear that (5.0.6) will not be satisfied if any two of the x_i are the same, since then two rows of the determinant would be identical.

Equations (5.0.1) and (5.0.5) can be regarded as $n + 2$ homogeneous equations in the $n + 2$ quantities $-1, \gamma_0, \gamma_1, \ldots, \gamma_n$. Hence their determinant must vanish:

$$\begin{vmatrix} \phi_0(x_0) & \phi_1(x_0) & \cdots & \phi_n(x_0) & y_0 \\ \phi_0(x_1) & \phi_1(x_1) & \cdots & \phi_n(x_1) & y_1 \\ \cdots & \cdots & \cdots & \cdots & \cdots \\ \phi_0(x_n) & \phi_1(x_n) & \cdots & \phi_n(x_n) & y_n \\ \phi_0(x) & \phi_1(x) & \cdots & \phi_n(x) & \Phi(x) \end{vmatrix} = 0. \tag{5.0.7}$$

This can be regarded as an equation in $\Phi(x)$. If we expand this determinant by elements of the last row, the last term will be $\Delta\Phi(x)$, and every other term will be equal to some $\phi_j(x)$ multiplied by its cofactor, which is a constant. Hence when we solve for $\Phi(x)$, we shall have $\Phi(x)$ expressed as a linear combination of the functions $\phi_j(x)$ in just the form (5.0.1), as required. Also if we set $x = x_i$ in (5.0.7) and subtract row $i + 1$ from the last row, we get

$$\Delta(\Phi(x_i) - y_i) = 0,$$

which shows that $\Phi(x_i) = y_i$, and the values of Φ at the points x_i agree with those of $f(x)$. We can write the solution of (5.0.7) in the form

$$\Delta\Phi(x) \equiv - \begin{vmatrix} \phi_0(x_0) & \phi_1(x_0) & \cdots & \phi_n(x_0) & y_0 \\ \cdots & \cdots & \cdots & \cdots & \cdots \\ \phi_0(x_n) & \phi_1(x_n) & \cdots & \phi_n(x_n) & y_n \\ \phi_0(x) & \phi_1(x) & \cdots & \phi_n(x) & 0 \end{vmatrix}. \tag{5.0.8}$$

If we expand the determinant on the right of (5.0.8) by elements of the last row, and divide by Δ, we obtain the form (5.0.1). Also we note that, if we expand along the last column and divide by Δ, we obtain a form

$$\Phi(x) \equiv \Sigma y_i \Lambda_i(x), \tag{5.0.9}$$

where each $\Lambda_i(x)$ is itself a particular linear combination of the $\phi_j(x)$ with coefficients depending only upon the x_k. Hence for a particular set of fundamental points the Λ_i can be calculated once and used for any $f(x)$. This would be useful, for example, if interpolations are to be made for each of several different functions, all of which are tabulated for the same values x_k.

Equation (5.0.9) exhibits an important property of interpolating functions: their linearity. Thus, to make explicit the fact that Φ is the interpolating function for $f(x)$, let us designate it $\Phi(f; x)$ and write (5.0.9) in the form

$$\Phi(f; x) \equiv \Sigma f(x_i)\Lambda_i(x). \tag{5.0.10}$$

By the same rule, if g is any other function, its interpolating function is

$$\Phi(g; x) \equiv \Sigma g(x_i)\Lambda_i(x).$$

But then if λ and μ are any constants, and $h(x) = \lambda f(x) + \mu g(x)$, it follows that

$$\begin{aligned} \Phi(h; x) &\equiv \Sigma[\lambda f(x_i) + \mu g(x_i)]\Lambda_i(x) \\ &\equiv \lambda\Phi(f; x) + \mu\Phi(g; x). \end{aligned} \tag{5.0.11}$$

It is understood that the basic functions ϕ_i and the fundamental points x_i are fixed throughout.

We may note further that, since by (5.0.7) $\Phi(\phi_j; x) \equiv \phi_j$, therefore

$$\phi_j(x) \equiv \sum_{i=0}^{n} \phi_j(x_i)\Lambda_i(x) \qquad (j = 0, 1, \ldots, n). \tag{5.0.12}$$

Suppose we require now that at x_n the derivatives of Φ and f shall be equal:

$$f'(x_n) = y'_n = \Phi'(x_n).$$

Then x_n may or may not coincide with one of the other x_i. We modify (5.0.7) in the next to the last line and write

$$\text{(5.0.13)} \qquad \begin{vmatrix} \phi_0(x_0) & \phi_1(x_0) & \cdots & \phi_n(x_0) & y_0 \\ \cdots & \cdots & \cdots & \cdots & \cdots \\ \phi'_0(x_n) & \phi'_1(x_n) & \cdots & \phi'_n(x_n) & y'_n \\ \phi_0(x) & \phi_1(x) & \cdots & \phi_n(x) & \Phi(x) \end{vmatrix} \equiv 0.$$

Again the function $\Phi(x)$ which satisfies this equation is a linear combination of the functions $\phi_i(x)$ which takes on the prescribed values y_j for $j = 0, 1, \ldots, n - 1$. Since (5.0.13) is an identity, we can differentiate, and only the last line is affected, all other elements being constant. Hence we see that Φ' takes on the prescribed value at x_n. If we require the derivatives of Φ and f to be equal for any other x_i, we replace also that row of (5.0.13) by a row of derivatives, and again the equation defines the required function Φ. This procedure can be used for derivatives of any order where the solution (5.0.8) is modified only by replacing appropriate rows in the two determinants by rows of derivatives of the order required. The form (5.0.1) still holds, and the form (5.0.9) is modified only by the replacement of certain y_i by the value of the derivative. The linearity property is unaffected.

5.01. *Some Expressions for the Remainder.* The determinant

$$\text{(5.01.1)} \qquad W(x) = \begin{vmatrix} \phi_0(x) & \cdots & \phi_n(x) \\ \cdots & \cdots & \cdots \\ \phi_0^{(n)}(x) & \cdots & \phi_n^{(n)}(x) \end{vmatrix}$$

is known as the Wronskian. If this remains different from zero everywhere on the interval of interpolation (a, b), one can define the linear operator L_{n+1} by the relation

$$\text{(5.01.2)} \quad L_{n+1}[\phi(x)] = W^{-1}(x) \begin{vmatrix} \phi_0(x) & \cdots & \phi_n(x) & \phi(x) \\ \cdots & \cdots & \cdots & \cdots \\ \phi_0^{(n)}(x) & \cdots & \phi_n^{(n)}(x) & \phi^{(n)}(x) \\ \phi_0^{(n+1)}(x) & \cdots & \phi_n^{(n+1)}(x) & \phi^{(n+1)}(x) \end{vmatrix}$$
$$\equiv \phi^{(n+1)} + a_1\phi^{(n)} + \cdots + a_{n+1}\phi,$$

and the linear differential equation of order $n + 1$

$$\text{(5.01.3)} \qquad L_{n+1}[\phi] = 0$$

is satisfied by each $\phi_i(x)$. Moreover every solution of (5.01.3) is expressible as a linear combination of the ϕ_i.

In general, for $\nu \leq n$ one can similarly define the linear operator $L_{\nu+1}$, and the differential equation of order $\nu + 1$

$$L_{\nu+1}[\phi] = 0 \tag{5.01.4}$$

is satisfied by $\phi_0, \phi_1, \ldots, \phi_\nu$, and every solution of (5.01.4) is expressible as a linear combination with constant coefficients of these ϕ's.

An equivalent definition of the operators $L_{\nu+1}$ can be obtained as follows: Define the differential operator

$$D = d/dx, \tag{5.01.5}$$

and select b_0 so that

$$(D - b_0)\phi_0(x) \equiv 0, \qquad b_0(x) = \phi_0'(x)/\phi_0(x). \tag{5.01.6}$$

Then

$$L_1[\phi] \equiv (D - b_0)\phi, \tag{5.01.7}$$

since the two differential equations $L[\phi] = 0$ and $(D - b_0)\phi = 0$ are both satisfied by ϕ_0. Again let b_1 satisfy

$$(D - b_1)(D - b_0)\phi_1 = 0, \qquad b_1 = (\phi_1' - b_0\phi_1)'/(\phi_1' - b_0\phi_1). \tag{5.01.8}$$

Hence

$$L_2[\phi] \equiv (D - b_1)(D - b_0)\phi \equiv (D - b_1)L_1[\phi(x)]. \tag{5.01.9}$$

Proceeding sequentially, we define $b_2, b_3, \ldots, b_n$ so that

$$L_{n+1}[\phi] \equiv (D - b_n) \cdots (D - b_0)\phi \equiv (D - b_n)L_n[\phi]. \tag{5.01.10}$$

A generalization of Rolle's theorem can now be stated: If the functions $b_i(x)$ are all analytic on the interval (a, b), and if $\phi(x)$ is analytic, and vanishes $n + 2$ times, counting multiplicities, then $L_{n+1}[\phi]$ vanishes at least once on the interval.

First consider any two consecutive zeros of ϕ, and define

$$\psi(x) = \phi(x) \exp\left[-\int_a^x b_0(x)\,dx\right].$$

Then

$$\psi'(x) = L_1[\phi] \exp\left[-\int_a^x b_0(x)\,dx\right].$$

Then ψ and ϕ vanish together, as do ψ' and $L_1[\phi]$. By Rolle's theorem ψ' vanishes at least once between consecutive zeros of ϕ. By a simple extension of the argument, $L_2[\phi]$ vanishes at least once between consecutive zeros of $L_1[\phi]$. Eventually we conclude that $L_{n+1}[\phi]$ vanishes at least once between consecutive zeros of $L_n[\phi]$, and hence at least once on the interval.

Define the function

$$(5.01.11)\quad g(x,\,s) = W^{-1}(s)\begin{vmatrix} \phi_0(s) & \cdots & \phi_n(s) \\ \cdots & \cdots & \cdots \\ \phi_0^{(n-1)}(s) & \cdots & \phi_n^{(n-1)}(s) \\ \phi_0(x) & \cdots & \phi_n(x) \end{vmatrix} = \Sigma g_i(s)\phi_i(x).$$

Then as a function of x, $g(x, s)$ satisfies (5.01.3), and

$$(5.01.12)\quad \partial^i g(x,\,s)/\partial x^i \Big|^{x=s} = \begin{matrix} 0, & i = 0, 1, \ldots, n-1, \\ 1, & i = n. \end{matrix}$$

Hence one can verify directly that

$$(5.01.13)\quad y(x) = \sum \alpha_i\phi_i + \int_a^x g(x,\,s)\psi(s)ds$$

satisfies the nonhomogeneous equation

$$(5.01.14)\quad L[y] = \psi$$

for any constants α_i.

Any solutions $\psi_i(x)$ of (5.01.3) with the nonvanishing Wronskian could replace the ϕ_i in (5.01.11), and the same $g(x,s)$ would result. This can be verified directly by writing

$$\phi_i = \Sigma\alpha_{ij}\psi_j$$

and substituting into (5.01.11), in which case the determinant $|\alpha_{ij}|$ appears as a factor which cancels out. Otherwise one can observe that the initial conditions (5.01.12) define the solution $g(x,s)$ of (5.01.3) uniquely. In particular the $\Lambda_i(x)$ are linear combinations of the ϕ_j and hence satisfy (5.01.3), together with the conditions

$$(5.01.15)\quad \Lambda_i(x_j) = \delta_{ij}.$$

If the Λ_i replace the ϕ_j in (5.01.11), one can write

$$(5.01.16)\quad g(x,s) = \Sigma\Gamma_i(s)\Lambda_i(x).$$

Note that

$$(5.01.17)\quad g(x_j,s) = \Sigma\Gamma_i(s)\Lambda_i(x_j) = \Gamma_j(s),$$

whence one can write

$$(5.01.18)\quad g(x,s) = \Sigma\Lambda_i(x)g(x_i,s).$$

From (5.01.12) and (5.01.16) we verify that

$$(5.01.19)\quad y = \sum \beta_i\Lambda_i(x) + \sum \Lambda_i(x) \int_{x_i}^x \Gamma_i(s)\psi(s)ds$$

satisfies (5.01.14) with $y(x_i) = \beta_i$. In particular

$$h(x) = \sum \Lambda_i(x) \int_{x_i}^{x} \Gamma_i(s)ds \tag{5.01.20}$$

satisfies

$$L_{n+1}[h] = 1, \qquad h(x_i) = 0. \tag{5.01.21}$$

With $h(x)$ defined by (5.01.20) or (5.01.21), we can obtain Petersson's form for the remainder

$$R(x) = f(x) - \Phi(x) = f(x) - \Sigma y_i \Lambda_i(x), \tag{5.01.22}$$

i.e., the error made in representing $f(x)$ by $\Phi(x)$. The function $R(x)$ vanishes at the $n + 1$ points x_i. For any $x' \neq x_i$, we can choose C so that

$$f(x') - \Phi(x') - Ch(x') = 0.$$

It is clear that $h(x') \neq 0$, since otherwise $h(x)$, which vanishes at every x_i, would have at least $n + 2$ zeros, whence $L_{n+1}[h]$ would vanish at least once, contrary to (5.01.21). When C is chosen so that

$$f(x) - \Phi(x) - Ch(x)$$

vanishes at x', that function has $n + 2$ zeros, whence

$$L_{n+1}[f(x) - \Phi(x) - Ch(x)] = L_{n+1}[f(x)] - C$$

vanishes at least once. Hence for some ξ on (a,b), $C = L_{n+1}[f(\xi)]$. Hence

$$R(x') = L_{n+1}[f(\xi)]h(x').$$

Hence if we drop the prime, we have

$$f(x) = \Sigma y_i \Lambda_i(x) + L_{n+1}[f(\xi)]h(x), \tag{5.01.23}$$

where ξ is some point on the interval, and $h(x)$ satisfies (5.01.20) and (5.01.21).

Since certainly $f(x)$ satisfies

$$L_{n+1}[y] = L_{n+1}[f], \tag{5.01.24}$$

we can apply (5.01.19) and assert that

$$\begin{aligned} f(x) &= \sum \Lambda_i(x) \left\{ y_i + \int_{x_i}^{x} \Gamma_i(s) L_{n+1}[f(s)]ds \right\} \\ &= \Phi(x) + \sum \Lambda_i(x) \int_{x_i}^{x} \Gamma_i(s) L_{n+1}[f(s)]ds. \end{aligned} \tag{5.01.25}$$

This can also be written

$$f(x) = \Phi(x) + \sum \Lambda_i(x) \int_{x_i}^{x} g(x_i,s) L_{n+1}[f(s)]ds, \tag{5.01.26}$$

because of (5.01.17).

Since $\int_{x_i}^{x} = \int_a^x - \int_a^{x_i}$, the remainder R can be written

(5.01.27) $$R(x) = \int_a^x g(x,s)L_{n+1}[f(s)]ds - \sum \Lambda_i(x) \int_a^{x_i} \Gamma_i(s)L_{n+1}[f(s)]ds$$

after applying (5.01.16) to the first integral. The formula remains equally valid, however, when the end point b of the interval of interpolation replaces the end point a in the limits of integration. When this replacement is made and the two equivalent expressions for $R(x)$ are combined, one obtains the more symmetric form,

(5.01.28) $$\begin{aligned} R(x) &= \int_a^b K(x,s)L_{n+1}[f(s)]ds, \\ 2K(x,s) &= g(x,s) \operatorname{sgn}(x - s) - \Sigma\Lambda_i(x)\Gamma_i(s) \operatorname{sgn}(x_i - s), \end{aligned}$$

where sgn u is the signum function whose value is $+1$ when the argument is positive and -1 when the argument is negative. Although this function is discontinuous where the argument vanishes, nevertheless the kernel $K(x,s)$ remains continuous, since $g(x,s)$ vanishes at $s = x$, and $\Gamma_i(s) = g(x_i, s)$ vanishes at $s = x_i$.

It is possible to generalize this development to cases where certain conditions $y_i = \Phi(x_i)$ are replaced by conditions of the form $f^{(\nu)}(x_i) = \Phi^{(\nu)}(x_i)$, which require the equality of derivatives of f and the approximating function Φ, rather than equality of their functional values. As one special case, consider the requirements that at some point α

$$f^{(\nu)}(\alpha) = \Phi^{(\nu)}(\alpha), \qquad \alpha = 0, 1, \ldots, n.$$

This gives the Taylor expansion when the functions are polynomials. Let the functions $\psi_\nu(x)$ be chosen so that

$$L_{\nu+1}[\psi_\nu(x)] = 0, \quad \psi_\nu(\alpha) = \psi_\nu'(\alpha) = \cdots = \psi_\nu^{(\nu-1)}(\alpha) = 0, \quad \psi_\nu^{(\nu)}(\alpha) = 1.$$

In (5.01.13) replace a by α, the ϕ_i by the equivalent set ψ_i, and let

$$\psi(s) = L_{n+1}[f(s)].$$

Hence $f(x)$ can be represented in the form

$$f(x) = \beta_0\psi_0(x) + \beta_1\psi_1(x) + \cdots + \beta_n\psi_n(x) + \int_\alpha^x g(x,s)L_{n+1}[f(s)]ds,$$

provided the β's are properly selected. On setting $x = \alpha$, we find

$$\beta_0 = f(\alpha).$$

Next, apply the operator L_1 and again set $x = \alpha$ to obtain

$$\beta_1 = L_1[f(\alpha)].$$

Proceeding thus we finally arrive at Petersson's generalized Taylor's expansion

$$f(x) = f(\alpha)\psi_0(x) + L_1[f(\alpha)]\psi_1(x) + \cdots + L_n[f(\alpha)]\psi_n(x) + \int_\alpha^x g(x,s)L_{n+1}[f(s)]ds. \tag{5.01.29}$$

5.1. Polynomial Interpolation. Consider now the case of polynomial interpolation. Equation (5.0.1) becomes

$$P(x) \equiv c_0 + c_1x + c_2x^2 + \cdots + c_nx^n. \tag{5.1.1}$$

The determinant Δ is the Vandermonde determinant,

$$\Delta = \begin{vmatrix} 1 & x_0 & \ldots & x_0^n \\ \cdots & \cdots & \cdots & \cdots \\ 1 & x_n & \ldots & x_n^n \end{vmatrix} = \prod_{i>j} (x_i - x_j), \tag{5.1.2}$$

which vanishes if and only if any two of the x_i coincide. Equation (5.0.8) takes the form

$$P(x) \equiv -\Delta^{-1} \begin{vmatrix} 1 & x_0 & x_0^2 & \ldots & x_0^n & y_0 \\ \cdots & \cdots & \cdots & \cdots & \cdots & \cdots \\ 1 & x_n & x_n^2 & \ldots & x_n^n & y_n \\ 1 & x & x^2 & \ldots & x^n & 0 \end{vmatrix}, \tag{5.1.3}$$

which coincides with (5.1.1) if we expand along the last row, but has the form

$$P(x) \equiv \Sigma y_i L_i(x) \tag{5.1.4}$$

when we expand along the last column. The L_i are themselves polynomials with coefficients which depend only upon the x_j. These polynomials are

$$L_i(x) = \prod_{j \neq i} [(x - x_j)/(x_i - x_j)]. \tag{5.1.5}$$

They can be obtained by direct expansion of the determinant, or we can verify that they satisfy the necessary conditions if we note that

$$L_i(x_j) = \delta_{ij} \tag{5.1.6}$$

with δ_{ij} the Kronecker δ. From this it follows that with $L_i(x)$ defined by (5.1.5) and $P(x)$ by (5.1.4) we have $P(x_i) = y_i$.

We can write $L_i(x)$ in another form if we define

$$\omega(x) \equiv \Pi(x - x_i), \tag{5.1.7}$$

for then

$$\omega'(x) = \sum_j \prod_{i \neq j} (x - x_i), \tag{5.1.8}$$

$$\omega'(x_j) = \prod_{i \neq j} (x_j - x_i). \tag{5.1.9}$$

Hence

$$L_i(x) \equiv \omega(x)/[(x - x_i)\omega'(x_i)], \tag{5.1.10}$$

and therefore

$$P(x) \equiv \omega(x) \sum_i y_i/[(x - x_i)\omega'(x_i)], \tag{5.1.11}$$

or

$$P(x) \equiv \omega(x) \sum_i f(x_i)/[(x - x_i)\omega'(x_i)]. \tag{5.1.12}$$

The form (5.1.4) with any of the equivalent representations of the $L_i(x)$ is the Lagrange interpolation formula.

If $f(x)$ is itself a polynomial of degree not greater than n, then $f(x)$ and $P(x)$ are identical. Hence the $L_i(x)$ satisfy the $n + 1$ identities

$$x^j \equiv \Sigma x_i^j L_i(x) \qquad (j = 0, \ldots, n). \tag{5.1.13}$$

This is the identity (5.0.12) for the case $\phi_j = x^j$.

The explicit forms for the $L_i(x)$ become rather complicated when derivatives of P and f are to be equated at some of the points x_j, but in any particular case they can be formed from the determinantal expression. However, an important special case arises when both conditions $P = f$ and $P' = f'$ are to be imposed at every x_i. For this case we have Hermite's interpolation formula for the polynomial $H(x)$ of degree $2n + 1$ satisfying

$$H(x_i) = f(x_i), \qquad H'(x_i) = f'(x_i) \qquad (i = 0, 1, \ldots, n). \tag{5.1.14}$$

We can surely express this in the form

$$H(x) \equiv \Sigma y_i h_i(x) + \Sigma y_i' H_i(x) \tag{5.1.15}$$

with suitable polynomials $h_i(x)$ and $H_i(x)$, each of degree $2n + 1$ or less. Instead of writing down the appropriate determinant (5.0.13) and expanding, it is easier to proceed indirectly. If in (5.1.15) we set $x = x_j$ and apply (5.1.14), it follows that

$$f(x_j) = \Sigma f(x_i)h_i(x_j) + \Sigma f'(x_i)H_i(x_j).$$

This relation must hold whatever may be the values of $f(x_i)$ and of $f'(x_i)$. Thus, we may have $f(x_j) = 1$ while $f(x_i) = 0$ for every $i \neq j$, and while $f'(x_i) = 0$ for all i. This implies that $h_j(x_j) = 1$. On the other hand, if, for some particular $k \neq j$, $f(x_k) = 1$ while $f(x_i) = 0$ for $i \neq k$, and $f'(x_i) = 0$ for all i, then we find that $h_k(x_j) = 0$ for $k \neq j$. Setting some $f'(x_k) = 1$ while all other $f'(x_i) = 0$ and all $f(x_i) = 0$ shows that every $H_i(x_k) = 0$.

This argument, with an analogous one applied to

$$H'(x) \equiv \Sigma y_i h_i'(x) + \Sigma y_i' H_i'(x), \tag{5.1.16}$$

shows that the polynomials $h_i(x)$ and $H_i(x)$ must satisfy

$$\begin{aligned} h_i(x_j) &= \delta_{ij}, \qquad & H_i(x_j) &= 0, \\ h_i'(x_j) &= 0, \qquad & H_i'(x_j) &= \delta_{ij}. \end{aligned} \tag{5.1.17}$$

We ask now whether with appropriately chosen linear polynomials $v_i(x)$ and $w_i(x)$ the polynomials h_i and H_i may have the form

$$h_i(x) = v_i(x)L_i^2(x), \qquad H_i(x) = w_i(x)L_i^2(x). \tag{5.1.18}$$

These are, indeed, of the necessary degree $2n + 1$. From (5.1.6) all conditions are satisfied for $h_i(x_j)$ and $H_i(x_j)$ with $j \neq i$. Before examining the case for $j = i$, note that, since by (5.1.10)

$$\omega(x) \equiv \omega'(x_i)(x - x_i)L_i(x),$$

therefore

$$\begin{aligned} \omega'(x) &\equiv \omega'(x_i)[L_i(x) + (x - x_i)L_i'(x)], \\ \omega''(x) &\equiv \omega'(x_i)[2L_i'(x) + (x - x_i)L_i''(x)]. \end{aligned}$$

Hence $\omega''(x_i) = 2\omega'(x_i)L_i'(x_i)$ or $2L_i'(x_i) = \omega''(x_i)/\omega'(x_i)$. Now

$$h_i(x_i) = v_i(x_i)L_i^2(x_i) = v_i(x_i)$$

by (5.1.6). Hence, by (5.1.17) $v_i(x_i) = 1$. Also, by differentiating (5.1.18)

$$h_i'(x) = v_i'(x)L_i^2(x) + 2v_i(x)L_i(x)L_i'(x),$$

so that, at x_i, $0 = v_i'(x_i) + \omega''(x_i)/\omega'(x_i)$, and since v_i is linear,

$$v_i(x) = 1 - (x - x_i)\omega''(x_i)/\omega'(x_i). \tag{5.1.19}$$

Next $0 = H_i(x_i) = w_i(x_i)L_i^2(x_i)$ so that $w_i(x_i) = 0$. Also

$$H_i'(x) = w_i'(x)L_i^2(x) + 2w_i(x)L_i(x)L_i'(x)$$

so that $1 = w_i'(x_i)$. Hence

$$w_i(x) \equiv x - x_i. \tag{5.1.20}$$

Hermite's formula is therefore

$$H(x) \equiv \Sigma[y_i v_i(x) + y_i' w_i(x)]L_i^2(x), \tag{5.1.21}$$

with v_i and w_i defined by (5.1.19) and (5.1.20).

Identities analogous to (5.1.13) hold for the h_i and H_i. In particular for $H(x) \equiv 1$ we have

$$1 \equiv \Sigma v_i(x)L_i^2(x). \tag{5.1.22}$$

5.11. *The Remainder Term.* While the polynomial $P(x)$ is determined to be equal to $f(x)$ at each of $n + 1$ distinct values of x, in general $P \neq f$ at all other values. Some expressions for the remainder

$$R(x) = f(x) - \Phi(x)$$

were derived in §5.01. A simpler derivation of one of these for the case of polynomial interpolation will be given now. Practically this provides only an upper bound for the error in terms of an upper bound for the derivative of order $n + 1$ of $f(x)$ on an interval which contains all the x_i. Even though we cannot, or do not wish to, evaluate f or any of its derivatives exactly, we may be able to set limits to the possible values of any of these derivatives, and in this case the error estimates will be helpful.

Let x_{n+1} be the point at which the error is to be evaluated. Define the function $g(x)$ by

$$\begin{vmatrix} 1 & x_0 & x_0^2 & \cdots & x_0^{n+1} & f(x_0) \\ \cdot & \cdot & \cdot & \cdots & \cdot & \cdot \\ 1 & x_{n+1} & x_{n+1}^2 & \cdots & x_{n+1}^{n+1} & f(x_{n+1}) \\ 1 & x & x^2 & \cdots & x^{n+1} & f(x) \end{vmatrix} \equiv g(x).$$

Then

$$g(x_0) = g(x_1) = \cdots = g(x_{n+1}) = 0.$$

By Rolle's theorem, therefore, $g'(x)$ must vanish at least once in each of the $n + 1$ intervals between consecutive values of the x_i. If we let x_i' designate the points at which g' vanishes, then again by Rolle's theorem $g''(x)$ must vanish at least once in each interval between consecutive values x_i'. Continuing thus, we conclude finally that $g^{(n+1)}(x)$ vanishes at least once at a point ξ which lies somewhere on the interval between the greatest and the least of the x_i. Hence for this ξ we have

$$\begin{vmatrix} 1 & x_0 & \cdots & x_0^{n+1} & f(x_0) \\ \cdot & \cdot & \cdots & \cdot & \cdot \\ 1 & x_{n+1} & \cdots & x_{n+1}^{n+1} & f(x_{n+1}) \\ 0 & 0 & \cdots & (n+1)! & f^{(n+1)}(\xi) \end{vmatrix} = 0.$$

This is exact, though we know nothing about ξ except the fact that it lies somewhere on the interval named. If we expand along the last row, we get

$$\begin{vmatrix} 1 & x_0 & \cdots & x_0^{n+1} \\ \cdot & \cdot & \cdots & \cdot \\ 1 & x_{n+1} & \cdots & x_{n+1}^{n+1} \end{vmatrix} f^{(n+1)}(\xi) = (n+1)! \begin{vmatrix} 1 & x_0 & \cdots & f(x_0) \\ \cdot & \cdot & \cdots & \cdot \\ 1 & x_{n+1} & \cdots & f(x_{n+1}) \end{vmatrix},$$

and if we solve this equation for $f(x_{n+1})$ and drop the subscript, we get

$$f(x) = P(x) + f^{(n+1)}(\xi)\omega(x)/(n+1)!, \tag{5.11.1}$$

where ω is the polynomial defined in (5.1.7). Hence the second term on the right represents the amount by which $P(x)$ deviates from $f(x)$. This corresponds to (5.01.23).

Now although we would not know the value of ξ, which is in any case a function of x, nevertheless we may know an upper bound for $f^{(n+1)}$ on the entire interval. Let us call this M_{n+1}. Then for any x on the interval containing all the x_i we have

$$|R(x)| \leq M_{n+1}|\omega(x)|/(n+1)!. \tag{5.11.2}$$

The right member of this inequality vanishes for $x = x_i$, as it should. Between successive x_i, $|\omega(x)|$ rises to a relative maximum. With uniform spacing of the x_i the maxima are highest near the ends of the interval, for if $x_0 < x_1 < \cdots < x_n$, then at one end the factor $|x - x_n|$ is large, and at the other $|x - x_0|$ is large. Hence in this case the approximation is best for values of x near the middle of the range. For values of x outside the range, the inequality (5.11.2) is still valid, provided we understand M_{n+1} to represent a bound for $f^{(n+1)}$ in the entire interval including also x. But outside the range, $\omega(x)$ itself becomes increasingly large, and this accounts for the high uncertainty of extrapolation.

5.12. *Chebyshev Polynomials and Optimum-interval Interpolation.* In the inequality (5.11.2) the factor M_{n+1} depends upon the particular function $f(x)$ but not upon the distribution within the interval of interpolation of the values x_i which determine the $P(x)$. On the other hand, $\omega(x)$ depends only upon the distribution of the x_i and not at all upon the function. Ordinarily, in short calculations one has available a set of tabulated values of $f(x)$ and must accept them as they are given. But when a table is being prepared, the location of the x_i can be chosen at will, and some choices might be better than others.

All that we know about the variation with x of the error in the interpolation is contained in the polynomial $\omega(x)$. The bounds of the error are least exact at those points x where $|\omega(x)|$ is greatest. Hence it is natural to prefer a selection of points x_i on the interval which reduces as far as possible the greatest maximum of $|\omega(x)|$. It is plausible to suppose that, since the relative maxima ordinarily vary in height from one to the next, a choice of the x_i that reduces the highest maximum will probably raise some of the others. Hence we might anticipate that the minimal maximum will be had, if at all, only in a case where all maxima are equal. And a succession of equal maxima suggests a trigonometric sine or cosine.

Introduce a change of scale and origin so that the interval over which the interpolations are to be made is the interval from -1 to $+1$. By a well-known trigonometric identity, $\cos n\theta$ is expressible as a polynomial in $\cos\theta$ of degree n. This is trivial for $n = 0$ and $n = 1$, while for $n = 2$,

$\cos 2\theta = 2\cos^2\theta - 1$. Suppose it verified that

$$\cos r\theta = P_r(\cos\theta) \tag{5.12.1}$$

is a polynomial of degree r for $r = 0, 1, \ldots, n$. Then

$$\cos(n+1)\theta + \cos(n-1)\theta = 2\cos n\theta \cos\theta$$

by a formula from elementary trigonometry, whence

$$P_{n+1} = 2P_nP_1 - P_{n-1} \tag{5.12.2}$$

is a polynomial of degree $n + 1$ in $\cos\theta$. Hence if

$$x = \cos\theta, \qquad \theta = \cos^{-1} x, \qquad 0 \le \theta \le \pi,$$

then $P_n(x)$ is a polynomial in x of degree n, and since $P_0 = 1$, $P_1 = x$, it follows from (5.12.2) that the coefficient of x^n in P_n is 2^{n-1}. Hence each polynomial

$$\begin{aligned} T_0 &= 1, \\ T_n &= 2^{1-n}\cos(n\cos^{-1}x) \qquad (n \ge 1), \end{aligned} \tag{5.12.3}$$

has leading coefficient 1, and since all its zeros lie on the interval, it is a possible $\omega(x)$. We can prove that no polynomial $R_n(x)$ exists with degree n and leading coefficient 1 whose maximum absolute values are all numerically less than those of $T_n(x)$.

Since $T_n(x) = 2^{1-n}\cos n\theta$, therefore $T_n(x)$ has the relative maxima and minima of $\pm 2^{1-n}$ for

$$\theta_j = j\pi/n \qquad (j = 0, 1, \ldots, n),$$

and hence for

$$x'_j = \cos(j\pi/n).$$

Now if R_n is of degree n with leading coefficient 1, and has no maximum or minimum numerically greater than those of T_n, then

$$T_n(x'_0) - R_n(x'_0) \ge 0, \qquad T_n(x'_1) - R_n(x'_1) \le 0, \ldots,$$

since the maximum of T_n at x'_0 cannot be less than the value of R_n at x'_0, the minimum of T_n at x'_1 cannot exceed the value of R_n at $x'_1, \ldots$. Hence the polynomial $T_n - R_n$ must vanish at least n times on the interval. But $T_n - R_n$ is of degree only $n - 1$, and therefore

$$T_n - R_n \equiv 0.$$

The application of this theorem is that, if one chooses the $n + 1$ values x_i to be the zeros of the polynomial T_{n+1}, making this the ω of (5.11.2), then the greatest possible error of interpolation anywhere on the interval is $2^{-n}M_{n+1}/(n + 1)!$ for any function whatever whose derivative of order $n + 1$ does not exceed M_{n+1} on the interval. Any other choice of fundamental points would replace the factor 2^{-n} by a larger one.

The polynomials defined by (5.12.3) are known as the Chebyshev polynomials. Their zeros are easily found, for they vanish when

$$n\theta = (2i + 1)\pi/2 \qquad (i = 0, 1, \ldots, n - 1),$$

and hence when

$$x_i = \cos\,[(2i + 1)\pi/(2n)]. \tag{5.12.4}$$

In these formulas it is understood that the range of interpolation has been transformed to the interval from -1 to $+1$.

Function tables may contain hundreds or thousands of entries, and for any particular interpolation one would expect to use only a few consecutive ones. When the table is to be printed in a book, ordinarily the abscissas x_i are uniformly spaced. When tabular entries are required for automatic computation, it is important to reduce to a minimum the number of entries to be recorded. The use of the Chebyshev points x_i may then be appropriate.

Suppose that we are willing to use an interpolation polynomial of degree n at most and that an error of magnitude ϵ can be tolerated. The entire range of the variable is to be broken up into subintervals within each of which at most $n + 1$ Chebyshev points x_i are to be selected at which to evaluate the entries $f(x_i)$ for the tabulation. The entries $f(x_i)$ on one of these subintervals will be used to determine the interpolation polynomial for that interval. We would like to use as few of these subintervals as possible, and hence we would like to make each subinterval as long as it can be made without allowing the interpolation error to exceed ϵ, or the degree of the polynomial to exceed n. In some circumstances an optimal solution is possible.

We consider here the problem of making a particular interval as long as possible. When the end points a and b are known, then the transformation

$$\begin{aligned} x &= [(b - a)u + (b + a)]/2, \\ u &= (2x - b - a)/(b - a), \end{aligned} \tag{5.12.5}$$

transforms the interval (a, b) in the variable x to the interval $(-1, 1)$ in the variable u. If

$$\phi(u) = \prod_0^n (u - u_i),$$

where u_i is given by substituting x_i in (5.12.5), then

$$\omega(x) = 2^{-n-1}(b - a)^{n+1}\phi(u),$$

as is verified directly.

In view of (5.11.1), the condition that the error shall nowhere exceed ϵ is that

$$|f^{(n+1)}(\xi)\omega(x)| \leq (n+1)!\epsilon.$$

Suppose $|f^{(n+1)}(x)|$ is monotonically decreasing. Then this inequality is surely satisfied if

$$|f^{(n+1)}(a)|M \leq (n+1)!\epsilon,$$

where M represents the maximum of $|\omega(x)|$ on the interval (a, b). If N is the maximum of $|\phi(u)|$ on the interval $(-1, 1)$, this is equivalent to

$$(b-a)^{n+1}N|f^{(n+1)}(a)| \leq (n+1)!2^{n+1}\epsilon.$$

Hence for a fixed a the longest admissible interval $b - a$ would be that for which the equality holds:

$$(b-a)^{n+1} = (n+1)!2^{n+1}\epsilon N^{-1}|f^{(n+1)}(a)|^{-1}.$$

If the Chebyshev points are used, $\phi(u) = T_{n+1}(u)$, then $N = 2^{-n}$, and therefore

$$b = a + 4[(n+1)!2^{-1}\epsilon|f^{(n+1)}(a)|^{-1}]^{1/(n+1)}. \tag{5.12.6}$$

When $|f^{(n+1)}(x)|$ is monotonically increasing, a and b can be interchanged.

If $|f^{(n+1)}(x)|$ remains monotonic over the entire range of the tabulation, the range can be divided into optimal intervals by starting at one end and working toward the other; if it has a single maximum, one can start with this and work toward the two ends. Other cases will require special treatment.

5.13. *Aitken's Method of Interpolation.* We turn now to computational procedures. Calculation of the $L_i(x)$ directly from (5.1.5) or (5.1.10) involves considerable labor if the degree is higher than one or two. Tables of the L_i are available for equally spaced x_i. If these are not at hand, or if the x_i are not equally spaced, then Aitken's method of computation is almost ideally simple.

We first obtain a generalization of the formula (5.1.3). Let P_i stand for $P(x_i)$; let P_{ij} stand for the linear interpolation polynomial determined by (x_i, y_i) and (x_j, y_j); let P_{ijk} stand for the quadratic interpolation polynomial determined by (x_i, y_i), (x_j, y_j), and (x_k, y_k); . . . ; and let $P_{01\ldots n}$ stand for P itself. As for P_i, we can regard it as the interpolation polynomial of zero degree determined by (x_i, y_i).

Note that

$$P_{ij} = P_{ji}$$

and that in general for any $P_{ij\ldots}$ permuting the subscripts leaves the polynomial unchanged. Note also that

$$P_{ij\ldots}(x_i) = y_i. \tag{5.13.1}$$

The generalization of (5.1.3) is that for any m, if M stands for the set of subscripts $m + 1, m + 2, \ldots, n$, then

$$\begin{vmatrix} 1 & x_0 & \ldots & x_0^m & P_{0M} \\ \cdot & \cdot & \cdot & \cdot & \cdot \\ 1 & x_m & \ldots & x_m^m & P_{mM} \\ 1 & x & \ldots & x^m & P \end{vmatrix} = 0. \tag{5.13.2}$$

If in place of the P_{iM} we were to write P_i, this equation would define the interpolation polynomial of degree m determined by $(x_0, y_0), \ldots, (x_m, y_m)$.

In proof we observe first that the polynomial P defined by (5.13.2) is of degree n at most, since in the expansion of the determinant x^m will multiply each of the interpolation polynomials P_{iM}, which are of degree $n - m$, and there are no terms of higher degree. Next we observe that, if in the determinant we set $x = x_i$ for $i \leq m$, then P must take the value assumed by P_{iM}, and this by (5.13.1) is y_i. This is true because all other elements of the last row are then identical with corresponding elements of the row $i + 1$. Finally, if in the determinant we set $x = x_j$ for $j > m$, then every P_{iM} becomes equal to y_j, making all elements but the last in the last column equal to y_j times the corresponding elements in the first column. Hence the determinant can vanish only if also P has the value y_j. Hence the P defined by (5.13.2) is in fact the polynomial P of (5.1.3), and the theorem is proved.

Aitken applies this principle in the following way: In application of (5.1.3) with $n = 1$, we have

$$\begin{vmatrix} 1 & x_0 \\ 1 & x_1 \end{vmatrix} P_{01} = - \begin{vmatrix} 1 & x_0 & y_0 \\ 1 & x_1 & y_1 \\ 1 & x & 0 \end{vmatrix},$$

and hence

$$P_{01} = \begin{vmatrix} x - x_0 & P_0 \\ x - x_1 & P_1 \end{vmatrix} \bigg/ (x_1 - x_0). \tag{5.13.3}$$

Likewise

$$P_{12} = \begin{vmatrix} x - x_1 & P_1 \\ x - x_2 & P_2 \end{vmatrix} \bigg/ (x_2 - x_1). \tag{5.13.4}$$

From the theorem then we can say that

$$P_{012} = \begin{vmatrix} x - x_0 & P_{01} \\ x - x_2 & P_{12} \end{vmatrix} \bigg/ (x_2 - x_0). \tag{5.13.5}$$

In like manner we can form P_{123}, and then

$$P_{0123} = \begin{vmatrix} x - x_0 & P_{012} \\ x - x_3 & P_{123} \end{vmatrix} \bigg/ (x_3 - x_0). \tag{5.13.6}$$

In a specific calculation x is a specific number, and the sequential computations yield the numerical values assumed by the various polynomials at that particular x. When different polynomials agree to a sufficient number of significant figures, one terminates the process.

In applying these formulas, one can choose at will the particular polynomials $P_{ij\ldots}$ to be evaluated, bearing in mind only that, when a pair $P_{ij\ldots}$ and $P_{kj\ldots}$ are used to evaluate $P_{ikj\ldots}$, they must agree in all subscripts but one, while the unlike subscripts i and k determine the x_i and x_k which are to appear explicitly. Aitken proposes a sequence and tabulation as follows:

$$\begin{array}{llllll} P_0 & & & & & x_0 - x \\ P_1 & P_{01} & & & & x_1 - x \\ P_2 & P_{02} & P_{012} & & & x_2 - x \\ P_3 & P_{03} & P_{013} & P_{0123} & & x_3 - x \\ P_4 & P_{04} & P_{014} & P_{0124} & P_{01234} & x_4 - x \end{array}$$

However, the best approximation can be expected at any stage when the abscissa x lies roughly in the middle of the interval containing the particular fundamental abscissas being utilized. Consequently it is advantageous in using this scheme to order the abscissas so that either $\ldots < x_4 < x_2 < x_0 < x < x_1 < x_3 < \cdots$, or else the reverse order holds.

5.14. *Divided Differences.* Aitken's method is disadvantageous when a number of interpolations must be carried out over the same range. An alternative to the computation of the Lagrange polynomials $L_i(x)$ is the use of divided differences in the construction of Newton's interpolation formula.

The polynomial

$$(5.14.1)\quad P(x) = a_0 + (x - x_0)a_1 + (x - x_0)(x - x_1)a_2 + \cdots + (x - x_0)(x - x_1) \cdots (x - x_{n-1})a_n$$

is of degree n and assumes the values $f(x_i)$ at x_i, provided

$$(5.14.2)\quad \begin{aligned} f(x_0) &= a_0, \\ f(x_1) &= a_0 + (x_1 - x_0)a_1, \\ f(x_2) &= a_0 + (x_2 - x_0)a_1 + (x_2 - x_0)(x_2 - x_1)a_2, \\ &\cdots\cdots\cdots\cdots\cdots\cdots \end{aligned}$$

and the coefficients a_i can be determined recursively from these relations. From these relations it is apparent that for any function $f(x)$ the polynomial $P_{01\ldots m}(x)$ as defined in the last section is

$$P_{012\ldots m}(x) = a_0 + (x - x_0)a_1 + \cdots + (x - x_0) \cdots (x - x_{m-1})a_m,$$

where the coefficients are the same as the first $m + 1$ coefficients in $P(x)$. Finally, since a_n is the coefficient of x^n in (5.14.1), this must be equal to the coefficient c_n of x^n in (5.1.1). Hence $a_n = c_n$ is expressible as the quotient of two determinants, where the denominator is the Vandermonde of order $n + 1$, and the numerator is the same except that the elements x_i^n are replaced by $f(x_i)$. Hence a_m is expressible as the quotient of two similar determinants of order $m + 1$. Thus, for the given function f each coefficient a_m in (5.14.1) is a function of the $m + 1$ variables x_0, $x_1, \ldots, x_m$. This function is called a divided difference of order m and is written

(5.14.3)

$$f(x_0, x_1, \ldots, x_m) = \begin{vmatrix} 1 & x_0 & \cdots & x_0^{m-1} & f(x_0) \\ \cdots & \cdots & \cdots & \cdots & \cdots \\ 1 & x_m & \cdots & x_m^{m-1} & f(x_m) \end{vmatrix} : \begin{vmatrix} 1 & x_0 & \cdots & x_0^m \\ \cdots & \cdots & \cdots & \cdots \\ 1 & x_m & \cdots & x_m^m \end{vmatrix}.$$

For the particular case when $f = x^r$ for $r < m$ this vanishes for any set of fundamental points, while for $r = m$

$$f(x_0, x_1, \ldots, x_m) = 1.$$

Hence the divided difference of order m for any polynomial of degree m is a constant, and for any polynomial of degree less than m it vanishes.

The notation $[x_0, x_1, \ldots, x_m]$ is often found in the literature in place of $f(x_0, x_1, \ldots, x_m)$, but this fails to place in evidence the function whose divided difference is being written.

Now consider the expansions

$$\begin{aligned} P_{01} &= f(x_0) + (x - x_0)f(x_0, x_1), \\ P_{02} &= f(x_0) + (x - x_0)f(x_0, x_2), \\ P_{012} &= f(x_0) + (x - x_0)f(x_0, x_1) + (x - x_0)(x - x_1)f(x_0, x_1, x_2). \end{aligned}$$

By a formula analogous to (5.13.5), however, it is also true that

$$P_{012} = \begin{vmatrix} x - x_1 & P_{01}(x) \\ x - x_2 & P_{02}(x) \end{vmatrix} \Big/ (x_2 - x_1).$$

By equating the coefficients of x^2 in the two expressions for $P_{012}(x)$, one obtains the identity

$$f(x_0, x_1, x_2) = [f(x_0, x_2) - f(x_0, x_1)]/(x_2 - x_1).$$

Since the divided difference is symmetric in all variables, it follows also that

$$\begin{aligned} f(x_0, x_1, x_2) &= [f(x_1, x_2) - f(x_0, x_1)]/(x_2 - x_0) \\ &= [f(x_0, x_2) - f(x_1, x_2)]/(x_0 - x_1). \end{aligned}$$

By (5.13.6) one finds likewise that

$$f(x_0, x_1, x_2, x_3) = [f(x_0, x_1, x_2) - f(x_1, x_2, x_3)]/(x_0 - x_3),$$

and in general

(5.14.4) $f(x_0, x_1, x_2, \ldots) = [f(x_0, x_2, \ldots) - f(x_1, x_2, \ldots)]/(x_0 - x_1),$

where the omitted variables are the same in all three places. This being the case, one can form divided differences of progressively higher order according to the scheme:

$$
(5.14.5) \qquad
\begin{array}{llllll}
x_0 & f(x_0) & & & & \\
 & & f(x_0, x_1) & & & \\
x_1 & f(x_1) & & f(x_0, x_1, x_2) & & \\
 & & f(x_1, x_2) & & f(x_0, x_1, x_2, x_3) & \\
x_2 & f(x_2) & & f(x_1, x_2, x_3) & \cdot & \\
 & & f(x_2, x_3) & \cdot & \cdot & \\
x_3 & f(x_3) & \cdot & \cdot & \cdot & \\
\cdot & \cdot & \cdot & \cdot & \cdot & \\
\cdot & \cdot & \cdot & \cdot & \cdot & \\
\cdot & \cdot & \cdot & \cdot & \cdot &
\end{array}
$$

where each f is equal to the difference of the two on its left, divided by the difference of the x's on the diagonals with it.

Another expression for the divided difference which can be obtained directly from (5.14.3) is of some theoretical interest. The coefficient of $f(x_i)$ on the right of this identity is the quotient of two Vandermonde determinants. When the common factors are canceled out, we are left with

$$
(5.14.6) \qquad f(x_0, x_1, \ldots, x_m) = \sum_{i=0}^{m} \left[f(x_i) \Big/ \prod_{j \neq i} (x_i - x_j) \right].
$$

This brings out again the fact that the divided difference of any order is symmetric in all the arguments x_i which appear in it.

5.141. *Integral form of the remainder.* Consider the function

$$\phi(t) = f[(1 - t)x_0 + tx_1]$$

for fixed x_0 and x_1. This satisfies

$$
\begin{aligned}
\phi'(t) &= (x_1 - x_0)f'[(1 - t)x_0 + tx_1], \\
\phi(0) &= f(x_0), \qquad \phi(1) = f(x_1).
\end{aligned}
$$

Hence integration of $\phi'(t)$ and division by $(x_1 - x_0)$ gives

$$
(5.141.1) \qquad f(x_0, x_1) = \int_0^1 f'[(1 - t)x_0 + tx_1]dt.
$$

Now consider

$$
\begin{aligned}
\phi(t_1, t_2) &= f[(1 - t_1)x_0 + (t_1 - t_2)x_1 + t_2x_2], \\
\phi_{t_1}(t_1, t_2) &= (x_1 - x_0)f'[(1 - t_1)x_0 + (t_1 - t_2)x_1 + t_2x_2], \\
\phi_{t_1,t_2}(t_1, t_2) &= (x_1 - x_0)(x_2 - x_1)f''[(1 - t_1)x_0 + (t_1 - t_2)x_1 + t_2x_2].
\end{aligned}
$$

Then

$$\int_0^{t_1} f''[(1 - t_1)x_0 + (t_1 - t_2)x_1 + t_2x_2]dt_2 = (x_1 - x_0)^{-1}$$
$$(x_2 - x_1)^{-1} \int_0^{t_1} \phi_{t_1t_2}(t_1, t_2)dt_2 = (x_1 - x_0)^{-1}(x_2 - x_1)^{-1}[\phi_{t_1}(t_1, t_1)$$
$$- \phi_{t_1}(t_1, 0)] = (x_2 - x_1)^{-1}\{f'[(1 - t_1)x_0 + t_1x_2] - f'[(1 - t_1)x_0 + t_1x_1]\}.$$

But if we now apply the original result (5.141.1), we find

$$\int_0^1 \int_0^{t_1} f''[(1 - t_1)x_0 + (t_1 - t_2)x_1 + t_2x_2]dt_2\, dt_1 = (x_2 - x_1)^{-1}[f(x_0, x_2) - f(x_0, x_1)],$$

whereas the right member of this is by (5.14.4) equal to $f(x_0, x_1, x_2)$. Hence

$$(5.141.2) \quad f(x_0, x_1, x_2) = \int_0^1 \int_0^{t_1} f''[(1 - t_1)x_0 + (t_1 - t_2)x_1 + t_2x_2]dt_2\, dt_1,$$

and by a simple induction we have in general

$$(5.141.3) \quad f(x_0, x_1, \ldots, x_n) = \int_0^1 \int_0^{t_1} \cdots \int_0^{t_{n-1}} f^{(n)}[(1 - t_1)x_0 + (t_1 - t_2)x_1 + \cdots + t_nx_n]dt_n \cdots dt_1.$$

In the special case when $x_0 = x_1$, the relation (5.141.1) gives

$$f(x_0, x_0) = f'(x_0).$$

Again when $x_0 = x_1 = x_2$, it follows from (5.141.2) that

$$f(x_0, x_0, x_0) = f''(x_0)/2,$$

and generally for $m + 1$ equal arguments

$$(5.141.4) \quad f(x_0, x_0, \ldots, x_0) = f^{(m)}(x_0)/m!.$$

Hence if the function and its derivatives up to and including the mth are known at some point x_0, and the derivatives of the interpolation polynomial are required to equal those of the function, the table of divided differences is formed by writing $f(x_0)$ $m + 1$ times, $f'(x_0)$ m times, $f'(x_0)/2!$ $m - 1$ times, . . . , and constructing the rest of the table as before.

In general, whether or not there are repeated arguments, the identity

$$f(x_0, x) = [f(x) - f(x_0)]/(x - x_0)$$

is valid for any x, including in the limit $x = x_0$, and it can be written

$$(5.141.5) \quad f(x) = f(x_0) + (x - x_0)f(x_0, x).$$

Again

$$f(x_0, x_1, x) = [f(x_0, x) - f(x_0, x_1)]/(x - x_1),$$

and the identity can be written

$$f(x_0, x) = f(x_0, x_1) + (x - x_1)f(x_0, x_1, x).$$

When this is substituted into (5.141.5), the result is

$$(5.141.6)\quad f(x) = f(x_0) + (x - x_0)f(x_0, x_1) + (x - x_0)(x - x_1)f(x_0, x_1, x).$$

In general,

$$\begin{aligned}(5.141.7)\quad f(x) &= f(x_0) + (x - x_0)f(x_0, x_1) + \cdots \\ &+ (x - x_0)(x - x_1)\cdots(x - x_{n-1})f(x_0, x_1, \ldots, x_n) \\ &+ (x - x_0)(x - x_1)\cdots(x - x_n)f(x_0, x_1, \ldots, x_n, x).\end{aligned}$$

If f were a polynomial of degree n or less, the last divided difference would vanish identically. For an arbitrary f, the last term represents the error made in replacing f by its interpolation polynomial of degree n as given by the preceding terms. On introducing (5.141.3), we can write

$$f(x) = P(x) + R(x),$$

$$(5.141.8)\quad R(x) = \omega(x)\int_0^1\int_0^{t_n}\cdots\int_0^{t_1} f^{(n+1)}\Big[x + t_0(x_0 - x) + \sum_1^n t_i(x_i - x_{i-1})\Big]dt_0 \cdots dt_n,$$

where $P(x)$ is the interpolation polynomial of degree n, and $R(x)$ is the remainder. The expression previously obtained for the remainder $R(x)$ involved the indeterminate quantity known only to lie somewhere on the interval containing $x_0, x_1, \ldots, x_n$ and x. Note that in case all the fundamental points $x_0, \ldots, x_n$ coincide the expression (5.141.7) becomes a Taylor expansion, and (5.141.8) is a well-known formula for the remainder.

5.15. *Operational Derivation of Equal-interval Formulas.* Naturally the Lagrangean formulas for interpolation can be specialized to the case of equally spaced abscissas x_i. However, a great many special forms exist, and most of these are readily derived directly by the use of an operational scheme.

Suppose that the entire tabulation is made at points $\ldots x_{-2}, x_{-1}, x_0, x_1, x_2, \ldots$ and that

$$x_{i+1} = x_i + h$$

for every integer i. Then

$$x_i = x_0 + ih$$

for every i. The interval width h is assumed fixed throughout. On making the change of variable

$$(5.15.1)\qquad x = x_0 + uh, \qquad u = (x - x_0)/h,$$

the function $f(x)$ becomes a function of u:

$$f(x_0 + uh) = g(u), \tag{5.15.2}$$

for which the interval width is unity.

We now define a set of operators with respect to the interval h. For any function $f(x)$, the displacement operator E, the forward-difference operator Δ, the backward-difference operator ∇, and the central-difference operator δ are defined as follows:

$$\begin{aligned} Ef(x) &= f(x+h), \\ \Delta f(x) &= f(x+h) - f(x), \\ \nabla f(x) &= f(x) - f(x-h), \\ \delta f(x) &= f(x+h/2) - f(x-h/2). \end{aligned} \tag{5.15.3}$$

Since it is natural to require that

$$E^2 f(x) = E[Ef(x)] = Ef(x+h) = f(x+2h),$$

and in general

$$E^u f(x) = f(x+uh) \tag{5.15.4}$$

for any integer u, we may indeed go a step further and accept (5.15.4) for all real u, integral or not. With this understanding we can write the following formal relations between pairs of operators:

$$\begin{aligned} \Delta &= E - 1 = E\nabla. \\ \delta &= \Delta E^{-1/2} = \nabla E^{1/2} = E^{1/2} - E^{-1/2}. \end{aligned} \tag{5.15.5}$$

In principle any of the four operators can be expressed in terms of any other, but the expressions are not all simple. Thus Δ is the "negative" root of the quadratic

$$\Delta^2 - \delta^2\Delta - \delta^2 = 0,$$

and ∇ is the "positive" root of

$$\nabla^2 + \delta^2\nabla - \delta^2 = 0.$$

In addition to the operators, we define also the factorials

$$\begin{aligned} u^{(r)} &= u(u-1) \cdots (u-r+1) \\ &= (x-x_0)(x-x_1) \cdots (x-x_{r-1})/h^r \end{aligned} \tag{5.15.6}$$

and the generalized binomial coefficient

$$u_{(r)} = u^{(r)}/r!. \tag{5.15.7}$$

Since $\Delta u(x) = 1$, we find for these quantities

$$\Delta u^{(r)} = ru^{(r-1)}, \qquad \Delta u_{(r)} = u_{(r-1)}. \tag{5.15.8}$$

Now if u is a positive integer, since $E = 1 + \Delta$, it follows that

$$E^u = 1 + u_{(1)}\Delta + u_{(2)}\Delta^2 + u_{(3)}\Delta^3 + \cdots, \tag{5.15.9}$$

and the series terminates after $u + 1$ terms. If these equivalent operators are applied to $f_0 = f(x_0)$, we have

$$f(x) = f_0 + u_{(1)}\Delta f_0 + u_{(2)}\Delta^2 f_0 + u_{(3)}\Delta^3 f_0 + \cdots, \tag{5.15.10}$$

where x is given by (5.15.1). If u is not a positive integer, the series does not terminate. But the polynomial in x which results from replacing u by its expression in terms of x on the right of

$$P(x) = f_0 + u_{(1)}\Delta f_0 + \cdots + u_{(n)}\Delta^n f_0 \tag{5.15.11}$$

is a polynomial of degree n in x, and for $i = 0, 1, \ldots, n$ it is true that $P(x_i) = f(x_i)$. Hence $P(x)$ defined by (5.15.11) is the interpolation polynomial determined at the points $x_0, \ldots, x_n$. This is Newton's formula for forward interpolation. In terms of x it is

$$P(x) = f_0 + \frac{x - x_0}{h}\Delta f_0 + \cdots + \frac{(x - x_0)\cdots(x - x_{n-1})}{n!h^n}\Delta^n f_0. \tag{5.15.12}$$

Note that the fundamental points which determine this polynomial are the points whose abscissas are $x_0, x_1, x_2, \ldots, x_n$.

Newton's formula for backward interpolation is obtained by expanding E^u in powers of ∇:

$$E^u = (1 + \nabla)^{-u} = 1 + u_{(1)}\nabla + (u + 1)_{(2)}\nabla^2 + (u + 2)_{(3)}\nabla^3 + \cdots. \tag{5.15.13}$$

On dropping terms of degree higher than n, one has

$$\begin{aligned} P(x) &= f_0 + u_{(1)}\nabla f_0 + (u + 1)_{(2)}\nabla^2 f_0 + \cdots + (u + n - 1)_{(n)}\nabla^n f_0 \\ &= f_0 + \frac{x - x_0}{h}\nabla f_0 + \frac{(x - x_0)(x - x_{-1})}{2!h^2}\nabla f_0 + \cdots + \frac{(x - x_0)\cdots(x - x_{-n+1})}{n!h^n}\nabla^n f_0. \end{aligned} \tag{5.15.14}$$

For this the fundamental points are the points whose abscissas are x_{-n}, $x_{-n+1}, \ldots, x_{-1}, x_0$. If these were the same as the points entering in (5.15.12), but with different designations, then the two polynomials (5.15.12) and (5.15.14) would be identical except in form. They would also be identical if f were itself a polynomial of degree n, whether or not the points were the same. In that case $\Delta^{n+1}f$ and $\nabla^{n+1}f$ would vanish, and both polynomials $P(x)$ would be identical with f.

There are many different forms in which the interpolation polynomial of degree n can be written, and these can be obtained, one from the others, by neglecting differences of order $n + 1$ and higher, and then by renaming the points. One simple scheme is based upon the "lozenge diagram." In the array

$$\begin{array}{ll} u_r\Delta^p f_q & \\ & \left.\begin{matrix}(u+1)_{r+1}\\ u_{r+1}\end{matrix}\right\}\Delta^{p+1}f_q, \\ u_r\Delta^p f_{q+1} & \end{array}$$

where parentheses are omitted from subscripts of u, the sum of the terms in the upper row is equal to the sum of the terms in the lower row:

$$u_r\Delta^p f_q + (u+1)_{r+1}\Delta^{p+1}f_q = u_r\Delta^p f_{q+1} + u_{r+1}\Delta^{p+1}f_q.$$

The identity follows directly from the relation

$$u_r(\Delta^p f_{q+1} - \Delta^p f_q) = [(u+1)_{r+1} - u_{r+1}]\Delta^{p+1}f_q,$$

which is easily verified. Consequently in the array

$$\begin{array}{ccccc}
\vdots & \vdots & & & \\
f_{-2} & & \vdots & & \\
& \left.\begin{matrix}(u+2)_1\\ (u+1)_1\end{matrix}\right\}\Delta f_{-2} & & \vdots & \\
f_{-1} & & \left.\begin{matrix}(u+2)_2\\ (u+1)_2\end{matrix}\right\}\Delta^2 f_{-2} & & \vdots \\
& \left.\begin{matrix}(u+1)_1\\ u_1\end{matrix}\right\}\Delta f_{-1} & & \left.\begin{matrix}(u+2)_3\\ (u+1)_3\end{matrix}\right\}\Delta^3 f_{-2} & \\
f_0 & & \left.\begin{matrix}(u+1)_2\\ u_2\end{matrix}\right\}\Delta^2 f_{-1} & & \left.\begin{matrix}(u+2)_4\\ (u+1)_4\end{matrix}\right\}\Delta^4 f_{-2} \\
& \left.\begin{matrix}u_1\\ (u-1)_1\end{matrix}\right\}\Delta f_0 & & \left.\begin{matrix}(u+1)_3\\ u_3\end{matrix}\right\}\Delta^3 f_{-1} & \vdots \\
f_1 & & \left.\begin{matrix}u_2\\ (u-1)_2\end{matrix}\right\}\Delta^2 f_0 & \vdots & \\
& \left.\begin{matrix}(u-1)_1\\ (u-2)_1\end{matrix}\right\}\Delta f_1 & \vdots & & \\
f_2 & \vdots & & & \\
\vdots & & & &
\end{array}$$

the sum of any two terms connected by a dash pointing down to the right is equal to the sum of the two terms connected by the dash just below. Now if we start with f_0 and proceed diagonally downward, summing the first $n + 1$ terms, we obtain the right member of (5.15.11). But the sum of any other sequence of terms obtained by proceeding to the right and ending with $\Delta^n f_0$ will have, according to the theorem, identically the same value. Hence we obtain different expressions for the same interpolation polynomial. By ending on $\Delta^n f_i$ for $i \neq 0$, we obtain an interpolation polynomial based upon a different set of fundamental points.

It has been remarked already that with uniform spacing the interpolation is most accurate for points near the middle of the range. For computational purposes it is convenient if the coefficients are small. Both conditions are satisfied if one designates as x_0 the fundamental point closest to the point x to be interpolated and if the series contains only terms as close as possible to the horizontal line through f_0. The two Newton-Gauss formulas result:

$$P(x) = f_0 + u_1\Delta f_0 + u_2\Delta^2 f_{-1} + (u+1)_3\Delta^3 f_{-1} + (u+1)_4\Delta^4 f_{-2} + (u+2)_5\Delta^5 f_{-2} + \cdots \tag{5.15.15}$$

and

$$P(x) = f_0 + u_1\Delta f_{-1} + (u+1)_2\Delta^2 f_{-1} + (u+1)_3\Delta^3 f_{-2} + (u+2)_4\Delta^4 f_{-2} + (u+2)_5\Delta^5 f_{-3} + \cdots, \tag{5.15.16}$$

the first following the lower, the second the upper broken line through f_0.

These last formulas are more neatly expressed by central differences. In this notation the same lozenge diagram appears as follows:

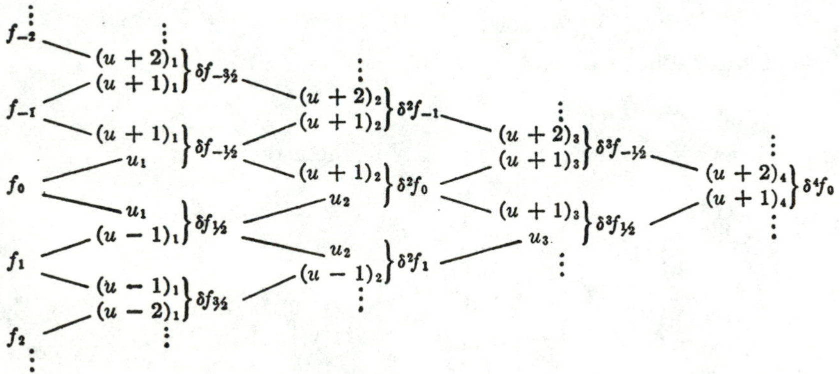

The two formulas (5.15.15) and (5.15.16) appear in these notations:

$$P(x) = f_0 + u_1\delta f_{1/2} + u_2\delta^2 f_0 + (u+1)_3\delta^3 f_{1/2} + (u+1)_4\delta^4 f_0 + \cdots \tag{5.15.17}$$

and

$$P(x) = f_0 + u_1\delta f_{-1/2} + (u+1)_2\delta^2 f_0 + (u+1)_3\delta^3 f_{-1/2} + (u+2)_4\delta^4 f_0 + \cdots. \tag{5.15.18}$$

These two formulas can be combined to give a single more symmetric formula, if we introduce the "central mean" operator μ defined by

$$2\mu = E^{1/2} + E^{-1/2}. \tag{5.15.19}$$

If we add the two formulas and divide by 2, the result is

$$P(x) = [1 + u\mu\delta + u^2\delta^2/2! + u(u^2 - 1^2)\mu\delta^3/3! + u^2(u^2 - 1^2)\delta^4/4! + u(u^2 - 1^2)(u^2 - 2^2)\mu\delta^5/5! + \cdots]f_0. \tag{5.15.20}$$

If between two differences of odd order in the table one writes their mean, the coefficients for the above formula all lie on the same horizontal line.

A great many other special formulas can be derived, but for these reference must be made to the copious literature on interpolation.

5.151. *The remainder in equal-interval interpolation.* All interpolation formulas that can be obtained from the lozenge diagram by following any broken-line path to a particular difference $\delta^n f_0$ are identical except in the arrangement of the terms. Hence they all have the same remainder term. If $n = 2p$ is an even number, the remainder is

$$(5.151.1)\quad R_{n+1} = h^{n+1}u(u^2 - 1^2) \cdots (u^2 - p^2)f^{(n+1)}(\xi)/(n+1)!,$$

where ξ is a point on the smallest interval containing $x, x_{-p}, x_{-p+1}, \ldots, x_0, \ldots, x_p$. If possible, for $|u| < 1$ a formula which ends in a difference $\delta^n f_0$, $\delta^n f_{-1}$, or $\delta^n f_1$ should be used, rather than one ending in some other $\delta^n f_i$, though this will not be possible when, for example, the tabulation begins with $u = 0$. The same expression (5.151.1) for the remainder holds for (5.15.20) if the series is terminated on a difference $\delta^{2p} f_0$ of even order.

If $0 < u < 1$, a formula (5.15.17) which ends on a difference of odd order is most symmetric with respect to that point. If $n = 2p - 1$, then in this case

$$(5.151.2)\quad R_{n+1} = \frac{h^{n+1}u(u^2 - 1^2) \cdots [u^2 - (p-1)^2](u - p)f^{(n+1)}(\xi)}{(n+1)!}.$$

To obtain an upper limit for the truncation error from R_{n+1} directly, one must have an upper limit for $f^{(n+1)}$ over the interval containing the x_i, and this is not necessarily easy to obtain. However, it may be known that each of certain consecutive derivatives retains a fixed sign over the interval, and these signs may be known. In this event, Steffenson's "error test" can be applied. This rests upon the simple observation that in any series

$$s = u_0 + u_1 + \cdots + u_n + R_{n+1},$$

where R_{n+1} is the error due to dropping terms beginning with u_{n+1}, if it is known that R_{n+1} and R_{n+2} have opposite signs, then since

$$R_{n+1} = u_{n+1} + R_{n+2},$$

it follows that

$$|R_{n+1}| \leq |u_{n+1}|$$

and the error is less than the first neglected term in absolute value.

5.2. Trigonometric and Exponential Interpolation. Given any set of constants α_i and β_i, among the possible choices of basic functions $\phi_i(x)$ for interpolation would be

$$\phi_i(x) = \sin(\alpha_i x + \beta_i)$$

or

$$\phi_i(x) = \exp(\alpha_i x + \beta_i).$$

The general methods described at the beginning of this chapter apply, but not much more can be said when the constants α_i and β_i are completely arbitrary and unrelated. Most often, however, the α_i will be in arithmetic progression, in which case certain special simplifications are possible.

If, possibly after changing scale, one can set

$$\alpha_k = k,$$

then for exponential interpolation one has

$$\Phi(x) = \Sigma\gamma_k e^{kx}. \tag{5.2.1}$$

However, if

$$z = e^x,$$

then

$$\Phi(x) = \Sigma\gamma_k z^k = \Psi(z),$$

and the exponential interpolation is the same as polynomial interpolation with a transformation on the independent variable.

Another exponential form is

$$\Phi(x) = \gamma_0 + \gamma_1 e^x + \gamma_{-1}e^{-x} + \gamma_2 e^{2x} + \gamma_{-2}e^{-2x} + \cdots + \gamma_n e^{nx} + \gamma_{-n}e^{-nx}. \tag{5.2.2}$$

Given now the $2n + 1$ points, $x_0, x_1, \ldots, x_{2n}$, let

$$E_i(x) = \prod_{j \neq i}^{0,2n} \{\exp[(x - x_j)/2] - \exp[-(x - x_j)/2]\}, \tag{5.2.3}$$

$$e_i(x) = E_i(x)/E_i(x_i). \tag{5.2.4}$$

Then

$$e_i(x_j) = \delta_{ij},$$

whence

$$\Phi(x) = \sum_0^{2n} y_i e_i(x) \tag{5.2.5}$$

satisfies

$$\Phi(x_i) = y_i$$

and expands into a polynomial of the form (5.2.2).

In the form

$$\Phi(x) = \gamma_0 + \gamma_1 e^{ix} + \gamma_{-1}e^{-ix} + \cdots + \gamma_n e^{inx} + \gamma_{-n}e^{-inx} \tag{5.2.6}$$

if γ_k and γ_{-k} are conjugate complex quantities, then (5.2.6) can be written in the form

$$\Phi(x) = \alpha_0 + \alpha_1 \cos x + \alpha_2 \cos 2x + \cdots + \alpha_n \cos nx + \beta_1 \sin x + \beta_2 \sin 2x + \cdots + \beta_n \sin nx. \tag{5.2.7}$$

Analogous to the functions $E_i(x)$ and $e_i(x)$ are the functions

$$S_i(x) = \prod_{j \neq i}^{0,2n} \sin [(x - x_j)/2], \tag{5.2.8}$$

$$s_i(x) = S_i(x)/S_i(x_i). \tag{5.2.9}$$

Then

$$\Phi(x) = \sum_0^{2n} y_i s_i(x). \tag{5.2.10}$$

5.3. Bibliographic Notes. The literature on interpolation, methods of approximate representation, and numerical quadrature is vast, and only a few elementary general principles are given here. On operational methods see Whittaker and Robinson, Steffenson (1927), Aitken (1929), McClintock (1895), Herget (1948), Jordan (1947), and the encyclopedias, both French and German.

The usual expression for the remainder in polynomial interpolation is (5.11.2); its generalization (5.01.23) is given by Petersson (1949). Steffenson (1927) gives (5.141.8). The form (5.01.25) for polynomials was given by Peano (1914; see also Mansion, 1914); Kowalewski (1932) gave this and (5.01.28), still for polynomials only. The more general form (5.01.25) for arbitrary interpolating functions was given by Sard (1948*a*) and Milne (1948*b*). See also Curry (1951*a*). These will reappear in the following two chapters. Attention is called to subsequent papers by Sard and collaborators.*

The discussion of optimal-interval interpolation is based on Harrison (1949).

The designation "Chebyshev polynomials" is sometimes applied to any set of orthogonal polynomials. The term "Chebyshev polynomial" is sometimes applied to that polynomial of degree n which among all polynomials of that degree most closely approximates the function under consideration, where the measure of departure is the maximum absolute difference between the polynomial and the function. This is not necessarily the interpolating polynomial which agrees with the function at the Chebyshev points (5.12.4), since in (5.11.1) it is not shown that ξ necessarily maximizes $f^{(n+1)}$ at the same time that x maximizes $\omega = T_n$. No

*In the Russian translation of this book it is pointed out that remainder formulas similar to those of Sard had been previously obtained by E. Ja. Remez: On some classes of linear functionals in spaces c_p and on remainder terms in formulas of approximate analysis, *Trudy Inst. Mat. AN U.S.S.R.* **3**(1939):21–62, **4**(1940):47–82; On remainder terms for some formulas of approximate analysis, *Doklady Akad. Nauk, SSSR* **26**(1940):130–134.

simple algorithm is available in general for constructing the closest polynomial approximation to a given function.

One might suppose that by increasing the degree of the interpolating polynomial one would necessarily improve the approximation to any given continuous function, at least when the fundamental abscissas are always uniformly spaced. However, this is not so. For discussion of these and related problems see Bernstein (1926), Ahiezer (1948), de la Vallée Poussin (1952), Fejér (1934), Jackson (1930), and Feldheim (1939).

Most treatments of interpolation discuss also the subject of "inverse interpolation," which is the evaluation of the independent variable at the point where the function takes on a prescribed value, not tabulated. An obvious possibility is to interchange the roles of the dependent and independent variables and interpolate in the ordinary manner. Otherwise one can equate the interpolating polynomial to the prescribed value and solve the resulting algebraic equation. In this connection see Kincaid (1948*b*).

Most standard texts discuss divided differences. See Aitken (1932) for the method called by his name and Neville (1934) for a similar method.

Tables of differences, ordinary or divided, are useful for detecting errors in computed tables of functions. See Miller (1950).

CHAPTER 6

MORE GENERAL METHODS OF APPROXIMATION

6. More General Methods of Approximation

In selecting a particular linear combination

$$\Phi(x) = \Sigma\gamma_i\phi_i(x) \tag{6.0.1}$$

of functions $\phi_i(x)$ to perform an interpolation, the aim is naturally to determine a combination Φ which approximates the given function f at whatever point or points the function f is to be evaluated. In the method of interpolation, so-called, certain values $y_i = f(x_i)$ of f, and possibly certain values $y_j^{(r)} = f^{(r)}(x_j)$, must be known already, or directly obtainable. Then the coefficients γ_i are determined by the requirement that each $y_i = \Phi(x_i)$ and that each $y_j^{(r)} = \Phi^{(r)}(x_j)$. However, this is only one of many possible schemes for determining the γ_i.

An expansion in powers of $x - x_0$ by Taylor's series up to and including the term in $(x - x_0)^n$ differs in appearance from interpolation, but can be made a special case in which the conditions for determining the coefficients are that

$$\Phi^{(r)}(x_0) = f^{(r)}(x_0), \qquad r = 0, \ldots, n. \tag{6.0.2}$$

Petersson's generalized expansion which satisfies conditions (6.0.2) but replaces polynomials by a more general set of functions $\phi_i(x)$ was given in §5.01. The well-known expansions in Fourier or other orthogonal series, with only a finite number of terms retained, illustrate other methods of obtaining approximate representations that are useful for some purposes.

In this chapter will be treated only certain fairly immediate extensions of the methods of interpolation.

6.1. Finite Linear Methods. In the expansion of a function as a series of orthogonal functions, the coefficients are determined by integration. The methods to be described here require summation rather than integration, and hence are called finite.

When the given quantities y_i are not exactly, but only approximately, equal to the $f(x_i)$, it is not necessarily worth while to require that the $\Phi(x_i)$ be exactly equal to the y_i. If there are $n + 1$ quantities y_i, approximately equal to f at $n + 1$ distinct points x_i, it may be saving in time, and quite adequate, to find some linear combination (6.0.1) of $m + 1 < n + 1$ functions ϕ_i, in such a way that the resulting $\Phi(x_i)$ will be as nearly as possible equal to the y_i, without however expecting strict equality at least at all of the points. This is especially the case when the y_i result from experimental measurements or result from computations in which the round-off is not negligible. But if strict equality can no longer be asked for, some other criterion must be found that will yield a system of only $m + 1$ equations in the $m + 1$ unknowns but that will in some sense treat all $n + 1$ of the points alike.

Let the index r run from 0 to m, to distinguish from $i, j, \ldots$ which run from 0 to $n > m$. We seek

$$\Phi(x) \equiv \Sigma\gamma_r\phi_r(x) \equiv f(x) - R(x), \tag{6.1.1}$$

where R does not necessarily vanish even at the x_i. The criterion will be given in the following form: Choose $m + 1$ functions $\psi_r(x)$ so that the matrix of the $\psi_r(x_i)$ has rank $m + 1$, and then determine the γ_r so that the $m + 1$ conditions

$$\sum_j \psi_r(x_j)R(x_j) = 0 \tag{6.1.2}$$

are satisfied. If the matrix of the $\phi_r(x_i)$ also has rank $m + 1$, then Eqs. (6.1.2) define the γ_i uniquely. Moreover, in the special case $m = n$, Φ becomes the ordinary interpolating function.

Let d, c, y, f_r, and p_r, respectively, represent the vectors whose elements are $R(x_j)$, γ_r, y_i, $\phi_r(x_j)$, and $\psi_r(x_j)$; let F and P represent the matrices whose columns are the vectors f_r and p_r, respectively. It is required to satisfy the equations

$$Fc = y + d, \tag{6.1.3}$$
$$P^{\mathsf{T}}d = 0. \tag{6.1.4}$$

Thus y is to be resolved into two components, one in the space of F and the other orthogonal to the space of P. On combining these two equations one obtains

$$P^{\mathsf{T}}Fc = P^{\mathsf{T}}y. \tag{6.1.5}$$

If each matrix P and F has rank $m + 1 \leq n + 1$, then the matrix $P^{\mathsf{T}}F$ is nonsingular, and there is a unique solution y of (6.1.5). Given a system of functions $\psi_r(x)$, the problem reduces to the multiplication of the matrices and solution of the linear system (6.1.5), and this is a problem already discussed at length in Chap. 2.

However, it may happen that at the outset one does not know how many functions ϕ_r one should use to obtain an adequate representation in the sense that the vector d has been made sufficiently small. This vector represents the amount by which the function Φ fails to agree even with the given values y_i of f. It is geometrically evident that, if the system (6.1.5) of order $m + 1$ has been solved, and then also a new system of order $m + 2$ obtained by adjoining a new function ϕ_{r+1}, the new residual vector d cannot be greater than the old one, and in general will actually be less. Moreover, if $n + 1$ functions are included, d will vanish since we are back to the case of interpolation. One would like, if possible, to start with a single function ϕ_0 and adjoin sequentially $\phi_1, \phi_2, \ldots,$ stopping only after the vector d has been made sufficiently small. It is possible to do this without having to solve a new system each time by forming a pair of biorthogonal systems of vectors.

The method can be developed by showing that unit triangular matrices U and V can be found such that for the two matrices

$$T = FV^{-1}, \qquad S = PU^{-1}, \tag{6.1.6}$$

the matrix product

$$S^{\mathsf{T}}T = D \tag{6.1.7}$$

is a diagonal matrix. This is a generalization of the theorem shown at the end of §2.201.

The theorem is trivial when P and F have but a single column each. An inductive proof can be given, which also exhibits the algorithm by assuming relations (6.1.6) and (6.1.7) and showing that, given vectors f, independent of the columns of F, and p, independent of the columns of P, it is possible to find vectors u, v, t, s, and a scalar δ, satisfying

$$\begin{aligned} (T, t)\begin{pmatrix} V & v \\ 0 & 1 \end{pmatrix} &= (F, f), \\ (S, s)\begin{pmatrix} U & u \\ 0 & 1 \end{pmatrix} &= (P, p), \\ (S, s)^{\mathsf{T}}(T, t) &= \begin{pmatrix} D & 0 \\ 0 & \delta \end{pmatrix}. \end{aligned} \tag{6.1.8}$$

The last relation gives δ as

$$s^{\mathsf{T}}t = \delta \tag{6.1.9}$$

and requires that

$$s^{\mathsf{T}}T = 0, \qquad S^{\mathsf{T}}t = 0. \tag{6.1.10}$$

The first two require that

$$Tv + t = f, \qquad Su + s = p. \tag{6.1.11}$$

Hence when the first of (6.1.11) is multiplied on the left by S^{T} and the second by T^{T}, s and t are eliminated, and equations in v alone and in u

alone result. Because of (6.1.7), and because D is diagonal and hence symmetric, v and u are given by

$$v = D^{-1}S^{\mathsf{T}}f, \qquad u = D^{-1}T^{\mathsf{T}}p, \tag{6.1.12}$$

provided only that no diagonal element in D is zero. After v and u are found from (6.1.12), t and s are obtained from (6.1.11), and δ from (6.1.9). This establishes the induction and provides an algorithm for finding the matrices S, T, U, V, and D.

Now observe that the columns of T are linear combinations of those of F, and the columns of S are linear combinations of those of P. Equivalent to (6.1.3) and (6.1.4) are the equations

$$TVc = y + d, \qquad S^{\mathsf{T}}d = 0.$$

Instead of solving for c, one solves for

$$b = Vc \tag{6.1.13}$$

by the simple relation

$$b = D^{-1}S^{\mathsf{T}}y. \tag{6.1.14}$$

To return to the induction, suppose that b is given and that the effect of adjoining a new function ϕ, and hence a new vector f, is to be investigated. Vectors s and t are found by the method already described. The vector b is then unaffected except by the adjunction of an additional element

$$\beta = \delta^{-1}s^{\mathsf{T}}y. \tag{6.1.15}$$

The equation

$$Tb = y + d \tag{6.1.16}$$

becomes

$$(T, t)\begin{pmatrix} b \\ \beta \end{pmatrix} = y + d',$$

where d' is the new residual, or on expanding and applying (6.1.16),

$$d + t\beta = d'.$$

The vector d is orthogonal to all columns of T; d' is orthogonal to all columns of the enlarged matrix (T, t); in particular d' and βt are orthogonal. Hence

$$d^{\mathsf{T}}d = \beta^2 t^{\mathsf{T}}t + d'^{\mathsf{T}}d'. \tag{6.1.17}$$

Thus the adjunction of the new function ϕ reduces the squared length of the residual d by the amount $\beta^2 t^{\mathsf{T}}t$.

Once it is decided how many functions ϕ are to be included in Φ, one can go to (6.1.13) and solve for c. Alternatively, one can form the functions $\tau_r(x)$ corresponding to the columns t_r of T. Thus let $\tau(x)$ represent the row vector whose elements are the first $m + 1$ functions $\tau_0(x), \ldots, \tau_m(x)$; let $\phi(x)$ represent the row vector whose elements are $\phi_0(x), \ldots, \phi_m(x)$. Then

$$\tau(x) = \phi(x)V^{-1}. \tag{6.1.18}$$

Then $\Phi(x)$ is given in either of the two ways:

$$\Phi(x) = \tau(x)b = \phi(x)c. \tag{6.1.19}$$

The matrix V and the functions $\tau(x)$ are entirely independent of the function to be approximated, and depend only upon the functions ϕ_r and ψ_r and upon the points x_i. Hence if the same basic functions and basic points are to be used in the representation of more than one function $f(x)$, it is advantageous to find the $\tau(x)$ and the matrices T, S, V, U once and for all.

6.101. *An expression for the remainder.* The solution c of Eqs. (6.1.5) can be written

$$c = (P^{\mathsf{T}}F)^{-1}P^{\mathsf{T}}y,$$

and hence the function Φ as

$$\Phi(x) = \phi(x)(P^{\mathsf{T}}F)^{-1}P^{\mathsf{T}}y. \tag{6.101.1}$$

If for the elements y_j of the vector y one substitutes the values $\phi_r(x_j)$ of any ϕ_r, the result is

$$\phi_r(x) = \phi(x)(P^{\mathsf{T}}F)^{-1}P^{\mathsf{T}}f_r, \tag{6.101.2}$$

since the quantities $\phi_r(x_j)$ are the elements of the column f_r of the matrix F. Thus (6.101.1) expresses $\Phi(x)$ as a linear combination of the values y_j of $f(x)$ the coefficients of which are certain functions $\Phi_j(x)$:

$$\Phi(x) = \sum_j \Phi_j(x)y_j, \tag{6.101.3}$$

and (6.101.2) implies that

$$\phi_r(x) \equiv \sum_j \Phi_j(x)\phi_r(x_j), \tag{6.101.4}$$

identically.

The Petersson expansion for $\alpha = a$ can be written

$$f(x) = \sum \alpha_r\phi_r(x) + \int_a^x g(x, s)L_{m+1}[f(s)]ds, \tag{6.101.5}$$

when the special functions denoted in §5.01 as ψ_r are replaced by linear combinations of the ϕ_r. Moreover,

$$y_j = \sum \alpha_r \phi_r(x_j) + \int_a^{x_j} g(x_j, s) L_{m+1}[f(s)]ds.$$

Now multiply this equation by $\Phi_j(x)$, sum over j, and subtract from (6.101.5). Because of (6.101.4), the result is

$$R(x) = \int_a^x g(x, s) L_{m+1}[f(s)]ds - \sum \Phi_j(x) \int_a^{x_j} g(x_j, s) L_{m+1}[f(s)]ds. \tag{6.101.6}$$

If, as in §5.01, we write $\int_a^{x_i} = \int_a^x - \int_{x_i}^x$, we obtain on applying (6.101.4)

$$R(x) = \sum \Phi_j(x) \int_{x_j}^x g(x_j, s) L_{m+1}[f(s)]ds, \tag{6.101.7}$$

similar to (5.01.26). For the symmetric form given in (5.01.28), we can again write (6.101.6) with the limit b instead of a and obtain

$$\begin{aligned} R(x) &= \int_a^b K(x, s) L_{m+1}[f(s)]ds, \\ 2K(x, s) &= g(x, s) \operatorname{sgn}(x - s) - \Sigma\Phi_j(x) g(x_j, s) \operatorname{sgn}(x_j - s). \end{aligned} \tag{6.101.8}$$

6.11. *Least-square Curve Fitting.* The residual vector d always vanishes when $m = n$, provided at each step the columns of F, as well as those of P, are kept linearly independent. Moreover, the vector c is then independent of the choice of the functions ψ, given only the linear independence of the sets of functional values. But for any given $m < n$, the vectors c and d and the length of d will depend upon the selection of the functions ψ. Thus, for example, one could choose functions ψ each of which vanishes at all but $m + 1$ of the points x_j. This would give the interpolating function $\Phi(x)$ which passes through these $m + 1$ points and would take no account of the other points.

It is natural to ask that for whatever $m < n$ one might fix upon the vector c be chosen so that the vector d is as small as possible. In all cases the vector y is resolved into two components one of which lies in the space of F and d being the other component. Then the shortest length possible for d is the length of the perpendicular from the point y to the space of F. Hence the component Fc of y in the space of F is to be chosen as the orthogonal projection of y upon this space. This is effected by making $\psi_r = \phi_r$, and hence by taking the matrix P to be the same as F. All equations in §6.1 can be specialized immediately to this case. It may be remarked in passing that in many cases there are good statistical grounds for minimizing the length (or the squared length) of d, but these will not be developed here.

Often experimental conditions may be such that certain values of the y_j may be less reliable than others. If so, greatest "weight" should be given to the measurements of highest reliability. This can be effected by associating a diagonal matrix W of order $n + 1$, with positive non-null diagonal elements, in which the magnitude of each diagonal element, say the jth, measures the degree of reliability attached to the value of the measurement y_j. The matrix W is then used as a metric for the space, and the orthogonality and lengths of the vectors are taken with reference to this metric. This can be achieved by setting

$$P = WF$$

in §6.1.

6.111. *Least-square fitting of polynomials.* A special simplification is possible when the function Φ is to be a polynomial, and

(6.111.1) $$\phi_r(x) = x^r.$$

Let X be the diagonal matrix whose diagonal elements are the abscissas x_j. Then

(6.111.2) $$f_r = X^r f_0.$$

The argument used in proving Lanczos's theorem (§2.22) can be applied here to show that each column t_{r+1} of the matrix T is expressible as a linear combination of Xt_r, t_r, and t_{r-1}. In fact,

(6.111.3)
$$\begin{aligned} t_0 &= f_0, \\ t_1 &= (X - \alpha_0\delta_0^{-1})t_0, \\ t_{r+1} &= (X - \alpha_r\delta_r^{-1})t_r - \delta_r\delta_{r-1}^{-1}t_{r-1}, \qquad r \geq 1, \end{aligned}$$

where

(6.111.4) $$\alpha_r = t_r^T W X t_r.$$

From $t_0 = f_0$ one calculates α_0 and δ_0, thence t_1, α_1, and δ_1, and so on, sequentially. The functions $\tau_r(x)$ are orthogonal on the set of points x_i and are given by

(6.111.5)
$$\begin{aligned} \tau_0(x) &= 1, \\ \tau_1(x) &= (x - \alpha_0\delta_0^{-1})\tau_0(x), \\ \tau_{r+1}(x) &= (x - \alpha_r\delta_r^{-1})\tau_r(x) - \delta_r\delta_{r-1}^{-1}\tau_{r-1}(x), \qquad r \geq 1. \end{aligned}$$

6.12. *Finite Fourier Expansions.* Let the abscissas $x_0, x_1, \ldots, x_{n-1}$ be uniformly spaced. After making a linear change of variable, we can assume that

(6.12.1) $$x_k = k/n,$$

and the function $f(x)$ is to be considered only on the range $0 \leq x < 1$, or is supposed periodic of period 1. Make a further change of variable by

$$\omega = \exp(2\pi i x), \qquad i = \sqrt{-1}. \tag{6.12.2}$$

Then

$$\omega_k = \exp(2\pi i k/n), \tag{6.12.3}$$

and

$$\omega_k^j = \exp(2\pi i jk/n) = \omega_j^k. \tag{6.12.4}$$

The function $f(x)$ becomes a function of ω, $F(\omega)$, and the interpolating polynomial in ω gives a representation of $f(x)$ of the form

$$f(x) = \beta_0 + \beta_1 \exp(2\pi i x) + \beta_2 \exp(4\pi i x) + \cdots + \beta_{n-1} \exp[2(n-1)\pi i x] + R(x).$$

If $f(x)$ is real, then either each term is real or else its conjugate complex also occurs, if it is understood that

$$\exp(2k\pi i x) = \cos(2k\pi x) + i \sin(2k\pi x). \tag{6.12.5}$$

Suppose $n = pm$ is some integral multiple of m, and consider a representation

$$f(x) = \beta_0 + \beta_1 \exp(2\pi i p x) + \cdots + \beta_{m-1}[2(m-1)\pi i p x] + R(x) \tag{6.12.6}$$

or

$$F(\omega) = \beta_0 + \beta_1 \omega^p + \cdots + \beta_{m-1}\omega^{p(m-1)} + P(\omega). \tag{6.12.7}$$

If (6.12.3) is taken as valid for all integral values of k, positive and negative, which amounts to so defining the ω_k for k negative, or exceeding $n-1$, then (6.12.4) is valid for all integral values of j and k. Hence

$$\sum_{k=0}^{n-1} \omega_k^{pj}\omega_k^{-ph} = \sum_{k=0}^{n-1} \omega_k^{p(j-h)} = \sum_{k=0}^{n-1} \omega_{p(j-h)}^k.$$

Any ω_k satisfies $\omega^n = 1$, but if $k \neq 0$, then $\omega_k \neq 1$. But since

$$\omega^n - 1 = (\omega - 1)(\omega^{n-1} + \omega^{n-2} + \cdots + \omega + 1),$$

it follows that for $k \neq 0$

$$\omega_k^{n-1} + \omega_k^{n-2} + \cdots + \omega_k + 1 = 0.$$

Hence

$$\sum_{k=0}^{n-1} \omega_k{}^j\omega_k^{-ph} = \begin{matrix} n & \text{if } j = h, \\ 0 & \text{if } j \neq h. \end{matrix} \tag{6.12.8}$$

Hence the functions ω^{pj} and ω^{-ph} are biorthogonal on the set ω_k in the sense of §6.1. Then if the functions ω^{pj} are taken as the τ_j, and ω^{-ph} as the σ_h of §6.1, the coefficients β are given by

$$\beta_j = n^{-1} \sum_k y_k \omega_k^{-pj}. \tag{6.12.9}$$

If $2j \neq m$, then

$$\beta_{m-j} = \sum_k y_k \omega_k^{-p(m-j)} = \sum_k y_k \omega_k^{pj},$$

or

$$\beta_j = n^{-1} \sum_k y_k[\cos(2\pi pjk/n) - i \sin(2\pi pjk/n)],$$

$$\beta_{m-j} = n^{-1} \sum_k y_k[\cos(2\pi pjk/n) + i \sin(2\pi pjk/n)].$$

Hence, if

$$\begin{aligned} A_j &= n^{-1} \sum_k y_k \cos(2\pi pjk/n), \\ B_j &= n^{-1} \sum_k y_k \sin(2\pi pjk/n), \end{aligned} \tag{6.12.10}$$

then

$$\begin{aligned} \beta_k &= A_k - iB_k, \\ \beta_{m-k} &= A_k + iB_k, \end{aligned}$$

and

$$\beta_k \omega^{pk} + \beta_{m-k}\omega^{-pk} = 2A_k \cos(2\pi pkx) + 2B_k \sin(2\pi pkx).$$

Hence if m is odd, the representation (6.12.6) can be written

$$\begin{aligned} f(x) = A_0 + 2 \sum_{k=1}^{(m-1)/2} A_k \cos(2\pi pkx) \\ + 2 \sum_{k=1}^{(m-1)/2} B_k \sin(2\pi pkx) + R(x), \end{aligned} \tag{6.12.11}$$

with the A's and B's given by (6.2.10). In case m is even, $B_{m/2} = 0$, while $A_{m/2} = 0$ if n is even, and $A_{m/2} = 1/n$ if n is odd.

6.2. Chebyshev Expansions. It was shown in the last chapter that if T_n is the Chebyshev polynomial of degree n, then T_n is that polynomial of degree n and leading coefficient unity whose maximum departure from 0 on the interval from -1 to $+1$ is least. This property was used there to obtain a particular set of points of interpolation that was optimal in the sense there described. The property may, however, be utilized in a different way to obtain an approximate representation of $f(x)$. If $f(x)$ is expanded in a series of polynomials T_n, and if the coefficients of these polynomials in the expansion do not themselves increase too rapidly, then the fact that each polynomial is small over the entire interval suggests that a small number of terms in the expansion might provide an approximation that is uniformly good over the entire interval. This is in contrast to a Taylor series expansion which requires more terms the farther one goes from the center of the expansion.

The analysis is simplified somewhat by introducing the functions

$$C_n(x) = 2\cos n\theta, \qquad x = 2\cos\theta. \tag{6.2.1}$$

Then

$$C_0(x) = 1, \qquad C_1(x) = x, \qquad C_2(x) = x^2 - 2, \ldots$$

and in general $C_n(x)$ is a polynomial in x of leading coefficient 1. We consider the representation of the function $f(x)$ on the range from -2 to $+2$ and seek an expansion

$$f(x) = \alpha_0 + \alpha_1 C_1(x) + \alpha_2 C_2(x) + \cdots, \tag{6.2.2}$$

or equivalently

$$f(2\cos\theta) = \alpha_0 + 2\sum_1^\infty \alpha_n \cos n\theta, \tag{6.2.3}$$

with θ ranging from 0 to π. Since

$$2\cos n\theta \cos m\theta = \cos(n+m)\theta + \cos(n-m)\theta,$$

it follows that

$$2\int_0^\pi \cos n\theta \cos m\theta \, d\theta = \begin{matrix} 0 & \text{for } n \neq m, \\ \pi & \text{for } n = m, \end{matrix} \tag{6.2.4}$$

and hence formally

$$\begin{aligned} \alpha_n &= \pi^{-1}\int_0^\pi f(2\cos\theta)\cos n\theta \, d\theta \\ &= (1/2\pi)\int_{-2}^2 f(x)\,[C_n(x)/\sqrt{4-x^2}]dx. \end{aligned} \tag{6.2.5}$$

Equations (6.2.4) express the orthogonality of the functions $\cos n\theta$ and $\cos m\theta$, $n \neq m$, on the interval from 0 to π. Equivalently, $C_n(x)$ and $C_m(x)$, $n \neq m$, are orthogonal with respect to the weight function $(4 - x^2)^{-1/2}$ on the interval from -2 to $+2$.

Any power x^n of x is expressible as a linear combination of C_n and polynomials of lower order:

$$x^n = C_n(x) + nC_{n-2}(x) + \binom{n}{2} C_{n-4}(x) + \cdots.$$

If we expand

$$f(x) = f(0) + xf'(0) + \tfrac{1}{2}x^2 f''(0) + \cdots$$

and substitute into the integrals (6.2.5), we obtain

$$\alpha_n = \frac{f^{(n)}(0)}{n!} + \frac{f^{(n+2)}(0)}{1!(n+1)!} + \frac{f^{(n+4)}(0)}{2!(n+2)!} + \cdots . \tag{6.2.6}$$

The first term alone on the right is the coefficient of x^n in the Taylor expansion. If this coefficient and α_n are comparable, the expansion in Chebyshev polynomials, carried to a given number of terms, may be expected to give a representation of the function that is uniformly good for all $|x| < 2$; for $2 > |x| > 1$ it is better than the same number of terms of the Taylor series, though the representation will be less good than the Taylor series for $|x| < 1$.

6.3. Bibliographic Notes. The remainder formulas given here are additional special cases of the general remainder formulas of Sard and Milne.

There is an extensive literature on methods of curve fitting, and only a few titles are given in the bibliography. Deming (1938) extends the method of least squares to cases where the parameters occur nonlinearly by employing a method of successive linearization. Remainder formulas in such a situation are not available. A "method of moments," common in the literature, leads to functions ψ_r which are different from ϕ_r. For the infinite case, reference is to be made to the literature on Fourier series, orthogonal series in general, and the problem of moments.

On the representation by Chebyshev polynomials see Lanczos (1938), Miller (1945), and Olds (1950). Nonlinear methods include the use of continued fractions (and reciprocal differences, described in the books on interpolation).

On methods of smoothing data subject to error see Schoenberg (1946, 1952), papers by Sard, and Lanczos (1952).

CHAPTER 7

NUMERICAL INTEGRATION AND DIFFERENTIATION

7. Numerical Integration and Differentiation

A function $\Phi(x)$ obtained by any of the methods outlined in the last two chapters, to the extent that it represents the function $f(x)$ it is intended to approximate, may replace f in any numerical operations required. Thus where a derivative or an integral of $f(x)$ is required, one may differentiate or integrate $\Phi(x)$. Unfortunately for finding the derivative, though $\Phi(x)$ might agree closely with $f(x)$ in value, even over the entire interval, it can nevertheless happen that the slopes might be radically different. With integration, however, the situation is more fortunate, for clearly if on the interval from a to b the deviation $R(x)$ of Φ from f nowhere exceeds ϵ in absolute value, then the error in the integral cannot exceed $\epsilon(a - b)$; and since ordinarily $R(x)$ will change signs at least once, one can expect the error to be much less than this.

7.1. The Quadrature Problem in General. With the $m + 1$ functions $\phi_r(x)$, consider in addition a fixed function $w(x)$ and the integrals

$$\int_a^b \phi_r(x)w(x)dx = \mu_r. \tag{7.1.1}$$

Suppose that values

$$y_i = f(x_i) \tag{7.1.2}$$

are known at $n + 1$ points x_i. We seek coefficients λ_i, independent of the values of y_i, so that in the expression

$$\int_a^b f(x)w(x)dx = \sum \lambda_i y_i + R \tag{7.1.3}$$

the remainder R vanishes whenever f is equal to any of the ϕ_r. Hence

$$\mu_r = \sum_i \lambda_i \phi_r(x_i). \tag{7.1.4}$$

Here are $m + 1$ equations for determining the $n + 1$ required coefficients λ_i. If the equations are consistent, they can be satisfied whenever $n \geq m$, and if $n > m$, it is possible to impose $n - m$ further conditions on the λ_i. If $n < m$, the equations can be satisfied if and only if the points x_i

are so placed that the matrices $(\mu_r, \phi_r(x_i))$ and $(\phi_r(x_i))$ have the same rank.

Consider again Petersson's expansion

$$(7.1.5)\qquad \begin{aligned} f(x) &= \sum \alpha_r\phi_r(x) + \int_a^x g(x, s)L_{m+1}[f(s)]ds, \\ y_i &= \sum \alpha_r\phi_r(x_i) + \int_a^{x_i} g(x_i, s)L_{m+1}[f(s)]ds. \end{aligned}$$

Assuming the multipliers λ_i to have been found, multiply the first equation here by $w(x)$ and integrate from a to b; the second by λ_i and sum, subtracting the result from the integral just obtained. By (7.1.3), (7.1.1), and (7.1.4), this gives

$$R = \int_a^b w(x) \int_a^x g(x, s)L_{m+1}[f(s)]ds\, dx - \sum \lambda_i \int_a^{x_i} g(x_i, s)L_{m+1}[f(s)]ds.$$

A similar expansion written for $x = b$ gives in the same way

$$R = -\int_a^b w(x) \int_x^b g(x, s)L_{m+1}[f(s)]ds\, dx + \sum \lambda_i \int_{x_i}^b g(x_i, s)L_{m+1}[f(s)]ds.$$

Hence by addition of the two expressions

$$(7.1.6)\qquad \begin{aligned} R &= \int_a^b M(s)L_{m+1}[f(s)]ds, \\ 2M(s) &= \int_a^b w(x)g(x, s) \operatorname{sgn}(x - s)dx - \sum \lambda_i g(x_i, s) \operatorname{sgn}(x_i - s). \end{aligned}$$

The function $M(s)$ is continuous since the discontinuity in the signum function occurs only where g vanishes.

As examples, let $m = n = 1$, $a = x_0 = 0$, and $b = x_1 = 1$. Let $w(x) = 1$, $\phi_0(x) = 1$, and $\phi_1(x) = x$. Then $g(x, s) = x - s$, and $\lambda_0 = \lambda_1 = \frac{1}{2}$. Then $M(s) = s(s - 1)/2$. This gives the common trapezoidal rule with remainder:

$$(7.1.7)\qquad \int_0^1 f(x)dx = (y_0 + y_1)/2 - \int_0^1 s(1 - s)f''(s)ds/2.$$

Next, let $w(x) = e^{\alpha x}$, $\phi_0(x) = 1$, and $\phi_1(x) = x$. Again $g(x, s) = x - s$. However,

$$\lambda_0 = (e^\alpha - 1 - \alpha)/\alpha^2, \qquad \lambda_1 = (\alpha e^\alpha - e^\alpha + 1)/\alpha^2.$$

Then $M(s) = [e^{\alpha s} - (e^\alpha - 1)s - 1]/\alpha^2$, and

$$(7.1.8)\qquad \begin{aligned} \int_0^1 f(x)e^{\alpha x}\, dx = {} & [y_0(e^\alpha - 1 - \alpha) + y_1(\alpha e^\alpha - e^\alpha + 1)]/\alpha^2 \\ & + \int_0^1 [e^{\alpha s} - (e^\alpha - 1)s - 1]f''(s)ds/\alpha^2. \end{aligned}$$

If $w(x) = 1$, $\phi_0(x) = 1$, and $\phi_1(x) = e^{\alpha x}$, then $g(x, s) = (e^{\alpha(x-s)} - 1)/\alpha$, $\lambda_0 = (e^\alpha - 1)^{-1}\alpha^{-1}(\alpha e^\alpha - e^\alpha + 1)$, and $\lambda_1 = (e^\alpha - 1)^{-1}\alpha^{-1}(e^\alpha - 1 - \alpha)$.

Then $M(s) = (e^\alpha - 1)^{-1}\alpha^{-1}[s(e^\alpha - 1) - e^\alpha(1 - e^{-\alpha s})]$. Hence

$$(7.1.9)\quad \int_0^1 f(x)dx = (e^\alpha - 1)^{-1}\alpha^{-1}[y_0(\alpha e^\alpha - e^\alpha + 1) + y_1(e^\alpha - 1 - \alpha)] + (e^\alpha - 1)^{-1}\alpha^{-1}\int_0^1 [s(e^\alpha - 1) - e^\alpha(1 - e^{-\alpha s})][f''(s) - \alpha f'(s)]ds.$$

Finally, let $w(x) = 1$, $\phi_0(x) = e^{\alpha x}$, and $\phi_1(x) = xe^{\alpha x}$. Then

$$\lambda_0 = (e^\alpha - \alpha - 1)/\alpha^2, \qquad \lambda_1 = (\alpha - 1 + e^{-\alpha})/\alpha^2,$$

and $g(x, s) = (x - s)e^{\alpha(x-s)}$. Then

$$M(s) = \{1 + [(1 - e^\alpha)s - 1]e^{-\alpha s}\}/\alpha^2.$$

Hence

$$(7.1.10)\quad \int_0^1 f(x)dx = [y_0(e^\alpha - \alpha - 1) + y_1(\alpha - 1 + e^{-\alpha})]/\alpha^2 + \int_0^1 \{1 + [(1 - e^\alpha)s - 1]e^{-\alpha s}\}[f''(s) - 2\alpha f'(s) + \alpha^2 f(s)]ds/\alpha^2.$$

When these formulas are applied repeatedly to intervals from $i - 1$ to i, for $i = 1, 2, \ldots, n$, then formulas (7.1.7), (7.1.9), and (7.1.10) give, neglecting remainders,

$$\begin{aligned}\int_0^n f(x)dx &\doteq \left(y_0/2 + \sum_{i=1}^{n-1} y_i + y_n/2\right) \\ &\doteq (e^\alpha - 1)^{-1}\alpha^{-1}[y_0(\alpha e^\alpha - e^\alpha + 1) + y_n(e^\alpha - 1 - \alpha)] + \sum_{i=1}^{n-1} y_i \\ &\doteq \alpha^{-2}[y_0(e^\alpha - \alpha - 1) + y_1(\alpha - 1 + e^{-\alpha})] \\ &\qquad + \alpha^{-2}(e^{\alpha/2} - e^{-\alpha/2})^2 \sum_{i=1}^{n-1} y_i.\end{aligned}$$

Equations (7.1.8) and (7.1.10) are identical if in the former $f(x)e^{\alpha x}$ is replaced by $f(x)$.

In (6.101.3) and (6.101.4), the $\Phi_i(x)$ are linear combinations with constant coefficients of the $\phi_r(x)$. On multiplying (6.101.4) by $w(x)$ and integrating, one finds

$$\mu_r = \sum \phi_r(x_i) \int_a^b w(x)\Phi_i(x)dx,$$

and the integral on the right is expressible as a linear combination of the μ_r. Thus the choice

$$(7.1.11)\qquad \lambda_i = \int_a^b w(x)\Phi_i(x)dx$$

satisfies (7.1.4) and determines the λ_i uniquely. Hence in case $n = m$, to use (7.1.4) to determine the λ_i for the expression (7.1.3) is equivalent to approximating the integral of fw by the integral of Φw, where Φ is the interpolating function; when $n > m$, one can always determine the λ_i by (7.1.4), together with suitable further conditions so that the sum on the right of (7.1.3), ignoring R, is the approximation given by integrating w with a function Φ determined by one of the methods of §6.1.

In case $n < m$, it may be possible to choose the points x_i at which the evaluations $f(x_i)$ are to be made so that the matrices $(\phi_r(x_i),\ \mu_r)$ and $(\phi_r(x_i))$ have the same rank. In this event Eq. (7.1.4) is consistent and has solutions λ_i; moreover, if the rank is $n + 1$, the solutions λ_i are unique.

Consider the polynomial case with

$$\phi_r(x) = x^r. \tag{7.1.12}$$

If the points x_i are all distinct, then the matrix $(\phi_r(x_i))$ has rank $n + 1$ for $m \geq n$. Hence Eq. (7.1.4) will be consistent if every square submatrix of $(x_i^r,\ \mu_r)$ of order $n + 2$ is singular. This implies that any matrix of the form

$$\begin{pmatrix} 1 & 1 & \cdots & 1 & \mu_r \\ x_0 & x_1 & \cdots & x_n & \mu_{r+1} \\ \cdots & \cdots & \cdots & \cdots & \cdots \\ x_0^{n+1} & x_1^{n+1} & \cdots & x_n^{n+1} & \mu_{r+n+1} \end{pmatrix}$$

must be singular, and therefore the matrix

$$\begin{pmatrix} 1 & 1 & \cdots & 1 & \mu_0 & \mu_1 & \mu_2 & \cdots \\ x_0 & x_1 & \cdots & x_n & \mu_1 & \mu_2 & \mu_3 & \cdots \\ \cdots & \cdots & \cdots & \cdots & \cdots & \cdots & \cdots & \cdots \\ x_0^{n+1} & x_1^{n+1} & \cdots & x_n^{n+1} & \mu_{n+1} & \mu_{n+2} & \mu_{n+3} & \cdots \end{pmatrix}$$

must have rank $n + 1$. In particular if $m = 2n + 1$, and the x_i are taken to satisfy the equation

$$\begin{vmatrix} 1 & \mu_0 & \mu_1 & \cdots & \mu_n \\ x & \mu_1 & \mu_2 & \cdots & \mu_{n+1} \\ \cdots & \cdots & \cdots & \cdots & \cdots \\ x^{n+1} & \mu_{n+1} & \mu_{n+2} & \cdots & \mu_{2n+1} \end{vmatrix} = 0, \tag{7.1.13}$$

then if these x_i are all distinct, the system (7.1.4) is consistent, and defines a set of coefficients uniquely.

With the x_i as determined by (7.1.13), let

$$\omega_n(x) \equiv \Pi(x - x_i) \equiv x^{n+1} + \omega_{n,0}x^n + \cdots + \omega_{n,n}. \tag{7.1.14}$$

Then $\omega_n(x)$ is equal to the determinant in (7.1.13) except for a constant factor. Hence

$$\mu_{n+1+i} + \omega_{n,0}\mu_{n+i} + \cdots + \omega_{n,n}\mu_i = 0. \tag{7.1.15}$$

But by (7.1.1), (7.1.13), and (7.1.14), this implies that

$$\int_a^b x^i \omega_n(x) w(x) dx = 0, \qquad i = 0, 1, \ldots, n. \tag{7.1.16}$$

Hence if $p(x)$ is any polynomial of degree $\leq n$,

$$\int_a^b p(x)\omega_n(x)w(x)dx = 0.$$

In particular if we consider the sequence of polynomials $\omega_0, \omega_1, \ldots, \omega_n$, it follows from this equation that

$$\int_a^b \omega_m(x)\omega_n(x)w(x)dx = 0, \qquad n \neq m. \tag{7.1.17}$$

The polynomials $\omega_m(x)$ are therefore said to be orthogonal on the interval (a, b) with respect to the weight function $w(x)$. The derivation of Lanczos's theorem can be paraphrased to show that by defining formally $\omega_{-1} = 1$,

$$\begin{aligned} \omega_{-1}(x) &= 1, \\ \omega_0(x) &= x - \alpha_0\delta_0^{-1}, \\ \omega_r(x) &= (x - \alpha_r\delta_r^{-1})\omega_{r-1}(x) - \delta_r\delta_{r-1}^{-1}\omega_{r-2}(x), \qquad r \geq 1, \end{aligned} \tag{7.1.18}$$

where

$$\begin{aligned} \alpha_r &= \int_a^b x\omega_{r-1}^2(x)w(x)dx, \\ \delta_r &= \int_a^b \omega_{r-1}^2(x)w(x)dx. \end{aligned} \tag{7.1.19}$$

Hence to obtain the polynomial $\omega_n(x)$, one can calculate recursively the $\omega_r(x)$, $r \leq n$, instead of expanding the determinant (7.1.13).

In this situation both the x_i and the λ_i were left completely arbitrary and determined so as to take account of a maximal set of the ϕ_r. Another possible procedure would be to restrict the x_i and λ_i by a limited number of conditions, and then impose as many of the conditions (7.1.4) as possible. As an example of this, it is sometimes desirable to select the x_i so that the λ_i are all equal:

$$\lambda_0 = \lambda_1 = \cdots = \lambda.$$

There are then to be determined the $n + 2$ quantities $x_0, \ldots, x_n$ and λ, and therefore in general $n + 2$ conditions (7.1.4) can be imposed. For polynomials ϕ_r these have the form

$$\begin{aligned} \lambda &= \mu_0/(n+1), \\ \Sigma x_i &= (n+1)\mu_1/\mu_0, \\ &\cdots\cdots\cdots\cdots \\ \Sigma x_i^{n+1} &= (n+1)\mu_{n+1}/\mu_0. \end{aligned}$$

Hence the sums of the powers up to the $(n + 1)$st are known. From these the elementary symmetric functions of the x_i can be formed by means of Eqs. (3.02.5), and hence the equation of degree $n + 1$ of which the x_i are the roots. Unfortunately the roots x_i do not necessarily always fall within the range of the integration, and when they do not, the method is unusable.

7.2. Numerical Differentiation. Equation (6.101.7) can be written

$$f(x) = \Phi(x) + \sum_j \Phi_j(x) \int_{x_j}^{x} g(x_j, s)L[f(s)]ds. \tag{7.2.1}$$

Hence

$$f'(x) = \Phi'(x) + \sum_j \Phi_j'(x) \int_{x_j}^{x} g(x_j, s)L[f(s)]ds + \sum_j \Phi_j(x)g(x_j, x)L[f(x)].$$

But

$$\begin{aligned}\sum_j \Phi_j(x)g(x_j, s) &= \sum_j \sum_r \Phi_j(x)\phi_r(x_j)g_r(s) \\ &= \sum_r \phi_r(x)g_r(s) = g(x, s),\end{aligned}$$

and

$$g(x, x) = 0.$$

Hence

$$f'(x) = \Phi'(x) + \sum_j \Phi_j'(x) \int_{x_j}^{x} g(x_j, s)L[f(s)]ds. \tag{7.2.2}$$

Again

$$\sum_j \Phi_j'(x)g(x_j, s) = \sum_r \phi_r'(x)g_r(s) = \partial g(x, s)/\partial x,$$

and

$$\partial g(x, s)/\partial x \Big|^{x=0} = 0,$$

so that

$$f''(x) = \Phi''(x) + \sum_j \Phi_j''(x) \int_{x_j}^{x} g(x_j, s)L[f(s)]ds. \tag{7.2.3}$$

Similar relations hold for $f^{(\nu)}$, $\nu \leq n$.

For polynomial interpolation in terms of divided differences the equation

$$\begin{aligned}f(x) = f(x_0) + (x - x_0)f(x_0, x_1) + (x - x_0)(x - x_1)f(x_0, x_1, x_2) \\ + \cdots + (x - x_0) \cdots (x - x_{n-1})f(x_0, x_1, \ldots, x_n) \\ + (x - x_0) \cdots (x - x_n)f(x_0, x_1, \ldots, x_n, x)\end{aligned} \tag{7.2.4}$$

is exact, the last term representing an expression for the remainder. Let

$$x_{(r)} = (x - x_0) \cdots (x - x_r).$$

Then

$$\begin{aligned}f'(x) = f(x_0, x_1) + x'_{(1)}f(x_0, x_1, x_2) + \cdots \\ + x'_{(n-1)}f(x_0, \ldots, x_n) + R',\end{aligned} \tag{7.2.5}$$

where

(7.2.6) $R' = x'_{(n)}f(x_0, x_1, \ldots, x_n, x) + x_{(n)}f(x_0, x_1, \ldots, x_n, x, x).$

When $x = x_i$, $x_{(n)} = 0$, and the second term drops out of the remainder. Continuing, one finds

(7.2.7) $f''(x) = 2f(x_0, x_1, x_2) + x''_{(2)}f(x_0, x_1, x_2, x_3) + \cdots + x''_{(n-1)}f(x_0, \ldots, x_n) + R'',$

(7.2.8) $R'' = x''_{(n)}f(x_0, x_1, \ldots, x_n, x) + 2x'_{(n)}f(x_0, x_1, \ldots, x_n, x, x) + x_{(n)}f(x_0, x_1, \ldots, x_n, x, x, x).$

All divided differences which occur in these remainders can be expressed in integral form as in §5.141.

7.3. Operational Methods. For polynomial interpolation with equally spaced fundamental points, formulas for numerical differentiation and integration can be derived by operational methods. The Taylor expansion of an analytic function can be written in the form

$$f(x_0 + uh) = f(x_0) + uhf'(x_0) + u^2h^2f''(x_0)/2! + \cdots .$$

Hence by introducing the differential operators

(7.3.1) $$D = d/dx, \qquad \theta = hD$$

and recalling the definitions of the displacement and difference operators, we can write formally

$$E^u = 1 + uhD + u^2h^2D^2/2! + \cdots = e^{u\theta},$$

and therefore set

(7.3.2) $$E = 1 + \Delta = (1 - \nabla)^{-1} = e^\theta,$$

whence

(7.3.3) $$\theta = \log E = \log (1 + \Delta) = -\log (1 - \nabla).$$

Hence

(7.3.4) $$\theta = \Delta - \Delta^2/2 + \Delta^3/3 - \Delta^4/4 + \cdots = \nabla + \nabla^2/2 + \nabla^3/3 + \nabla^4/4 + \cdots .$$

These expansions can be used to obtain the derivative at any point x_0 in terms of forward or backward differences. To obtain $f'(x) = f'(x_0 + uh)$, write

(7.3.5) $E^u\theta = (1 + \Delta)^u \log (1 + \Delta) = \Delta + (2u - 1)\Delta^2/2 + (3u^2 - 6u + 2)\Delta^2/6 + \cdots$

or the corresponding expression in terms of ∇.

For the derivative in terms of central differences, we have

$$\delta = E^{1/2} - E^{-1/2} = e^{\theta/2} - e^{-\theta/2} = 2 \sinh (\theta/2),$$
$$\mu = (E^{1/2} + E^{-1/2})/2 = (e^{\theta/2} + e^{-\theta/2})/2 = \cosh (\theta/2).$$

From this it follows that formally

$$d\delta/d\theta = \mu = (1 + \delta^2/4)^{1/2},$$

and therefore

$$\begin{aligned} \theta &= \int_0^\delta (1 + \tau^2/4)^{-1/2} d\tau \\ &= \delta - \frac{1^2 \cdot \delta^3}{3! \cdot 2^2} + \frac{1^2 \cdot 3^2 \cdot \delta^5}{5! \cdot 2^4} - \cdots . \end{aligned} \tag{7.3.6}$$

This formula gives the value of $f'(x_0)$ in terms of the values of f at points $x_{\pm n/2}$. To obtain $f'(x_0)$ in terms of values of f at points $x_{\pm n}$, proceed as follows:

It can be verified directly that θ/μ satisfies the differential equation

$$(1 + \delta^2/4) d(\theta/\mu)/d\delta + (\delta/4)(\theta/\mu) = 1. \tag{7.3.7}$$

Also θ/μ is an odd function of δ. Hence assume the expansion

$$\theta/\mu = \delta + \alpha_3\delta^3 + \alpha_5\delta^5 + \alpha_7\delta^7 + \cdots ,$$

substitute into the differential equation, and equate coefficients of like powers of δ. By this means one finds that

$$\theta = \mu \left[\delta - \frac{1}{3!}\delta^3 + \frac{(2!)^2}{5!}\delta^5 - \frac{(3!)^2}{7!}\delta^7 + \cdots \right]. \tag{7.3.8}$$

To continue to higher derivatives, we have next that

$$d(\theta^2)/d\delta = 2\theta\, d\theta/d\delta = 2\theta/\mu.$$

Hence

$$\theta^2 = 2 \int_0^\delta \left[\tau - \frac{1}{3!}\tau^3 + \frac{(2!)^2}{5!}\tau^5 \cdots \right] d\tau,$$

or

$$\theta^2 = 2 \left[\frac{1}{2!}\delta^2 - \frac{1}{4!}\delta^4 + \frac{(2!)^2}{6!}\delta^6 - \frac{(3!)^2}{8!}\delta^8 + \cdots \right]. \tag{7.3.9}$$

It can now be shown inductively that

$$(1 + \delta^2/4) d(\theta^{2\nu+1}/\mu)/d\delta + (\delta/4)(\theta^{2\nu+1}/\mu) = (2\nu + 1)\theta^{2\nu}$$

and

$$d(\theta^{2\nu})/d\delta = 2\nu\theta^{2\nu-1}/\mu.$$

From this one can proceed sequentially to find θ^3/μ, θ^4, θ^5/μ, ...

Some estimate of the error can be had by noting the magnitude of the first neglected term. It is possible to obtain an exact expression for the error in any particular case by a method that will be used for numerical integration, now to be described.

Ordinarily in numerical integrations the integral is to be evaluated between two of the interpolation points, say x_i and x_j. At any rate it is no restriction to assume that the lower limit is such a point. Let $F(x)$ be any function satisfying

$$DF(x) = f(x). \tag{7.3.10}$$

One wishes to evaluate $(E^u - E^i)F(x_0)$ where i is an integer. If $i = 0$, we have

$$\begin{aligned}\int_{x_0}^{x_0+uh} f(x)dx &= (E^u - 1)F(x_0) \\ &= (u\theta + u^2\theta^2/2 + u^3\theta^3/3! + \cdots)F(x_0) \\ &= uh(1 + u\theta/2 + u^2\theta^2/3! + \cdots)f(x_0).\end{aligned} \tag{7.3.11}$$

If the powers of θ are replaced by their expansions in terms of any of the difference operators up to whatever power may be desired, the result is a formula for numerical integration. If one so desires, he can replace all difference operators retained in the formula by their expressions in terms of the displacement operator E, so obtaining a formula directly in terms of the $f(x_j)$.

7.31. *The Trapezoidal Rule.* Next to a Riemann sum this is the simplest rule of all. It is given in (7.1.7), but will now be derived operationally. In (7.3.11) set $u = 1$; we carry the expansion to the first power only of the difference operator, and to this degree of approximation $\theta = \Delta$. Hence (7.3.11) gives

$$\begin{aligned}\int_{x_0}^{x_1} f(x)dx &\doteq h(1 + \Delta/2)f(x_0) = h(1 + E)f(x_0)/2 \\ &= h(y_0 + y_1)/2.\end{aligned} \tag{7.31.1}$$

To evaluate the remainder one can proceed as follows: The formulas

$$\begin{aligned}e^\theta &= 1 + \theta + \theta^2 \int_0^1 \tau e^{(1-\tau)\theta}\, d\tau \\ &= 1 + \theta + \tfrac{1}{2}\theta^2 + \tfrac{1}{2}\theta^3 \int_0^1 \tau^2 e^{(1-\tau)\theta}\, d\tau\end{aligned} \tag{7.31.2}$$

are exact and can be verified directly by integrating by parts. If e^θ is replaced by E, and $e^\theta - 1$ by Δ, one has by the second of these

$$\Delta F(x_0) = h\left(1 + \tfrac{1}{2}\theta + \tfrac{1}{2}\theta^2 \int_0^1 \tau^2 E^{1-\tau}\, d\tau\right)f(x_0)$$

and by the first therefore

$$\Delta F(x_0) = h\left[1 + \tfrac{1}{2}\Delta - \tfrac{1}{2}\theta^2 \int_0^1 \tau(1 - \tau)E^{1-\tau}\, d\tau\right]f(x_0).$$

Thus the integral operator provides the desired remainder. This can be transformed

$$h\theta^2 \int_0^1 \tau(1-\tau)E^{1-\tau}\,d\tau f(x_0) = h^3 \int_0^1 \tau(1-\tau)f''[x_0 + (1-\tau)h]d\tau$$
$$= \int_{x_0}^{x_1} (x_1 - v)(v - x_0)f''(v)dv$$

by introducing the new variable of integration

$$v = x_0 + (1-\tau)h = x_1 - \tau h.$$

Hence we obtain finally the trapezoidal rule with remainder

(7.31.3) $$\int_{x_0}^{x_1} f(x)dx = \tfrac{1}{2}h(y_0 + y_1) - \tfrac{1}{2}\int_{x_0}^{x_1} (x_1 - v)(v - x_0)f''(v)dv.$$

By the law of the mean, since

$$\int_{x_0}^{x_1} (x_1 - v)(v - x_0)dv = (x_1 - x_0)^3/6 = h^3/6,$$

we can write

(7.31.4) $$\int_{x_0}^{x_1} f(x)dx = \tfrac{1}{2}h(y_0 + y_1) - \tfrac{1}{12}h^3 f''(\xi), \qquad x_0 \le \xi \le x_1.$$

Upper and lower limits for the true value of the integral can be had by introducing minimum and maximum values of f''. If $f''(x)$ does not change sign on the interval, the law of the mean can be invoked again to give

$$\int_{x_0}^{x_1} (x_1 - v)(v - x_0)dv = (x_1 - \xi)(\xi - x_0)\int_{x_0}^{x_1} f''(v)dv$$
$$= (x_1 - \xi)(\xi - x_0)(y_1' - y_0').$$

Since the quadratic factor cannot exceed $h^2/4$, it follows that, when f'' does not change sign, then for some positive $\epsilon \le 1$ it is true that

(7.31.5) $$\int_{x_0}^{x_1} f(x)dx = \tfrac{1}{2}h(y_0 + y_1) - \tfrac{1}{8}\epsilon h^2(y_1' - y_0').$$

7.32. *The Maclaurin Quadrature Formula.* The identities (7.31.2) are special cases of the more general form

(7.32.1) $$e^{u\theta} = 1 + u\theta + \cdots + u^m\theta^m/m! + (u^{m+1}\theta^{m+1}/m!)\int_0^1 \tau^m e^{(1-\tau)u\theta}\,d\tau.$$

On setting $u = 1$ and $u = -1$, subtracting, and applying to $F(x_0)$, one obtains

(7.32.2) $$\int_{x_{-1}}^{x_1} f(x)dx = 2h\left[f(x_0) + \frac{h^2}{3!}f''(x_0) + \cdots + \frac{h^{2m}}{(2m+1)!}f^{(2m)}(x_0)\right] + R,$$

where

$$(7.32.3)\quad R = \frac{h^{2m+2}}{(2m+1)!}\int_0^1 \tau^{2m+1}\{f^{(2m+1)}[x_0 + (1-\tau)h] - f^{(2m+1)}[x_0 - (1-\tau)h]\}d\tau$$

$$= \frac{1}{(2m+1)!}\int_0^h (h-v)^{2m+1}[f^{(2m+1)}(x_0+v) - f^{(2m+1)}(x_0-v)]dv.$$

This is the Maclaurin quadrature formula. For $m = 0$,

$$(7.32.4)\quad \int_{x_{-1}}^{x_1} f(x)dx = 2hf(x_0) + \int_0^h (h-v)[f'(x_0+v) - f'(x_0-v)]dv.$$

The integral on the right expresses the difference between the area under the curve and that under the tangent to the curve at the point (x_0, y_0).

7.33. *Simpson's Rule.* When $m = 1$, the formulas (7.32.2) and (7.32.3) in operational form are

$$E - E^{-1} = 2\theta + \tfrac{1}{3}\theta^3 + (\theta^4/6)\int_0^1 \tau^3(E^{1-\tau} - E^{-(1-\tau)})d\tau,$$

with operators applied to $F(x_0)$. One factor θ on the right applied to $F(x_0)$ replaces it by $hf(x_0)$. Consider the term θ^2. Since

$$2 + \delta^2 = 2\cosh\theta = e^\theta + e^{-\theta}$$

by (7.3.5), we can apply (7.32.1) to obtain

$$\delta^2 = \theta^2 + \tfrac{1}{2}\theta^3\int_0^1 \tau^2(E^{1-\tau} - E^{-(1-\tau)})d\tau.$$

Hence

$$(7.33.1)\quad E - E^{-1} = 2\theta + \tfrac{1}{3}\theta\delta^2 - \tfrac{1}{6}\theta^4\int_0^1 \tau^2(1-\tau)(E^{1-\tau} - E^{-(1-\tau)})d\tau$$

$$= \tfrac{1}{3}\theta(E + 4 + E^{-1}) - \tfrac{1}{6}\theta^4\int_0^1 \tau^2(1-\tau)(E^{1-\tau} - E^{-(1-\tau)})d\tau.$$

This gives Simpson's rule with remainder when applied in the usual way to $F(x_0)$. After changing the variable of integration on the right, one has

$$(7.33.2)\quad \int_{x_{-1}}^{x_1} f(x)dx = \tfrac{1}{3}h(y_{-1} + 4y_0 + y_1) - \tfrac{1}{6}\int_0^h (h-v)^2v[f'''(x_0+v) - f'''(x_0-v)]dv.$$

It is customary to point out that, whereas Simpson's rule utilizes only second differences, δ^2, and should therefore have a vanishing remainder when f is a quadratic polynomial, the remainder vanishes in fact even when f is a cubic polynomial. This follows from the fact that for a cubic polynomial f the third derivative is constant, and the integrand therefore vanishes in the remainder term.

In case f''' is monotonic (but not necessarily constant), the integrands which occur on the right in (7.33.2) and in (7.32.3) for $m = 1$ are both positive or both negative. However, one integral appears as an additive, and one as a subtractive, term. Hence in case f''' is monotonic, of the two approximations $h(y_{-1} + 4y_0 + y_1)/3$ and $2h(y_0 + h^2y_0''/6)$, one is an overestimate and one an underestimate of the integral which appears on the left.

Another "enclosure" theorem can be obtained when f''' is monotonic, a theorem, that is, which provides an upper and a lower bound to the true value of the integral. When f''' is monotonic, then the difference between the derivatives which appears in the integral on the right of (7.33.2) is bounded by 0 and $f'''(x_1) - f'''(x_{-1})$. Hence for some positive $\epsilon < 1$,

$$\int_0^h (h - v)^2v[f'''(x_0 + v) - f'''(x_0 - v)]dv = \epsilon(y_1''' - y_{-1}''') \int_0^h (h - v)^2v\, dv = \epsilon(y_1''' - y_{-1}''')h^4/12.$$

Hence

$$\text{(7.33.3)} \quad \int_{x_{-1}}^{x_1} f(x)dx = \tfrac{1}{3}h(y_{-1} + 4y_0 + y_1) - \epsilon h^4(y_1''' - y_{-1}''')/72.$$

The bounds are obtained by setting ϵ equal to 0 and 1.

7.34. *Newton's Three-eighths Rule.* To obtain a formula using third differences requires the expansion of e^θ to the fourth power of θ. In order to gain the advantages of symmetry in the expressions, consider the evaluation of $(E^{3/2} - E^{-3/2})F(x_0)$ in terms of $E^{\pm 3/2}f(x_0)$ and $E^{\pm 1/2}f(x_0)$, assuming the fundamental abscissas to be $x_{\pm 3/2}$ and $x_{\pm 1/2}$. Write

$$e^{3\theta/2} = 1 + \tfrac{3}{2}\theta + \tfrac{9}{8}\theta^2 + \tfrac{9}{16}\theta^3 + R_{3/2} = 1 + \tfrac{3}{2}\theta + \tfrac{9}{8}\theta^2 + \tfrac{9}{16}\theta^3 + \tfrac{27}{128}\theta^4 + 3\theta R_{3/2}'$$

and the corresponding expansions of $e^{-3\theta/2}$ with remainders $R_{-3/2}$ and $R_{-3/2}'$. We find then that

$$\text{(7.34.1)} \quad e^{3\theta/2} - e^{-3\theta/2} = E^{3/2} - E^{-3/2} = 3\theta(1 + \tfrac{3}{8}\theta^2 + R_{3/2}' + R_{-3/2}').$$

The expression $1 + 3\theta^2/8$ is required in terms of μ, μ^3, and remainders. For this we have

$$E^{3/2} + E^{-3/2} = 8\mu^3 - 6\mu = 2 + 9\theta^2/4 + R_{3/2} + R_{-3/2}$$

and

$$E^{1/2} + E^{-1/2} = 2\mu = 2 + \theta^2/4 + R_{1/2} + R_{-1/2}.$$

Hence

$$\text{(7.34.2)} \quad 1 + \tfrac{3}{8}\theta^2 = \mu^3 - \tfrac{1}{8}(R_{3/2} + R_{-3/2}) - \tfrac{3}{8}(R_{1/2} + R_{-1/2}).$$

The required formula is obtained by combining (7.34.1) and (7.34.2). Write this result:

$$\text{(7.34.3)} \quad \int_{x_{-3/2}}^{x_{3/2}} f(x)dx = \tfrac{3}{8}h(y_{-3/2} + 3y_{-1/2} + 3y_{1/2} + y_{3/2}) + R,$$

where R is obtained by combining the various remainder terms which appear above. Consider the terms in $\theta R_{3/2}$ and $\theta R_{3/2}'$. When either of these operators is applied to $F(x_0)$, they yield integrals with respect to τ between the limits 0 and 1 of a polynomial in τ multiplied by

$$f^{IV}[x_0 + 3h(1 - \tau)/2].$$

By introducing a new variable of integration

$$v = x_0 + 3h(1 - \tau)/2,$$

the integrals are taken from x_0 to $x_{3/2}$. After a little algebraic manipulation, when the coefficient is included, they combine to give the integral

$$1/24 \int_{x_0}^{x_{3/2}} (x_{3/2} - v)^3(x_0 - v)f^{IV}(v)dv.$$

A similar manipulation of $\theta R_{-3/2}$ and $\theta R_{-3/2}'$ gives

$$1/24 \int_{x_{-3/2}}^{x_0} (x_{-3/2} - v)^3(x_0 - v)f^{IV}(v)dv,$$

a result that can be written down immediately from symmetry.

There remain the terms in $\theta R_{1/2}$ and $\theta R_{-1/2}$. For the first of these introduce the variable of integration

$$v = x_0 + h(1 - \tau)/2,$$

the integral being taken then from x_0 to $x_{1/2}$. The result is

$$-3/16h \int_{x_0}^{x_{1/2}} (x_{1/2} - v)^3 f^{IV}(v)dv.$$

The same integral from $x_{-1/2}$ to x_0 results from the term in $\theta R_{-1/2}$. These three integrals can be rearranged to give finally

$$R = 1/24 \left\{ \int_{x_{-3/2}}^{x_{-1/2}} (x_{-3/2} - v)^3(x_0 - v)f^{IV}(v)dv \right. \tag{7.34.4}$$
$$\left. + \int_{x_{-1/2}}^{x_{1/2}} [(x_0 - v)^4 - 9/16]f^{IV}(v)dv + \int_{x_{1/2}}^{x_{3/2}} (x_{3/2} - v)^3(x_0 - v)f^{IV}(v)dv \right\}.$$

Equations (7.34.3) and (7.34.4) together give Newton's three-eighths rule with remainder.

The remainder just given is identical in form with that which one would obtain by using (7.1.6), and the polynomials in v which multiply $f^{IV}(v)$ in the several integrands define the function $M(v)$ which appears in (7.1.6). More generally, the operational method is merely a device that may be applied for calculating a quadrature formula and remainder in the special case when the intervals are equal, the base functions are polynomials, and $w(x) = 1$, whereas the method of §7.1 can be applied in general.

It may be noted that $M(v)$ is of fixed sign throughout the interval of integration in (7.34.4), and hence the law of the mean can be applied as in the previous cases.

7.35. *Open Formulas.* The trapezoidal rule, Simpson's rule, and the three-eighths rule are said to be of closed type, since the integration extends from the first to the last of the fundamental abscissas. A formula of open type is one which extends beyond. Such a formula is less exact but is often required, in particular in most methods of solving ordinary differential equations. In the most common methods for doing this, the dependent variable y is evaluated sequentially at successive points $x_i, x_{i+1}, x_{i+2}, \ldots$. When y_i has been evaluated, then one proceeds to evaluate y_{i+1}, first approximately by an open integration of y', thereafter by successive approximations by closed integrations, each closed integration employing the currently available approximation to y'_{i+1}. Under quite general conditions this process converges.

There are many possible formulas of open type, just as there are many of closed type. The three of the latter type which have already been given by no means exhaust the list. Of open formulas we shall give only one, which Milne uses in conjunction with Simpson's rule as the closed formula, for solving a differential equation. The formula in question gives the integral of $y = f(x)$ from x_{-2} to x_{+2} in terms of the three middle ordinates y_{-1}, y_0, and y_1. Hence it is a quadratic formula.

In (7.32.1) for $m = 3$, set $u = 2$ and $u = -2$ and subtract:

$$E^2 - E^{-2} = \tfrac{4}{3}\theta\left[3 + 2\theta^2 + 2\theta^3 \int_0^1 \tau^3(E^{2(1-\tau)} - E^{-2(1-\tau)})d\tau\right].$$

Again in (7.32.1) for $m = 1$, set $u = 2$ and $u = -2$ and add:

$$E + E^{-1} = 2 + \theta^2 + \tfrac{1}{2}\theta^3 \int_0^1 \tau^2(E^{1-\tau} - E^{-(1-\tau)})d\tau.$$

The left member of this equation is $\delta^2 + 2$; on eliminating the θ^2 from the first and applying the operator to $F(x_0)$ as usual, we have

$$\int_{x_{-2}}^{x_2} f(x)dx = \tfrac{4}{3}h(3 + 2\delta^2) + R,$$

where

$$R = \tfrac{4}{3}h^4\left\{-\int_0^1 \tau^2[f'''(x_1 - \tau h) - f'''(x_{-1} + \tau h)]d\tau + 2\int_0^1 \tau^3[f'''(x_2 - 2\tau h) - f'''(x_{-2} + 2\tau h)]d\tau\right\}.$$

After a change of the variables of integration this can be written

$$(7.35.1) \quad R = \tfrac{1}{6}\int_{x_1}^{x_2} (x_2 - v)^3 f'''(v)dv + \tfrac{1}{6}\int_{x_0}^{x_1} (v - x_0)[(v - x_{-1})^2 - 5h^2]f'''(v)dv + \tfrac{1}{6}\int_{x_{-2}}^{x_{-1}} (v - x_{-2})^3 f'''(v)dv + \tfrac{1}{6}\int_{x_{-1}}^{x_0} (x_0 - v)[(x_1 - v)^2 - 5h^2]f'''(v)dv.$$

This is the remainder R in the required formula

$$\int_{x_{-2}}^{x_2} f(x)dx = \tfrac{4}{3}h(2y_{-1} - y_0 + 2y_1) + R. \tag{7.35.2}$$

7.36. *The Euler-Maclaurin Summation Formula.* If the polynomials $\psi(\tau)$ satisfy

$$\psi_0(\tau) = 1, \qquad d\psi_{\nu+1}(\tau)/d\tau = \psi_\nu(\tau), \tag{7.36.1}$$

then repeated integration by parts yields the formula

$$e^\theta - 1 = \sum_{\nu=1}^{n} \theta^\nu[\psi_\nu(1) - e^\theta\psi_\nu(0)] + \theta^{n+1} \int_0^1 \psi_n(\tau)e^{(1-\tau)\theta}\, d\tau. \tag{7.36.2}$$

If e^θ is replaced by E, and the operators applied to $f(x_0)$, one has an "expansion" of $f(x_0 + h)$ in powers of h, in which however the coefficients involve derivatives evaluated at both x_0 and $x_0 + h$. Thus the generalization of Taylor's expansion of Hummel and Seebeck comes from choosing

$$\psi_{n+m} = \tau^m(1 - \tau)^n/(n + m)!.$$

In general, it is easily verified that for any set of polynomials satisfying (7.36.1) if we write

$$\psi_\nu(\tau) = a_0\tau^\nu/\nu! + a_1\tau^{\nu-1}/(\nu - 1)! + \cdots + a_\nu, \tag{7.36.3}$$

then each a_i is the same for any ψ_ν with $\nu \geq i$. Let

$$a_i = B_i/i!. \tag{7.36.4}$$

Then

$$\nu!\psi_\nu(\tau) = B_0\tau^\nu + \binom{\nu}{1} B_1\tau^{\nu-1} + \binom{\nu}{2} B_2\tau^{\nu-2} + \cdots + B_\nu.$$

This can be written symbolically in the form

$$\psi_\nu(\tau) = (B + \tau)^\nu/\nu! \tag{7.36.5}$$

if we understand that in the expansion B^i is to be replaced by B_i. If we now require that

$$B_0 = 1, \qquad (B + 1)^\nu = B^\nu, \qquad \nu > 1, \tag{7.36.6}$$

then we have a recursion that defines the B_i, and hence the $\psi_i(\tau)$, uniquely, and (7.36.1) takes a particularly simple form:

$$e^\theta - 1 = \tfrac{1}{2}\theta(1 + E) - \sum_{\nu=2}^{n} B_\nu\Delta\theta^\nu/\nu! + \theta^{n+1} \int_0^1 \psi_\nu(\tau)e^{(1-\tau)\theta}d\tau. \tag{7.36.7}$$

When this is applied to $F(x_0)$, the result expresses the integral of $f(x)$ from x_0 to x_1 in terms of f and its derivatives evaluated at x_0 and x_1:

$$(7.36.8)\quad \int_{x_0}^{x_1} f(x)dx = \tfrac{1}{2}h(y_0 + y_1) - \sum_{\nu=2}^{n} B_\nu h^\nu (y_1^{(\nu-1)} - y_0^{(\nu-1)})/\nu! + R,$$

and R can be evaluated by applying the integral operator in (7.36.7). If corresponding expansions are written for the integral from x_1 to x_2, . . . , x_{m-1} to x_m, and the results added, then the derivatives drop out everywhere except at x_0 and x_m:

$$(7.36.9)\quad \int_{x_0}^{x_m} f(x)dx = h(y_0/2 + y_1 + \cdots + y_{m-1} + y_m/2) - \sum_{\nu=2}^{n} B_\nu h^\nu (y_m^{(\nu-1)} - y_0^{(\nu-1)})/\nu! + R'.$$

This is the Euler-Maclaurin summation formula, and it is often used for approximating the value of a sum $\Sigma y_i = \Sigma f(x_i)$ in case the function $f(x)$ is readily integrated.

The constants B_ν are the Bernoulli numbers, and the polynomials $(B + \tau)^\nu$ the Bernoulli polynomials. It turns out that

$$B_{2\nu+1} = 0, \qquad \nu > 0.$$

For if n is allowed to approach infinity in (7.36.7), the expansion becomes

$$e^\theta - 1 = \tfrac{1}{2}\theta(1 + e^\theta) - (e^\theta - 1)\sum_{\nu=2}^{\infty} B_\nu \theta^\nu/\nu!.$$

Considering θ a real number, we can write

$$1 + \sum_{\nu=2}^{\infty} B_\nu \theta^\nu/\nu! = \frac{\theta(e^\theta + 1)}{2(e^\theta - 1)} = \frac{\theta(e^{\theta/2} + e^{-\theta/2})}{2(e^{\theta/2} - e^{-\theta/2})} = \frac{\theta}{2}\coth\frac{\theta}{2}.$$

Since the right member of this identity is an even function of θ, only even powers can appear on the left, and this proves the assertion.

7.4. Bibliographic Notes. The developments in Eqs. (7.2.4) and following and in (7.3.6) and following are based on Steffenson (1927). An interesting and suggestive development of quadrature formulas based on polynomial interpolation is to be found in Kowalewski (1932).

The selection of points so as to minimize the number of points leads to the Gaussian quadrature formulas, and that for equalizing the coefficients leads to the Chebyshev formulas. See Sard (1948*b*, 1949*a*, and 1949*b*) for other criteria for a "good" formula and (1951) for extension to more than one variable. The introductory treatment here follows Kneschke (1949*a* and 1949*b*) in the main.

CHAPTER 8

THE MONTE CARLO METHOD

8. The Monte Carlo Method

In §1.6, it was pointed out that in many, if not most, computations the occurrence of the maximum possible error may be extremely improbable and that it may be sufficient for practical purposes to be able to say that the probability is p that the error in the result will exceed some quantity δ, where p is small, perhaps one-tenth or one-hundredth of 1 per cent, and δ is within the limits of tolerance. It may well happen that computational labor of astronomical proportions would be required for a result which is certain to be in error less than δ, if indeed such a result is attainable at all, whereas with only moderate labor the probability of an error greater than δ can be made extremely small. Thus though the computation is strictly deterministic, it may be both possible and advantageous to employ nondeterministic, *i.e.*, statistical, methods to appraise the result.

The result of the computation is therefore treated as an estimate rather than a true approximation. Actually physical measurements are often of this sort. A physical measurement may be the average of many measurements, all differing from one another though taken under conditions as nearly identical as it is possible to make them. Along with the mean, the experimenter will then compute the probable error, which is not the maximum error that could have been made, since that is poorly defined or entirely undefined. Instead, the probable error is the error which one expects will be exceeded half the time. In more precise terms this means that, if one follows the practice of asserting with respect to any given measured quantity that the mean of the measurements does not differ from the true value by more than the probable error, then he may expect to be wrong in the case of about half the quantities under consideration. But the point of primary interest here is the fact that, if the maximal error of measurement is not defined, then neither can the maximal propagated error be specified for any computation which makes use of these measurements as data.

This being granted, it is reasonable to consider the feasibility of using nondeterministic methods for the computations themselves. This would mean obtaining an estimate of the desired quantity by means of some random sampling process, rather than obtaining an approximation by a rigorous computation. This is known as the Monte Carlo method. In a few situations it is the only feasible method known for "solving" the problem, though it is by no means a general method. A few examples will be given to explain and illustrate it, but any reasonably complete treatment would have to make rather extensive use of the theory of probability.

8.1. Numerical Integration. Consider first the problem of evaluating a multiple integral,

$$\Phi = \int \phi(x)dv. \tag{8.1.1}$$

The variable x is taken to represent a vector in the space of the coordinates $\xi_1, \ldots, \xi_n$ (where we could have $n = 1$, in particular); dv represents an element of volume in the n-dimensional space; and the integral is to be taken between fixed and finite limits. The assumption of finiteness for the range of integration is no restriction in itself, since this can always be achieved by a transformation of variables if necessary. However, it is assumed that ϕ is everywhere finite and bounded in this region. Hence the integral represents a hypervolume in the space of $n + 1$ dimensions with coordinates $\xi_1, \ldots, \xi_n, \eta$; therefore we can introduce scale factors and translations and assume that in the region

$$0 \leq \phi(x) \leq 1 \tag{8.1.2}$$

and that the integration extends over the region

$$0 \leq \xi_i \leq 1. \tag{8.1.3}$$

It follows that, if points were drawn at random from the entire unit $(n + 1)$-dimensional hypercube with uniform probability, then the probability is Φ that any particular one of the randomly selected points has a coordinate η satisfying

$$\eta \leq \phi(x), \tag{8.1.4}$$

if $x = (\xi_1, \ldots, \xi_n)$ represents the other n coordinates of this point. Hence if one made random drawings of a large number of points N, testing inequality (8.1.4) each time a drawing is made, and if the inequality is satisfied for N' of these points, then N'/N provides an estimate of Φ.

This is the essential idea underlying the Monte Carlo method of numerical integration. However, in any digital computation one cannot draw arbitrary points from the cube, but only points whose coordinates have a digital representation.

Suppose it has been determined that for representing the coordinate ξ_i in the base β it is sufficient to use σ_i places and that the computation of

$\phi(x)$ will then be accurate to τ places. Suppose, further, one has available some process for drawing digits $0, 1, \ldots, \beta - 1$ at random with equal probability. One therefore makes $\sigma_1 + \sigma_2 + \cdots + \sigma_n + \tau$ drawings. The first σ_1 digits in order provide the representation of the coordinate ξ_1; the next σ_2 provide the representation of the coordinate $\xi_2, \ldots$; the last τ provide the representation of the coordinate η. With these representations one tests the inequality (8.1.4) to decide whether the selected point in $(n + 1)$ space lies inside or outside the volume. There is a question whether the equality sign should be allowed in (8.1.4), or only the strict inequality. If the equalities do not arise in sufficient numbers to make a significant difference, then it is immaterial whether these points are counted as inside or outside or are neglected altogether. If they make a significant contribution, the decision must be based upon a consideration of the routine for computing ϕ.

If one could really make random selections from all points $(\xi_1, \ldots, \xi_n, \eta)$ in the unit hypercube, rather than from those points only whose coordinates are digital numbers, and could obtain the strict mathematical value of $\phi(x)$ for any point, one would be repeating the occurrences of an event with two possible outcomes: "success" and "failure" with probabilities Φ and $1 - \Phi$. By standard statistical formulas one can determine the probability that in N trials the number of actual successes N_1 will differ from $N\Phi$ by more than any given amount.

But since only points x are drawn for which the ξ_i are digital, one is at best not estimating the volume Φ, but a slightly different volume Φ', and the statistical formula gives the probability of deviations from $N\Phi'$, rather than from $N\Phi$. The nature of the volume Φ' is best illustrated for the case $n = 2$.

Each of the σ_i digits of ξ_i can have any one of β possible values. Hence there are $\beta^{\sigma_1+\sigma_2}$ possible points x. Associated with each x is a computed $\phi^*(x)$, defined by the computational routine. This is a digital number with τ places. For each x let η' represent a quantity differing from $\phi^*(x)$ by not more than $\beta^{-\tau}/2$, and whose exact value depends upon the error $\phi^*(x) - \phi(x)$ and upon the rule for including or excluding the equality in (8.1.4). Then Φ' is equal to $\beta^{-\sigma_1-\sigma_2}$ times the sum of the quantities η' for all possible x. Hence the quantity Φ' being estimated is essentially that approximation to Φ that would be obtained by employing a Riemann sum of $\beta^{\sigma_1+\sigma_2}$ terms for the integral.

The total error in the entire computation is therefore

$$\Phi - N_1/N = (\Phi - \Phi') + (\Phi' - N_1/N). \tag{8.1.5}$$

The second parentheses is the so-called sampling error, which is the deviation of the estimate from the quantity being estimated. In the assertion that the probability is p that this error does not exceed δ, if δ

is regarded as a function of N for fixed p, then δ is inversely proportional to the square root of N. Thus to cut δ in one-half one must quadruple the size N of the sample.

The first parentheses, $\Phi - \Phi'$, represents the usual computational error, generated and residual. There is no initial error, since each $\phi^*(x)$ is to be computed for exactly that x that has been drawn by the random process. For a given ϕ, the error $\phi - \phi^*$ generated in computing any $\phi^*(x)$ depends upon the values of the σ_i. Increasing these, which is to say, employing more places in the computation, will decrease the generated error, though at the expense of the additional labor involved in carrying along the extra places. But increasing the σ_i will also decrease the residual error, which is the deviation of the Riemann sum from the true value Φ of the integral. And the decrease in residual error comes about without the need for actually computing additional terms in the sum. Thus if the function ϕ is quite irregular, so that many subdivisions of the range of integration would be required to make the residual error sufficiently small, the direct computation of the Riemann sum might require a prohibitively large number of terms, hence ϕ computed for a prohibitively large number of x's, whereas in Monte Carlo computations the fineness of the subdivision has no effect whatever upon the number of values of x for which ϕ must be computed. This may be extremely important when the space is of high dimensionality. Thus, if $n = 6$, and each $\sigma_i = 10$, the Riemann sum contains 10^6 terms; if $\sigma_i = 20$, it contains 20^6 terms.

There are known techniques for making Monte Carlo estimates of an element of an inverse matrix, and solutions of certain functional equations, but it is not clear that the method is useful unless variables occur which are to be integrated out in the solution that is required.

It should be mentioned in passing that (8.1.4) can be replaced by any relation equivalent to it. Thus if ϕ is a square root, the relation $\eta^2 \leq \phi^2$ is equivalent and more easily examined.

8.2. Random Sequences. To employ the Monte Carlo method one must be able somehow to obtain random sequences of digits, or at least sequences which resemble random sequences in all essential aspects. What constitutes an adequate "resemblance" is not altogether clear, but at least the digits used must be in roughly equal proportions, and no digit may show a marked tendency to follow any other particular digit. Printed tables of randomly selected decimal digits are available, and the Rand Corporation has prepared punched-card tables of random decimal digits.

For high-speed machines neither printed tables nor punched cards are suitable sources. If one selects a 2ν-digit number for ν sufficiently large (say, 5 or more) squares, extracts the middle 2ν digits, and repeats, one is bound eventually to return again to the original number. Thus the process is cyclic. If the cycle is sufficiently long, then the extracted digits form a sequence that resembles a random sequence, and the operation is easily programed for a computing machine. Similar processes are multiplying by a fixed multiplier and extracting middle digits; and squaring and reducing modulo some prime.

8.3. Bibliographic Notes. The Monte Carlo method achieved its first popularity among the atomic-energy laboratories, following some successes in its use by von Neumann and Ulam. For general discussions see Metropolis and Ulam (1949) and proceedings of the Monte Carlo Symposium edited by Householder, Forsythe, and Germond (1951). More recently a series of papers and reports have come out from the Institute for Numerical Analysis: Kac and Donsker (1950), Kac (1951), Wasow (1950, 1951*a*, and 1951*b*), Fortet (1952*a* and 1952*b*), Curtiss (1952), Forsythe and Liebler (1950, 1951), and Cutkosky (1951). See also the proceedings of the several Endicott symposia, and the Quarterly Progress Reports of the National Applied Mathematics Laboratories of the National Bureau of Standards.

For problems other than numerical quadrature, the method has been used for matrix inversion, for solving functional equations of various types, but more especially for problems associated with physical processes that are essentially stochastic in character. In this last connection see the proceedings of the Monte Carlo Symposium, and also Nelson (1949) and Kahn (1949, 1950). On integral equations, which specialize directly to linear algebraic systems, see Albert (1951–1952) and Nygaard (1952), in addition to references already mentioned. It is this author's opinion that the method has proved and will prove most useful for the intrinsically stochastic physical problems.

BIBLIOGRAPHY

Shmuel Agmon (1951): "The Relaxation Method for Linear Inequalities," National Bureau of Standards, NAML Report 52-27.

N. I. Ahiezer (1947): "Lectures on the Theory of Approximation" (Russian), Moscow and Leningrad, 323 pp.

Franz Aigner and Ludwig Flamm (1912): Analyse von Abklingungskurven, *Physik. Z.*, **13**:1151–1155.

A. C. Aitken (1926): On Bernoulli's Numerical Solution of Algebraic Equations, *Proc. Roy. Soc. Edinburgh*, **46**:289–305.

——— (1929): A General Formula of Polynomial Interpolation, *Proc. Edinburgh Math. Soc.*, **1**:199–203.

——— (1931): Further Numerical Studies in Algebraic Equations and Matrices, *Proc. Roy. Soc. Edinburgh*, **51**:80–90.

——— (1932*a*): On Interpolation by Iteration of Proportional Parts, without the Use of Differences, *Proc. Edinburgh Math. Soc.* (2), **3**:56–76.

——— (1932*b*): On the Evaluation of Determinants, the Formation of Their Adjugates, and the Practical Solution of Simultaneous Linear Equations, *Proc. Edinburgh Math. Soc.* (2), **3**:207–219.

——— (1932*c*): On the Graduation of Data by the Orthogonal Polynomials of Least Squares, *Proc. Roy. Soc. Edinburgh*, **53**:54–78.

——— (1933): On Fitting Polynomials to Data with Weighted and Correlated Errors, *Proc. Roy. Soc. Edinburgh*, **A54**:12–16.

——— (1934): On Least Squares and Linear Combination of Observations, *Proc. Roy. Soc. Edinburgh*, **A55**:42–48.

——— (1936–1937*a*): Studies in Practical Mathematics. I. The Evaluation with Application of a Certain Triple Product Matrix, *Proc. Roy. Soc. Edinburgh*, **57**:172–181.

——— (1936–1937*b*): Studies in Practical Mathematics. II. The Evaluation of the Latent Roots and Latent Vectors of a Matrix, *Proc. Roy. Soc. Edinburgh*, **57**:269–304.

——— (1937–1938): Studies in Practical Mathematics. III. The Application of Quadratic Extrapolation to the Evaluation of Derivatives, and to Inverse Interpolation, *Proc. Roy. Soc. Edinburgh*, **58**:161–175.

——— (1945): Studies in Practical Mathematics. IV. On Linear Approximation by Least Squares, *Proc. Roy. Soc. Edinburgh*, **A62**:138–146.

G. E. Albert (1951–1952): "A General Approach to the Monte Carlo Estimation of the Solutions of Certain Fredholm Integral Equations," I–III, Oak Ridge National Laboratory Internal Memorandum.

Franz L. Alt (1952): Almost-triangular Matrices, *Proc. Intern. Congr. Math.*, 1950, **1**:657.

H. Andoyer (1906): Calcul des différences et interpolation, *Encyclopédie sci. math.*, I, **21**:47–160.

R. V. Andree (1951): Computation of the Inverse of a Matrix, *Am. Math. Monthly*, **58**:87–92.

V. A. Bailey (1941): Prodigious Calculations, *Australian J. Sci.*, **3**:78–80.

L. Bairstow (1914): "Investigations Relating to the Stability of the Aeroplane," Reports and Memoranda No. 154 of Advisory Committee for Aeronautics.

T. Banachiewicz (1937): Zur Berechnung der Determinanten, wie auch der Inversen, und zur durauf basierten Auflösung der Systeme linearen Gleichungen, *Acta Astron.*, **3**:41–72.

V. Bargmann, D. Montgomery, and J. von Neumann (1946): "Solution of Linear Systems of High Order," Princeton, N.J., Institute for Advanced Study Report, BuOrd, Navy Dept.

M. S. Bartlett (1951): An Inverse Matrix Adjustment Arising in Discriminant Analysis, *Ann. Math. Stat.*, **22**:107–111.

Julius Bauschinger (1904): Interpolation, *Enc. Math. Wiss.*, I D, **3**:799–820.

Edmund C. Berkeley (1949): "Giant Brains, or Machines That Think," John Wiley & Sons, Inc., New York, xvi + 270 pp.

Serge Bernstein (1926): "Leçons sur les propriétés extrémales et la meilleure approximation des fonctions analytiques d'une variable réelle," Gauthier-Villars & Cie, Paris, x + 207 pp.

Raymond T. Birge and J. W. Weinberg (1947): Least Squares Fitting of Data by Means of Polynomials, *Revs. Mod. Phys.*, **19**:298–360.

M. Š. Birman (1950): Some Estimates for the Method of Steepest Descent (Russian), *Uspekhi Mat. Nauk* 5, **3**(37):152–155.

D. R. Blaskett and H. Schwerdtfeger (1945): A Formula for the Solution of an Arbitrary Analytic Equation, *Quart. Appl. Math.*, **3**:266–268.

E. Bodewig (1935): Über das Euler'sche Verfahren zur Auflösung numerischer Gleichungen, *Comment. Math. Helv.*, **8**:1–4.

——— (1946*a*): On Graeffe's Method of Solving Algebraic Equations, *Quart. Appl. Math.*, **4**:177–190.

——— (1946*b*): Sur la méthode de Laguerre pour l'approximation des racines de certaines équations algébriques et sur la critique d'Hermite, *Koninkl. Ned. Akad. Wetenschap. Proc.*, **49**:910–921.

——— (1947): Comparison of Some Direct Methods for Computing Determinants and Inverse Matrices, *Koninkl. Ned. Akad. Wetenschap. Proc.*, **50**:49–57.

——— (1947–1948): Bericht über die verschiedenen Methoden zur Lösung eines Systems linearer Gleichungen mit reellen Koeffizienten, *Koninkl. Ned. Akad. Wetenschap. Proc.*, **50**:930–941, 1104–1166, 1285–1295; **51**:53–64, 211–219.

——— (1949): On Types of Convergence and on the Behavior of Approximations in the Neighborhood of a Multiple Root of an Equation, *Quart. Appl. Math.*, **7**:325–333.

O. Bottema (1950): A Geometrical Interpretation of the Relaxation Method, *Quart. Appl. Math.*, **7**:422–423.

O. L. Bowie (1951): Practical Solution of Simultaneous Linear Equations, *Quart. Appl. Math.*, **8**:369–373.

Alfred Brauer (1946): Limits for the Characteristic Roots of a Matrix, *Duke Math. J.*, **13**:387–395.

——— (1947): Limits for the Characteristic Roots of a Matrix, II, *Duke Math. J.*, **14**:21–26.

——— (1948): Limits for the Characteristic Roots of a Matrix, III, *Duke Math. J.*, **15**:871–877.

Otto Braunschmidt (1943): Über Interpolation, *J. reine angew. Math.*, **185**:14–55.

P. Brock and F. J. Murray (1952): "The Use of Exponential Sums in Step by Step Integration," unpublished manuscript.

S. Brodetsky and G. Smeal (1924): On Graeffe's Method for Complex Roots of Algebraic Equations, *Proc. Cambridge Phil. Soc.*, **22**:83–87.

E. T. Browne (1930): On the Separation Property of the Roots of the Secular Equation, *Am. J. Math.*, **52**:843–850.

E. M. Bruins (1951): "Numerieke Wiskunde," Servire, Den Haag, 127 pp.

Joseph G. Bryan (1950): "A Method for the Exact Determination of the Characteristic Equation and Latent Vectors of a Matrix with Applications to the Discriminant Function for More Than Two Groups," thesis, Harvard University.

Hans Buckner (1948): A Special Method of Successive Approximations for Fredholm Integral Equations, *Duke Math. J.*, **15**:197–206.

H. Burkhardt (1904): Trigonometrische Interpolation, *Enc. Math. Wiss.*, II A, **9a**:642–693.

Gino Cassinis (1944): I metodi di H. Boltz per la risoluzione dei sistemi di equazioni lineari e il loro impiego nella compenzazione delle triangolazione, *Riv. catasto e servici tecnici erariali*, No. 1.

L. Cesari (1931): Sulla risoluzione dei sistemi di equazioni lineari per approssimazioni successive, *Rass. poste, telegrafi e telefoni*, Anno 9.

——— (1937): Sulla risoluzione dei sistemi di equazioni lineari per approssimazioni successive, *Atti accad. nazl. Lincei Rend., Classe sci. fis., mat. e nat.* (6*a*), **25**:422–428.

F. Cohn (1894): Ueber die in recurrirender Weise gebildeten Grössen und ihren Zusammenhang mit den algebraischen Gleichungen, *Math. Ann.*, **44**:473–538.

A. R. Collar (1948): Some Notes on Jahn's Method for the Improvement of Approximate Latent Roots and Vectors of a Square Matrix, *Quart. J. Mech. Appl. Math.*, **1**:145–148.

L. Collatz (1950*a*): Iterationsverfahren für komplexe Nullstellen algebraischer Gleichungen, *Z. angew. Math. u. Mech.*, **30**:97–101.

——— (1950*b*): Über die Konvergenzkriterien bei Iterationsverfahren für lineare Gleichungssysteme, *Math. Z.*, **53**:149–161.

Computation Laboratory (1946): "A Manual of Operation for the Automatic Sequence Controlled Calculator," Harvard University Press, Cambridge, 561 pp.

——— (1949): "Description of a Relay Calculator," Harvard University Press, Cambridge, 366 pp.

J. L. B. Cooper (1948): The Solution of Natural Frequency Equations by Relaxation Methods, *Quart. Appl. Math.*, **6**:179–182.

A. F. Cornock and J. M. Hughes (1943): The Evaluation of the Complex Roots of Algebraic Equations, *Phil. Mag.* (7), **34**:314–320.

J. G. van der Corput (1946): Sur l'approximation de Laguerre des racines d'une équation qui a toutes ses racines réelles, *Koninkl. Ned. Akad. Wetenschap. Proc.*, **49**:922–929.

Charles L. Critchfield and John Beck, Jr. (1935): A Method for Finding the Roots of the Equation $f(x) = 0$ Where f Is Analytic, *J. Research Nat. Bur. Standards*, **14**:595–600.

L. L. Cronvich (1939): On the Graeffe Method of Solution of Equations, *Am. Math. Monthly*, **46**:185–190.

Prescott D. Crout (1941): A Short Method for Evaluating Determinants and Solving Systems of Linear Equations with Real or Complex Coefficients, *Trans. AIEE*, **60**:1235–1240.

Haskell B. Curry (1944): The Method of Steepest Descent for Non-linear Minimization Problems, *Quart. Appl. Math.*, **2**:258–261.

——— (1951*a*): Abstract Differential Operators and Interpolation Formulas, *Portugaliae Math.*, **10**:135–162.

——— (1951*b*): Note on Iterations with Convergence of Higher Degree, *Quart. Appl. Math.*, **9**:204–205.

J. H. Curtiss (1952): "A Unified Approach to the Monte Carlo Method," report presented at meeting of Association for Computing Machinery, Pittsburgh, May 2–3, 1952.

R. E. Cutkosky (1951): A Monte Carlo Method for Solving a Class of Integral Equations, *J. Research Nat. Bur. Standards*, **47**:113–115.

W. Edwards Deming (1938): "Statistical Adjustment of Data," John Wiley & Sons, Inc., New York, x + 261 pp.

Bernard Dimsdale (1948): On Bernoulli's Method for Solving Algebraic Equations, *Quart. Appl. Math.*, **6**:77–81.

C. Domb (1949): On Iterative Solutions of Algebraic Equations, *Proc. Cambridge Phil. Soc.*, **45**:237–240.

Paul S. Dwyer (1951): "Linear Computations," John Wiley & Sons, Inc., New York, xi + 344 pp.

Engineering Research Associates, Inc. (1950): "High-speed Computing Devices," McGraw-Hill Book Company, Inc., New York, xiii + 440 pp.

I. M. H. Etherington (1932): On Errors in Determinants, *Proc. Edinburgh Math. Soc.*, **3**:107–117.

G. Faber (1910): Über die Newton'sche Näherungsformel, *J. reine angew. Math.*, **138**:1–21.

V. N. Faddeeva (1950): "Computational Methods of Linear Algebra" (Russian), Moscow and Leningrad (Chap. 1, "Basic Material from Linear Algebra," translated by Curtis D. Benster), National Bureau of Standards Report 1644.

Leopold Fejér (1934): On the Characterization of Some Remarkable Systems of Points of Interpolation by Means of Conjugate Points, *Am. Math. Monthly*, **41**:1–14.

Ervin Feldheim (1939): Théorie de la convergence des procédés d'interpolation et de quadrature mécanique, *Mém. sci. math. acad. sci. Paris*, No. 95.

William Feller and George E. Forsythe (1951): New Matrix Transformations for Obtaining Characteristic Vectors, *Quart. Appl. Math.*, **8**:325–331.

Henry E. Fettis (1950): A Method for Obtaining the Characteristic Equation of a Matrix and Computing the Associated Modal Columns, *Quart. Appl. Math.*, **8**:206–212.

Donald A. Flanders and George Shortley (1950): Numerical Determination of Fundamental Modes, *J. Appl. Phys.*, **21**:1326–1332.

A. Fletcher, J. C. P. Miller, and L. Rosenhead (1946): "Index of Mathematical Tables," McGraw-Hill Book Company, Inc., New York, 450 pp.

L. R. Ford (1925): The Solution of Equations by the Method of Successive Approximations, *Am. Math. Monthly*, **32**:272–287.

George E. Forsythe (1951): "Tentative Classification of Methods and Bibliography on Solving Systems of Linear Equations," National Bureau of Standards, INA 52-7 (internal memorandum).

——— (1952): "Bibliographic Survey of Russian Mathematical Monographs, 1930–1951," National Bureau of Standards Report 1628.

——— and Richard A. Leibler (1950): Matrix Inversion by a Monte Carlo Method, *MTAC*, **4**:127–129.

——— and ——— (1951): Correction to the article, "Matrix Inversion by a Monte Carlo Process," *MTAC*, **5**:55.

——— and Theodore S. Motzkin (1952): An Extension of Gauss' Transformation for Improving the Condition of Systems of Linear Equations, *MTAC*, **6**:9–17.

Tomlinson Fort (1948): "Finite Differences and Difference Equations in the Real Domain," Oxford University Press, New York.

R. Fortet (1952): On the Estimation of an Eigenvalue by an Additive Functional of a

Stochastic Process, with Special Reference to the Kac-Donsker Method, *J. Research Nat. Bur. Standards*, **48**:68–75.

L. Fox (1950): Practical Methods for the Solution of Linear Equations and the Inversion of Matrices, *J. Roy. Stat. Soc.*, **B12**:120–136.

——— and J. C. Hayes (1951): More Practical Methods for the Inversion of Matrices, *J. Roy. Stat. Soc.*, **B13**:83–91.

———, H. D. Huskey, and J. H. Wilkinson (1948): Notes on the Solution of Algebraic Linear Simultaneous Equations, *Quart. J. Mech. Appl. Math.*, **1**:149–173.

J. S. Frame (1944): A Variation of Newton's Method, *Am. Math. Monthly*, **51**:36–38.

——— (1949): A Simple Recursion Formula for Inverting a Matrix (Abstract), *Bull. Am. Math. Soc.*, **55**:1045.

R. A. Frazer (1947): Note on the Morris Escalator Process for the Solution of Linear Simultaneous Equations, *Phil. Mag.*, **38**:287–289.

——— and W. J. Duncan (1929): On the Numerical Solution of Equations with Complex Roots, *Proc. Roy. Soc. (London)*, **A125**:68–82.

———, ———, and A. R. Collar (1946): "Elementary Matrices and Some Applications to Dynamics and Differential Equations," The Macmillan Company, New York, xvi + 416 pp.

G. F. Freeman (1943): On the Iterative Solution of Linear Simultaneous Equations, *Phil. Mag.* (7), **34**:409–416.

B. Friedman (1949): Note on Approximating Complex Zeros of a Polynomial, *Communs. Pure Appl. Math.*, **2**:195–208.

Thornton C. Fry (1945): Some Numerical Methods for Locating Roots of Polynomials, *Quart. Appl. Math.*, **3**:89–105.

Eduard Fürstenau (1860): Neue Methode zur Darstellung und Berechnung der imaginären Wurzeln algebraischer Gleichungen durch Determinanten der Coeffizienten, *Ges. Bef. ges. Naturw., Marburg*, **9**:19–48.

A. de la Garza (1951): "An Iterative Method for Solving Systems of Linear Equations," Oak Ridge, K-25 Plant, Report K-731.

N. K. Gavurin (1950): Application of Polynomials of Best Approximation to Optimal Convergence of Iterative Processes (Russian), *Uspekhi Mat. Nauk* 5, **3**(37):156–160.

Wallace Givens (1951): "Computation of Eigenvalues," Oak Ridge National Laboratory (internal memorandum).

——— (1952): Fields of Values of a Matrix, *Proc. Am. Math. Soc.*, **3**:206–209.

James W. Glover (1924): Quadrature Formulae When Ordinates Are Not Equidistant, *Proc. Intern. Math. Congr., Toronto*, 831–835.

Herman H. Goldstine and John von Neumann (1951): Numerical Inverting of Matrices of High Order, II, *Proc. Am. Math. Soc.*, **2**:188–202.

Michael Golomb (1943): Zeros and Poles of Functions Defined by Taylor Series, *Bull. Am. Math. Soc.*, **49**:581–592.

E. T. Goodwin (1950): Note on the Evaluation of Complex Determinants, *Proc. Cambridge Phil. Soc.*, **46**:450–452.

Lawrence M. Graves (1946): "The Theory of Functions of Real Variables," McGraw-Hill Book Company, Inc., New York, x + 300 pp.

Robert E. Greenwood (1949): Numerical Integration for Linear Sums of Exponential Functions, *Ann. Math. Stat.*, **20**:608–611.

——— and Masil B. Danford (1949): Numerical Integration with a Weight Function x, *J. Math. Phys.*, **28**:99–106.

D. P. Grossman (1950): On the Problem of the Numerical Solution of Systems of Simultaneous Linear Algebraic Equations (Russian), *Uspekhi Mat. Nauk* 5, **3**(37):87–103.

P. G. Guest (1950): Orthogonal Polynomials in the Least Squares Fitting of Observations, *Phil. Mag.*, **41**(7):124–137.

——— (1951): The Fitting of Polynomials by the Method of Weighted Grouping, *Ann. Math. Stat.*, **22**:537–548.

Jules Haag (1949): Sur la stabilité des points invariant d'une transformation, *Bull. sci. math.*, **73**:123–134.

J. Hadamard (1892): Essai sur l'étude des fonctions données par leur développement de Taylor, *J. Math.* (4), **8**:101–186.

Hugh J. Hamilton (1946): Roots of Equations by Functional Iteration, *Duke Math. J.*, **13**:113–121.

——— (1950): A Type of Variation on Newton's Method, *Am. Math. Monthly*, **57**:517–522.

J. M. Hammersley (1949): The Numerical Reduction of Nonsingular Matrix Pencils, *Phil. Mag.*, **40**(7):783–807.

Joseph O. Harrison, Jr. (1949): Piecewise Polynomial Approximation for Large-scale Digital Calculators, *MTAC*, **3**:400–407.

——— (1951): "On the Growth of Error in the Numerical Integration of Differential Equations," thesis, Columbia University.

Philip Hartman (1949): Newtonian Approximations to a Zero of a Function, *Comment. Math. Helv.*, **21**:321–326.

D. R. Hartree (1949): Notes on Iterative Processes, *Proc. Cambridge Phil. Soc.*, **45**:230–236.

Paul Herget (1948): "The Computation of Orbits," Edwards Bros., Inc., Ann Arbor, Mich., ix + 177 pp.

M. Herzberger (1949): The Normal Equations of the Method of Least Squares and Their Solution, *Quart. Appl. Math.*, **7**:217–223.

——— and R. H. Morris (1947): A Contribution to the Method of Least Squares, *Quart. Appl. Math.*, **5**:354–357.

Magnus R. Hestenes and William Karush (1951*a*): A Method of Gradients for the Calculation of the Characteristic Roots and Vectors of a Real Symmetric Matrix, *J. Research Nat. Bur. Standards*, **47**:45–61.

——— and ——— (1951*b*): Solutions of $Ax = \lambda Bx$, *J. Research Nat. Bur. Standards*, **47**:471–478.

——— and Marvin L. Stein (1951): "The Solution of Linear Equations by Minimization," National Bureau of Standards, NAML Report 52-45.

——— and Eduard Stiefel (1952): "Method of Conjugate Gradients for Solving Linear Systems," National Bureau of Standards Report 1659.

T. J. Higgins, R. P. Agnew, J. Barkley Rosser, and R. J. Walker (1942): Note on Whittaker's Method for the Roots of a Power Series, *Am. Math. Monthly*, **49**:462–465.

T. H. Hildebrandt and L. M. Graves (1927): Implicit Functions and Their Differentials in General Analysis, *Trans. Am. Math. Soc.*, **29**:127–153.

Jerome Hines (1951): On Approximating the Roots of an Equation by Iteration, *Math. Mag.*, **24**:123–127.

Frank L. Hitchcock (1938): Finding Complex Roots of Algebraic Equations, *J. Math. Phys.*, **17**:55–58.

——— (1939): Algebraic Equations with Complex Coefficients, *J. Math. Phys.*, **18**:202–210.

——— (1944): An Improvement on the G. C. D. Method for Complex Roots, *J. Math. Phys.*, **23**:69–74.

P. G. Hoel (1941): On Methods of Solving Normal Equations, *Ann. Math. Stat.*, **12**:354–359.

——— and D. D. Wall (1947): The Accuracy of the Root-squaring Method for Solving Equations, *J. Math. Phys.*, **26**:156–164.

Hans Hornich (1950): Zur Auflösung von Gleichungssystemen, *Monatsch. Math.*, **54**:130–134.

Paul Horst (1935): A Method of Determining the Coefficients of a Characteristic Equation, *Ann. Math. Stat.*, **6**:83–84.

Harold Hotelling (1936): Simplified Calculation of Principal Components, *Psychometrika*, **1**:27–35.

——— (1943*a*): Some New Methods in Matrix Calculation, *Ann. Math. Stat.*, **14**:1–34.

——— (1943*b*): Further Points on Matrix Calculation and Simultaneous Equations, *Ann. Math. Stat.*, **14**:440–441.

——— (1949): Practical Problems of Matrix Calculation, *Proc. Symposium Math. Stat. Prob. Berkeley*, 275–294.

A. S. Householder (1950): Some Numerical Methods for Solving Systems of Linear Equations, *Am. Math. Monthly*, **57**:453–459.

——— (1951): Polynomial Iterations to Roots of Algebraic Equations, *Proc. Am. Math. Soc.*, **2**:718–719.

———, G. E. Forsythe, and H. H. Germond (eds.) (1951): "Monte Carlo Method," National Bureau of Standards, Applied Mathematics Series 12, vii + 42 pp.

——— and Gale Young (1938): Matrix Approximation and Latent Roots, *Am. Math. Monthly*, **45**:165–171.

P. M. Hummel (1946): The Accuracy of Linear Interpolation, *Am. Math. Monthly*, **53**:364–366.

——— and C. L. Seebeck (1949): A Generalization of Taylor's Expansion, *Am. Math. Monthly*, **56**:243–247.

——— and ——— (1951): A New Interpolation Formula, *Am. Math. Monthly*, **58**:383–389.

Harry D. Huskey and Douglas R. Hartree (1949): On the Precision of a Certain Procedure of Numerical Integration, *J. Research Nat. Bur. Standards*, **42**:57–62.

C. A. Hutchinson (1935): On Graeffe's Method for the Numerical Solution of Algebraic Equations, *Am. Math. Monthly*, **42**:149–161.

S. Inman (1950): The Probability of a Given Error Being Exceeded in Approximate Computation, *Math. Gaz.*, **34**:99–113.

C. Isenkrahe (1888): Ueber die Anwendung iterirter Funktionen zur Darstellung der Wurzeln algebraischer und transcendenter Gleichungen, *Math. Ann.*, **31**:309–317.

V. K. Ivanov (1939): On the Convergence of Iterative Processes for the Solution of Systems of Linear Algebraic Equations (Russian), *Izvest. Akad. Nauk SSSR* (1939):477–483.

Dunham Jackson (1921): The General Theory of Approximation by Polynomials and Trigonometric Sums, *Bull. Am. Math. Soc.*, **27**:415–431.

——— (1930): The Theory of Approximation, *Am. Math. Soc. Colloquium Publs.*, Vol. 11.

C. G. J. Jacobi (1835): Observatiunculae ad theoriam equationum pertinentes, *J. reine angew. Math.*, **13**:340–352.

——— (1846): Über ein leichtes Verfahren die in der Theorie der Säcularstörungen vorkommenden Gleichungen numerisch Aufzulösen, *J. reine angew. Math.*, **30**:51–94.

H. A. Jahn (1948): Improvement of an Approximate Set of Latent Roots and Modal Columns of a Matrix by Methods Akin to Those of Classical Perturbation Theory, *Quart. J. Mech. Appl. Math.*, **1**:131–144.

H. Jensen (1944): Attempt at a Systematic Classification of Some Methods for the Solution of Normal Equations, *Geodaet. Insts. Meddelelsa*, No. 18.

C. W. Jones, J. C. P. Miller, J. F. C. Conn, and R. C. Pankhurst (1945): Tables of Chebyshev Polynomials, *Proc. Roy. Soc. Edinburgh*, **A62**:187–203.

Charles Jordan (1947): "Calculus of Finite Differences," 2d ed., Chelsea Publishing Company, New York, xxi + 652 pp.

W. B. Jordan (1951): An Iterative Process, *MTAC*, **5**:183.

F. Jossa (1940): Risoluzione progressiva di un sistema di equazioni lineari, *Rend. accad. sci. fis. mat. e nat. soc. reale Napoli* (4), **10**:346–352.

Mark Kac (1951): On Some Connections between Probability Theory and Differential and Integral Equations, *Proc. Symposium Math. Stat. Prob., 2d Symposium Berkeley*, 1950:189–215.

——— and Michael Cohen (1952): "A Statistical Method for Determining the Lowest Eigenvalue of Schrödinger's Equation," National Bureau of Standards Report 1553.

——— and M. D. Donsker (1950): A Sampling Method for Determining the Lowest Eigenvalue and the Principal Eigenfunction of Schrödinger's Equation, *J. Research Nat. Bur. Standards*, **44**:551–557.

Herman Kahn (1949): "Modification of the Monte Carlo Method," The RAND Corporation, Report P-132.

——— (1950): Random Sampling (Monte Carlo) Techniques in Neutron Attenuation Problems, *Nucleonics*, **6**: May, 27–33; June, 60–65.

L. V. Kantorovich (1939): The Method of Successive Approximations for Functional Equations, *Acta Math.*, **71**:62–97.

——— (1947): On the Method of Steepest Descent (Russian), *Doklady Akad. Nauk SSSR*, **56**:233–236.

——— (1948*a*): On a General Theory of Methods of Analytical Approximations (Russian), *Doklady Akad. Nauk SSSR*, **60**:957–960.

——— (1948*b*): On the Method of Newton for Functional Equations (Russian), *Doklady Akad. Nauk SSSR*, **59**:1237–1240.

Truman L. Kelley (1935): "Essential Traits of Mental Life," Harvard University Press, Cambridge, Mass., 145 pp.

W. M. Kincaid (1947): Numerical Methods for Finding Characteristic Roots and Vectors of Matrices, *Quart. Appl. Math.*, **5**:320–345.

——— (1948*a*): Note on the Error in Interpolation of a Function of Two Independent Variables, *Ann. Math. Stat.*, **19**:85–88.

——— (1948*b*): Solution of Equations by Interpolation, *Ann. Math. Stat.*, **19**:207–219.

Anna Klingst (1941): Eine Verallgemeinerung der Euler-Maclaurin'schen Reihe und der Bernoulli'schen Zahlen, *Sitzber. Akad. Wiss. Wien.* II*a*, **150**:221–256.

A. Kneschke (1949*a*): Theorie der genäherten Quadratur, *J. reine angew. Math.*, **187**:115–128.

——— (1949*b*): Zur Theorie der Interpolation, *Math. Z.*, **52**:137–149.

Julius König (1876): Ein allgemeiner Ausdruck für die ihren absoluten Betrage nach kleinste Wurzel der Gleichung *n*ten Grades, *Math. Ann.*, **9**:530–540.

——— (1884): Ueber eine Eigenschaft der Potenzreihen, *Math. Ann.*, **23**:447–449.

Walter Kohn (1949): A Variational Iteration Method for Solving Secular Equations, *J. Chem. Phys.*, **17**:670.

Zdenek Kopal, Pierre Carrus, and Katherine E. Kavanagh (1951): A New Formula for Repeated Mechanical Quadratures, *J. Math. Phys.*, **30**:44–48.

Gerhard Kowalewski (1932): "Interpolation und genäherte Quadratur," B. G. Teubner, Leipzig, v + 146 pp.

——— (1938): Entwicklung einer Funktion nach Lagrangeschen Polynomen und ihren Integralen, *Deut. Mathematik*, **3**:275–280.

——— (1948): "Einführung in die Determinantentheorie," 3d ed., Chelsea Publishing Company, New York, 320 pp.

A. N. Krylov (1950): "Lectures on Approximate Computations" (Russian), 5th ed., Moscow, Leningrad, 400 pp.

A. C. Kuroš, A. I. Markusevič, and P. R. Raševskiĭ (ed.) (1948): "Mathematics in SSSR in the 30 years 1917–1947" (Russian), Gosudarstvennoe Izdatel'stvo Tehniko-Teoreticeskiĭ Literatury, Moscow-Leningrad, 1044 pp.

Jack Laderman (1948): The Square Root Method for Solving Simultaneous Linear Equations, *MTAC*, **3**:13–16.

Cornelius Lanczos (1938): Trigonometric Interpolation of Empirical and Analytic Functions, *J. Math. Phys.*, **17**:123–199.

——— (1950): An Iteration Method for the Solution of the Eigenvalue Problem of Linear Differential and Integral Operators, *J. Research Nat. Bur. Standards*, **45**:255–282.

——— (1951): "Solution of Systems of Linear Equations by Minimized Iterations," National Bureau of Standards, NAML Report 52-13.

——— (1952): "Analytical and Practical Curve Fitting of Equidistant Data," National Bureau of Standards Report 1591.

D. H. Lehmer (1945): The Graeffe Process As Applied to Power Series, *MTAC*, **1**:377–383.

D. C. Lewis (1947): Polynomial Least Square Approximations, *Am. Jour. Math.*, **69**:273–278.

V. B. Lidskiĭ (1950): On Proper Values of Sums and Products of Symmetric Matrices (Russian), *Doklady Akad. Nauk SSSR*, **75**:769–772.

H. Liebmann (1918): Die angenäherte Ermittlung harmonischer Funktionen und konformer Abbildung, *Sitzber. math.-naturw. Kl. bayer. Akad. Wiss. München*, 1918:385–416.

Shih-nge Lin (1941): A Method of Successive Approximation of Evaluating the Real and Complex Roots of Cubic and Higher-order Equations, *J. Math. Phys.*, **20**:231–242.

——— (1943): A Method for Finding Roots of Algebraic Equations, *J. Math. Phys.*, **22**:60–77.

A. T. Lonseth (1947): The Propagation of Error in Linear Problems, *Trans. Am. Math. Soc.*, **62**:193–212.

Y. L. Luke and Dolores Ufford (1951): On the Roots of Algebraic Equations, *J. Math. Phys.*, **30**:94–101.

Cyrus Colton MacDuffee (1943): "Vectors and Matrices," The Mathematical Association of America, Menasha, Wis., xi + 192 pp.

——— (1946): "The Theory of Matrices," Chelsea Publishing Company, New York, v + 110 pp.

P. Mansion (1914): Théorème général de Peano sur le reste dans les formules de quadrature, *Mathesis*, **34**:169–174.

Emory McClintock (1895): Theorems in the Calculus of Enlargement, *Am. Jour. Math.*, **17**:69–80.

James McMahon (1894): On the General Term in the Reversion of Series, *Bull. NY Math. Soc.*, **3**:170–172.

N. N. Meiman (1949): Some Questions Relating to the Location of the Zeros of Polynomials (Russian), *Uspekhi Mat. Nauk*, **4**:154–188.

Nicholas Metropolis and S. Ulam (1949): The Monte Carlo Method, *J. Am. Stat. Assoc.*, **44**:335–341.

Franz Meyer (1889): Zur Auflösung der Gleichungen, *Math. Ann.*, **33**:511–524.

Leroy F. Meyers and Arthur Sard (1950*a*): Best Approximate Integration Formulas, *J. Math. Phys.*, **29**:118–123.

——— (1950*b*): Best Interpolation Formulas, *J. Math. Phys.*, **29**:198–206.

J. C. P. Miller (1945): Two Numerical Applications of Chebyshev Polynomials, *Proc. Roy. Soc. Edinburgh*, **A62**:204–210.

——— (1950): Checking by Differences, *MTAC*, **4**:3–11.

William Edmund Milne (1949*a*): "Numerical Calculus. Approximations, Interpolation, Finite Differences, Numerical Integration, and Curve Fitting," Princeton University Press, Princeton, N.J., x + 393 pp.

——— (1949*b*): The Remainder in Linear Methods of Approximation, *J. Research Nat. Bur. Standards*, **43**:501–511.

L. M. Milne-Thompson (1933): "Calculus of Finite Differences," Macmillan & Co., Ltd., London, xxiii + 558 pp.

R. von Mises (1936): Über allgemeine Quadraturformeln, *J. reine angew. Math.*, **174**:56–67.

——— and Hilda Pollaczek-Geiringer (1929): Praktische Verfahren der Gleichungsauflösung, *Z. angew. Math. u. Mechan.*, **9**:58–77, 152–164.

Edward F. Moore (1949): A New General Method for Finding Roots of Polynomial Equations, *MTAC*, **3**:486–488.

Oskar Morgenstern and Max A. Woodbury (1950): The Stability of Inverses of Input-Output Matrices, *Econometrica*, **18**:190–192.

Joseph Morris (1946): An Escalator Process for the Solution of Linear Simultaneous Equations, *Phil. Mag.* (7), **37**:106–120.

——— (1947): "The Escalator Method in Engineering Vibration Problems," John Wiley & Sons, Inc., New York, xv + 270 pp.

——— and J. W. Head (1942): "Lagrangian Frequency Equations. An 'Escalator' Method for Numerical Solution" (appendix by G. Temple), *Aircraft Eng.*, **14**: 312–316.

T. S. Motzkin and J. L. Walsh (1951): "On the Derivative of a Polynomial and Chebyshev Approximation," National Bureau of Standards Report 1444.

M. Müller (1948): Über ein Eulersches Verfahren zur Wurzelberechnung, *Math. Z.*, **51**:474–496.

Thomas Muir (1906, 1911, 1920, 1923): "The Theory of Determinants in the Historical Order of Development," Macmillan & Co., Ltd., London, 1:xi + 491 pp., 2:xvi + 475 pp., 3:xxvi + 503 pp., 4:xxvi + 508 pp.

——— (1930): "Contributions to the History of Determinants, 1900–1920," Blackie & Son, Ltd., Glasgow, xxiii + 408 pp.

——— and W. H. Metzler (1930): "Theory of Determinants," Albany, 606 pp.

Francis J. Murray (1947): "The Theory of Mathematical Machines," King's Crown Press, New York, vii + 116 pp.

——— (1949): Linear Equation Solvers, *Quart. Appl. Math.*, **7**:263–274.

——— (1951): Error Analysis for Mathematical Machines, *Trans. NY Acad. Sci.*, **13**(II):168–174.

H. Nägelsbach (1876, 1877): Studien zur Fürstenau's neuer Methode, *Arch. Math. Phys.*, **59**:147–192; **61**:19–85.

M. Lewis Nelson (1949): "A Monte Carlo Computation Being Made for Neutron Attenuation in Water," Oak Ridge National Laboratory, ORNL-439.

Eugen Netto (1887): Ueber einen Algorithmus zur Auflösung numerischer algebraischer Gleichungen, *Math. Ann.*, **29**:141–147.

——— (1887): Zur Theorie der iterirten Funktionen, *Math. Ann.*, **29**:148–153.

John von Neumann and H. H. Goldstine (1947): Numerical Inverting of Matrices of High Order, *Bull. Am. Math. Soc.*, **53**:1021–1099.

E. H. Neville (1934): Iterative Interpolation, *Ind. Math. Soc. J.*, **20**:87–120.

N. E. Nörlund (1926): "Leçons sur les séries d'interpolation," Gauthier-Villars & Cie, Paris, vii + 236 pp.

Kristen Nygaard (1952): "On the Solution of Integral Equations by Monte-Carlo Methods," Norwegian Defence Research Establishment, Rapport Nr. F-R94.

C. D. Olds (1950): The Best Polynomial Approximation of Functions, *Am. Math. Monthly*, **57**:617–621.

Ascher Opler (1951): Monte Carlo Matrix Calculation with Punched Card Machines, *MTAC*, **5**:115–120.

Alexander Ostrowski (1936): Konvergenzdiskussion und Fehlerabschätzung für die Newton'sche Methode bei Gleichungssystemen, *Comment. Math. Helv.*, **9**:79–103.

——— (1937*a*): Sur la détermination des bornes inférieures pour une classe des déterminants, *Bull. sci. math.* (2), **61**:19–32.

——— (1937*b*): Über die Determinanten mit überwiegender Hauptdiagonale, *Comment. Math. Helv.*, **10**:69–96.

——— (1937*c*): Über die Konvergenz und die Abrundungsfestigkeit des Newtonschen Verfahrens, *Rec. Math.*, **2**:1073–1095.

——— (1938): Über einen Fall der Konvergenz des Newtonschen Näherungsverfahrens, *Rec. Math.*, **3**:254–258.

——— (1940): Recherches sur la méthode de Graeffe et les zéros des polynomes et des series de Laurent, *Acta Math.*, **72**:99–155.

——— (1952): Note on Bounds for Determinants with Dominant Principal Diagonal, *Proc. Am. Math. Soc.*, **3**:26–30.

——— and Olga Taussky (1951): On the Variation of the Determinant of a Positive Definite Matrix, *Koninkl. Ned. Akad. Wetenschap. Proc.*, **54**:383–385.

W. V. Parker (1948): Characteristic Roots and the Field of Values of a Matrix, *Duke Math. J.*, **15**:439–442.

——— (1948): Sets of Complex Numbers Associated with a Matrix, *Duke Math. J.*, **15**:711–715.

——— (1951): Characteristic Roots and Field of Values of a Matrix, *Bull. Am. Math. Soc.*, **57**:103–108.

Maurice Parodi (1949): Sur les limites des modules des racines des équations algébriques, *Bull. sci. math.*, **73**:135–144.

——— (1951): Sur les familles de matrices auxquelles est applicable une méthode d'itération, *Compt. rend.*, **232**:1053–1054.

——— (1952): Sur quelques propriétés des valeurs caractéristiques des matrices carrées, *Mém. sci. math. acad. sci. Paris*, No. 118.

G. Peano (1914): Residuo in formulas de quadratura, *Mathesis*, **34**:1–10.

Karl Pearson (1920): "On the Construction of Tables and on Interpolation, Tracts for Computers II," Cambridge University Press, London, 70 pp.

Hans Petersson (1949): Über Interpolation durch Lösungen linearer Differential Gleichungen, *Abhandl. Math. Sem. Hamburg*, **16**:40–55.

S. Pincherle (1915): Funktional Operationen und Gleichungen, *Enc. Math. Wiss.*, II A, **11**:761–817.

R. Plunkett (1950): On the Convergence of Matrix Iteration Processes, *Quart. Appl. Math.*, **7**:419–421.

G. Polya (1915): Über das Graeffesche Verfahren, *Z. Math. u. Phys.*, **63**:275–290.

A. Porter and C. Mack (1949): New Methods for the Numerical Solution of Algebraic Equations, *Phil. Mag.* (7), **40**:578–585.

G. Baley Price (1951): Bounds for Determinants with Dominant Principal Diagonal, *Proc. Am. Math. Soc.*, **2**:497–502.

"Proceedings of a Second Symposium on Large-scale Digital Calculating Machinery" (1949), Harvard University Press, Cambridge, Mass., 393 pp.

R. Prony (An IV): Essai expérimentale et analytique, *J. polytechnique Cah.*, **2**:24–76.

Hans A. Rademacher (1947): On the Accumulation of Errors in Processes of Integration on High-speed Calculating Machines, *Proc. Symposium Large-scale Digital Calculating Machinery*, 176–185, Harvard University Press, Cambridge, Mass., 1948.

——— and I. J. Schoenberg (1947): An Iteration Method for Calculation with Laurent Series, *Quart. Appl. Math.*, **4**:142–159.

Johann Radon (1935): Restausdrücke bei Interpolations- und Quadraturformeln durch bestimmte Integrale, *Monatsh. Math. u. Phys.*, **42**:389–396.

Raymond Redheffer (1948): Errors in Simultaneous Linear Equations, *Quart. Appl. Math.*, **6**:342–343.

Edgar Reich (1949): On the Convergence of the Classical Iterative Method of Solving Linear Simultaneous Equations, *Ann. Math. Stat.*, **20**:448–451.

Lewis F. Richardson and J. Arthur Gaunt (1927): The Deferred Approach to the Limit, *Phil. Trans. Roy. Soc. (London)*, **A226**:299–361.

H. W. Richmond (1944): On Certain Formulae for Numerical Approximation, *J. London Math. Soc.*, **19**:31–38.

R. D. Richtmyer (1952): "The Evaluation of Definite Integrals, and a Quasi-Monte-Carlo Method Based on the Properties of Algebraic Numbers," Los Alamos Scientific Laboratory Report, LA-1342.

Mari Sofia Roma (1947): Il metodo dell'ortogonalizzazione per la risoluzione numerica dei sistemi di equazioni algebriche, *Riv. Catasto e serviti tecnici Erariali*, No. 1.

Arnold E. Ross (1950): "A General Theory of the Iterative Methods of Solution of Linear Systems," University of Notre Dame, Office of Air Research Contract, Technical Report No. 1.

J. Barkley Rosser (1951): Transformations to Speed the Convergence of Series, *J. Research Nat. Bur. Standards*, **46**:56–64.

———, C. Lanczos, M. R. Hestenes, and W. Karush (1951): The Separation of Close Eigenvalues of a Real Symmetric Matrix, *J. Research Nat. Bur. Standards*, **47**:291–297.

Carl Runge (1885): Entwicklung der Wurzeln einer algebraischen Gleichung in Summen von rationalen Funktionen der Coefficienten, *Acta Math.*, **6**:305–318.

——— and H. König (1924): "Vorlesungen über numerisches Rechnen," Springer-Verlag, Berlin.

——— and F. A. Willers (1915): Numerische und graphische Quadratur und Integration gewohnlicher und partieller Differentialgleichungen, *Enc. Math. Wiss.*, II C, **2**:47–176.

S. Rushton (1951): On Least Squares Fitting by Orthonormal Polynomials Using the Choleski Method, *J. Roy. Stat. Soc.*, **B13**:92–99.

D. H. Sadler (1950): Maximum-interval Tables, *MTAC*, **4**:129–132.

Herbert E. Salzer (1951): Checking and Interpolation of Functions Tabulated at Certain Irregular Logarithmic Intervals, *J. Research Nat. Bur. Standards*, **46**:74–77.

Paul A. Samuelson (1942): A Method of Determining Explicitly the Coefficients of the Characteristic Equation, *Ann. Math. Stat.*, **13**:424–429.

——— (1945): A Convergent Iterative Process, *J. Math. Phys.*, **24**:131–134.

——— (1949): Iterative Computation of Complex Roots, *J. Math. Phys.*, **28**:259–267.

L. Sancery (1862): De la méthode des substitutions successives pour le calcul des racines des équations, *Nouvelles ann. math.* (2), **1**:305–315.

Arthur Sard (1948*a*): Integral Representations of Remainders, *Duke Math. J.*, **15**:333–345.

——— (1948*b*): The Remainder in Approximations by Moving Averages, *Bull. Am. Math. Soc.*, **54**:788–792.

——— (1949*a*): Best Approximate Integration Formulas; Best Approximation Formulas, *Am. J. Math.*, **71**:80–91.

——— (1949*b*): Smoothest Approximation Formulas, *Ann. Math. Stat.*, **20**:612–615.

——— (1951): Remainders: Functions of Several Variables, *Acta Math.*, **84**: 319–346.

James B. Scarborough (1950): "Numerical Mathematical Analysis," 2d ed., Johns Hopkins Press, Baltimore, xviii + 511 pp.

Erhard Schmidt (1908): Über die Auflösung linearer Gleichungen mit unendlich vielen Unbekannten, *Rend. circ. mat. Palermo*, **25**:53–77.

Hermann Schmidt (1950): Über Wurzelapproximation nach Euler und Fixgebilde linearer Transformationen, *Math. Z.*, **52**:547–556.

Robert J. Schmidt (1941): On the Numerical Solution of Linear Simultaneous Equations by an Iterative Method, *Phil. Mag.* (7), **32**:369–383.

——— (1935): Die Allgemeine Newtonsche Quadraturformel und Quadraturformeln für Stieltjesintegrale, *J. reine angew. Math.*, **173**:52–59.

I. J. Schoenberg (1946): Contributions to the Problem of Approximation of Equidistant Data by Analytic Functions, *Quart. Appl. Math.*, **4**:45–99, 112–141.

——— (1952): "On Smoothing Operations and Their Generating Functions," National Bureau of Standards Report 1734.

Ernst Schröder (1870): Über unendliche viele Algorithmen zur Auflösung der Gleichungen, *Math. Ann.*, **2**:317–365.

——— (1871): Ueber iterirte Funktionen, *Math. Ann.*, **3**:296–322.

Gunther Schulz (1933): Iterative Berechnung der reziproken Matrix, *Z. angew. Math. u. Mech.*, **13**:57–59.

——— (1942): Über die Lösung von Gleichungssystemen durch Iteration, *Z. angew. Math. u. Mech.*, **22**:234–235.

Hans Schwerdtfeger (1950): "Introduction to Linear Algebra and the Theory of Matrices," P. Noordhoff N. V., Groningen, 280 pp.

L. Seidel (1874): Über ein Verfahren die Gleichungen, auf Welche die Methode der kleinsten Quadrate führt, sowie lineare Gleichungen überhaupt, durch successive Annäherung aufzulösen, *Abhandl. bayer. Akad. Wiss., Math.-physik. Kl.*, **11**:81–108.

K. A. Semendiaev (1943): "The Determination of Latent Roots and Invariant Manifolds of Matrices by Means of Iterations" (translated by Curtis D. Benster), National Bureau of Standards Report 1402.

W. F. Sheppard (1924): Interpolation with Least Mean Square of Error, *Proc. Intern. Math. Congr. Toronto*, 821–830.

J. Sherman and W. J. Morrison (1949): Adjustment of an Inverse Matrix Corresponding to Changes in the Elements of a Given Column or a Given Row of the Original Matrix, *Ann. Math. Stat.*, **20**:621.

——— (1950): Adjustment of an Inverse Matrix Corresponding to a Change in One Element of a Given Matrix, *Ann. Math. Stat.*, **21**:124.

J. A. Shohat (1929): On a Certain Formula of Mechanical Quadratures with Non-equidistant Ordinates, *Trans. Am. Math. Soc.*, **31**:448–463.

——— and A. V. Bushkovitch (1942): On Some Applications of the Tchebycheff Inequality for Definite Integrals, *J. Math. Phys.*, **21**:211–217.

——— and C. Winston (1934): On Mechanical Quadratures, *Rend. circ. math. Palermo*, **58**:153–165.

Y. A. Šreĭder (1951): The Solution of Systems of Linear Consistent Algebraic Equations (Russian), *Doklady Akad. Nauk SSSR*, **76**:651–654.

J. F. Steffenson (1927): "Interpolation," The Williams & Wilkins Company, Baltimore.

Marvin L. Stein (1952): Gradient Methods in the Solution of Systems of Linear Equations, *J. Research Nat. Bur. Standards*, **48**:407–413.

P. Stein (1951*a*): A Note on Inequalities for the Norm of a Matrix, *Am. Math. Monthly*, **58**:558–559.

——— (1951*b*): The Convergence of Seidel Iterants of Nearly Symmetric Matrices, *MTAC*, **5**:237–240.

——— (1952*a*): A Note on Bounds of Multiple Characteristic Roots of a Matrix, *J. Research Nat. Bur. Standards*, **48**:59–60.

——— (1952*b*): Some General Theorems on Iterants, *J. Research Nat. Bur. Standards*, **48**:82–83.

J. K. Stewart (1951): Another Variation of Newton's Method, *Am. Math. Monthly*, **58**:331–334.

Eduard Stiefel (1952): Über einige Methoden der Relaxationsrechnung, *Z. angew. Math. u. Phys.*, **3**:1–33.

J. L. Synge (1944): A Geometrical Interpretation of the Relaxation Method, *Quart. Appl. Math.*, **2**:87–89.

Olga Taussky (1949): A Recurring Theorem on Determinants, *Am. Math. Monthly*, **56**:672–676.

——— (1950): Notes on Numerical Analysis—2. Note on the Condition of Matrices, *MTAC*, **4**:111–112.

F. Theremin (1855): Recherches sur la résolution des équations de tous les dégrés, *J. reine angew. Math.*, **49**:187–243.

T. N. Thiele (1909): "Interpolationsrechnung," B. G. Teubner, Leipzig, xii + 175 pp.

John Todd (1949*a*): The Condition of a Certain Matrix, *Proc. Cambridge Phil. Soc.*, **46**:116–118.

——— (1949*b*): The Condition of Certain Matrices, I, *Quart. J. Mech. Appl. Math.*, **2**:469–472.

L. B. Tuckerman (1941): On the Mathematically Significant Figures in the Solution of Simultaneous Linear Equations, *Ann. Math. Stat.*, **12**:307–316.

A. M. Turing (1948): Rounding-off Errors in Matrix Processes, *Quart. J. Mech. Appl. Math.*, **1**:287–308.

H. W. Turnbull (1929): "The Theory of Determinants, Matrices, and Invariants," Blackie & Son, Ltd., Glasgow, xvi + 338 pp.

——— and A. C. Aitken (1930): "An Introduction to the Theory of Canonical Matrices," Blackie & Son, Ltd., Glasgow, xiii + 192 pp.

C. de la Vallée Poussin (1952): "Leçons sur l'approximation des fonctions d'une variable réelle," Gauthier-Villars & Cie, Paris, vi + 151 pp.

C. C. Van Orstrand (1910): Reversion of Power Series, *Phil. Mag.* (6), **19**:366–376.

Frank M. Verzuh (1949): The Solution of Simultaneous Linear Equations with the Aid of the 602 Calculating Punch, *MTAC*, **3**:453–462.

Bernard Vinograde (1950): Note on the Escalator Method, *Proc. Am. Math. Soc.*, **1**:162–164.

H. S. Wall (1948): A Modification of Newton's Method, *Am. Math. Monthly*, **55**:90–94.

Wolfgang Wasow (1950): "On the Duration of Random Walks," National Bureau of Standards prepublication copy.

——— (1951*a*): "A Note on the Inversion of Matrices by Random Walks," National Bureau of Standards, NAML Report 52-15.

——— (1951*b*): Random Walks and the Eigenvalues of Elliptic Difference Equations, *J. Research Nat. Bur. Standards*, **46**:65–73.

Harold Wayland (1945): Expansion of Determinantal Equations into Polynomial Form, *Quart. Appl. Math.*, **2**:277–306.

F. Wenzl (1952): Iterationsverfahren zur Berechnung komplexer Nullstellen von Gleichungen, *Z. angew. Math. u. Mech.*, **32**:85–87.

E. T. Whittaker and G. Robinson (1940): "The Calculus of Observations. A Treatise on Numerical Mathematics," 3d ed., Blackie & Son, Ltd., Glasgow, xvi + 395 pp.

Helmut Wielandt (1944): Das Iterationsverfahren bei nicht selbstadjungierten linearen Eigenwertaufgaben, *Math. Z.*, **50**:93–143.

Albert Wilansky (1951): The Row-sums of the Inverse Matrix, *Am. Math. Monthly*, **58**:614–615.

Maurice V. Wilkes, David J. Wheeler, and Stanley Gill (1951): "The Preparation of Programs for an Electronic Digital Computer," Addison-Wesley Press, Inc., Cambridge, 170 pp.

Fr. A. Willers (1950): "Methoden der praktischen Analysis," 2d ed., Walter De Gruyter & Company, Berlin, 410 pp.

C. Winston (1934): On Mechanical Quadratures Formulae Involving the Classical Orthogonal Polynomials, *Ann. Math.*, **35**:658–677.

H. Wittmeyer (1936*a*): Einfluss der Änderung einer Matrix auf die Lösung des zugehörigen Gleichungssystems, sowie auf die characteristischen Zahlen und die Eigenvektoren, *Z. angew. Math. u. Mech.*, **16**:287–300.

——— (1936*b*): Über die Lösung von linearen Gleichungssysteme durch Iteration, *Z. angew. Math. u. Mech.*, **16**:301–310.

J. R. Womersley (1946): Scientific Computing in Great Britain, *MTAC*, **2**:110–117.

Y. K. Wong (1935): An Application of Orthogonalization Process to the Theory of Least Squares, *Ann. Math. Stat.*, **6**:53–75.

Max Woodbury (1950): Inverting Modified Matrices, Memorandum Report 42, Statistical Research Group, Princeton.

F. L. Wren (1937): Neo-Sylvester Contractions and the Solution of Systems of Linear Equations, *Bull. Am. Math. Soc.*, **43**:823–834.

Rudolf Zurmühl (1950): "Matrizen. Eine Darstellung für Ingenieure," Springer-Verlag, Berlin, xv + 427 pp.

PROBLEMS

CHAPTER 1

1. Suppose that a, b, c, and x are digital numbers and that $|ax + b| < |c|$. Assume a machine forming pseudoproducts and pseudoquotients with maximum error ϵ. For calculating

$$y = (ax + b)/c$$

(*a*) If $|a| < |c|$ and $|b| < |c|$, what routine is optimal, and what is the error?
(*b*) If $|a| \geq |c|$ and $|x| \geq |c|$, what error may occur?

2. If

$$y = \sum_{0}^{n} a_i x^{n-i}, \qquad |a| \leq 2^{-1},\ |x| \leq 2^{-1},$$

a_i and x digital, describe a routine producing only digital intermediate quantities and obtain limits for the final error.

3. A table of values of $f(x)$ is to be prepared at equally spaced values of x with values of Δf given to facilitate interpolation. Should one give

$$\Delta f_i^* = f^*(x_{i+1}) - f^*(x_i)$$

or

$$\Delta^* f_i = [f(x_{i+1}) - f(x_i)]^*?$$

That is, should one give the rounded difference of the f's or the difference of the rounded f's?

4. Find the error in the evaluation of

$$2^{-1}(2^{-1} + 2^{-1/2})^{1/2}$$

if the computation makes use of the routine described above for square roots.

5. Obtain formulas for the errors Δy and relative errors $\Delta y/\Delta x$ due to errors Δx for

(*a*) $y = \sin x$,
(*b*) $y = \tan x$,
(*c*) $y = \sec x$,
(*d*) $y = \exp(\alpha x)$,
(*e*) $y = \log x$.

6. Obtain a formula for the error Δx in the solution of the quadratic equation $ax^2 + bx + c = 0$ if the coefficients may be in error by amounts Δa, Δb, Δc.

7. If $f = 1 - x^2/a$, the iteration

$$x_{i+1} = x_i[1 + f(x_i)/2]$$

converges to the square root of a. Devise a routine based upon this iteration, assuming a machine with the same characteristics as described in §1.5, and analyze the errors.

CHAPTER 2

1. Solve in four distinct ways, using at least one direct and at least one iterative method:

$$\begin{aligned}3.2x - 2.0y + 3.9z &= 13.0,\\ 2.1x + 5.1y - 2.9z &= 8.6,\\ 5.9x + 3.0y + 2.2z &= 6.9.\end{aligned}$$

2. If $\mathbf{a}$, $\mathbf{b}$, $\mathbf{c}$, and $\mathbf{d}$ are arbitrary vectors in a plane, show that

$$[\mathbf{b}, \mathbf{c}][\mathbf{a}, \mathbf{d}] + [\mathbf{c}, \mathbf{a}][\mathbf{b}, \mathbf{d}] + [\mathbf{a}, \mathbf{b}][\mathbf{c}, \mathbf{d}] = 0,$$

and write the corresponding determinantal identity.

3. If $\mathbf{a}, \ldots, \mathbf{e}$ are vectors in 3 space, show that

$$[\mathbf{b}, \mathbf{c}, \mathbf{e}][\mathbf{a}, \mathbf{d}, \mathbf{e}] + [\mathbf{c}, \mathbf{a}, \mathbf{e}][\mathbf{b}, \mathbf{d}, \mathbf{e}] + [\mathbf{a}, \mathbf{b}, \mathbf{e}][\mathbf{c}, \mathbf{d}, \mathbf{e}] = 0.$$

4. If $\mathbf{a}_1, \ldots, \mathbf{a}_4$ are linearly independent, then the vectors $\mathbf{a}^i$ satisfying

$$\mathbf{a}^i \cdot \mathbf{a}_j = \delta_{ij}$$

are said to form a set reciprocal to the initial one. Show that

$$\begin{aligned}[\mathbf{a}_1, \ \mathbf{a}_2, \mathbf{a}_3, \ \mathbf{a}_4]:[\mathbf{e}_1, \ \mathbf{e}_2, \mathbf{e}_3, \ \mathbf{e}_4] &= [\mathbf{e}_1, \ \mathbf{e}_2, \mathbf{e}_3, \ \mathbf{e}_4]:[\mathbf{a}^1, \ \mathbf{a}^2, \mathbf{a}^3, \ \mathbf{a}^4]\\ &= [\mathbf{a}_1, \ \mathbf{a}_2, \mathbf{a}_3, \ \mathbf{a}_4]:[\mathbf{e}_1, \ \mathbf{e}_2, \mathbf{e}_3, \ \mathbf{a}^4]\\ &= \cdots\end{aligned}$$

and write the corresponding determinantal identities.

5. If the $\mathbf{e}_i$ form an arbitrary set of (linearly independent) reference vectors, let $\mathbf{e}^i$ represent the reciprocal set. Show that, if G is the matrix of $\mathbf{e}_i \cdot \mathbf{e}_j$, then G^{-1} is the matrix of $\mathbf{e}^i \cdot \mathbf{e}^j$. Also if a' is the covariant representation of $\mathbf{a}$ in the system $\mathbf{e}_i$, then it is the contravariant representation in the system $\mathbf{e}^i$, and conversely.

6. With the system of equations $Ax = y$ of Prob. 1, form the equivalent system $A^{\mathsf{T}}Ax = A^{\mathsf{T}}y$. Solve using (*i*) Seidel iterations, (*ii*) relaxations, (*iii*) triangular factorizations, and (*iv*) the method of Stiefel and Hestenes.

7. Suppose it required to evaluate $a^{\mathsf{T}}x$, where a is a known vector and x satisfies $Ax = y$. Show that in the factorization of the bordered matrix

$$\begin{pmatrix} A & y \\ -a & 0 \end{pmatrix} = L'W',$$

where L' is unit lower triangular and W' upper triangular, $a^{\mathsf{T}}x$ is the element in the lower right-hand corner of W'.

8. In the process of making a triangular factorization of a matrix A, certain quantities may vanish and necessitate a reordering of rows or columns, or both, in order to proceed. Show that the process will go through without such rearrangements if and only if all the following determinants are non-null:

$$\alpha_{11}, \quad \begin{vmatrix} \alpha_{11} & \alpha_{12} \\ \alpha_{21} & \alpha_{22} \end{vmatrix}, \quad \begin{vmatrix} \alpha_{11} & \alpha_{12} & \alpha_{13} \\ \alpha_{21} & \alpha_{22} & \alpha_{23} \\ \alpha_{31} & \alpha_{32} & \alpha_{33} \end{vmatrix}, \quad \ldots .$$

CHAPTER 3

1. If $f(x)$ is of degree $r \leq n$, and

$$g(x) = \prod_{0}^{n} (x - x_i)$$

has no repeated factors, then

$$\frac{f(x)}{g(x)} = \begin{vmatrix} 1 & \cdots & 1 \\ x_0 & \cdots & x_n \\ \cdots & \cdots & \cdots \\ x_0^{n-1} & \cdots & x_n^{n-1} \\ \dfrac{f(x_0)}{x - x_0} & \cdots & \dfrac{f(x_n)}{x - x_n} \end{vmatrix} : \begin{vmatrix} 1 & \cdots & 1 \\ x_0 & \cdots & x_n \\ \cdots & \cdots & \cdots \\ x_0^n & \cdots & x_n^n \end{vmatrix}.$$

2. From Eqs. (3.02.9) express S_i as a determinant in the σ_j and σ_j as a determinant in the S_i.

3. Obtain equations similar to (3.02.5) and (3.02.9) relating the S_i and the s_j. From these obtain determinantal expressions for members of each set in terms of those of the others.

4. If the x_i are distinct, show that

$$\begin{vmatrix} 1 & 1 & 1 \\ x_1 & x_2 & x_3 \\ x_1^3 & x_2^3 & x_3^3 \end{vmatrix} : \begin{vmatrix} 1 & 1 & 1 \\ x_1 & x_2 & x_3 \\ x_1^2 & x_2^2 & x_3^2 \end{vmatrix} = x_1 + x_2 + x_3,$$

$$\begin{vmatrix} 1 & 1 & 1 \\ x_1^2 & x_2^2 & x_3^2 \\ x_1^3 & x_2^3 & x_3^3 \end{vmatrix} : \begin{vmatrix} 1 & 1 & 1 \\ x_1 & x_2 & x_3 \\ x_1^2 & x_2^2 & x_3^2 \end{vmatrix} = x_2x_3 + x_3x_1 + x_1x_2.$$

Generalize for $x_1, x_2, \ldots, x_n$.

5. Evaluate π as the smallest root of $x^{-1} \sin x = 0$ by applying (*i*) Bernoulli's method and (*ii*) Graeffe's method, either to the equation itself or to a suitable transform.

6. Accelerate the convergence of the Bernoulli sequence for the preceding problem by applying the δ^2 process.

7. Form a third-order iteration for $x^3 + x - 1 = 0$.

8. Form a second-order polynomial iteration for the same equation.

CHAPTER 4

1. Find the characteristic equation of the matrix

$$\begin{pmatrix} 3 & 1 & 2 & 4 \\ 7 & 1 & 0 & 1 \\ 2 & 1 & 2 & 3 \\ 4 & 1 & 2 & 2 \end{pmatrix}.$$

2. Evaluate the largest proper value (*s*) of the above matrix by iteration.

3. Apply Lanczos's method (§4.23) to obtain the proper values and proper vectors of the same matrix.

4. Diagonalize by the method of §4.115

$$\begin{pmatrix} 3 & 2 & 1 & 0 \\ 2 & 3 & 2 & 1 \\ 1 & 2 & 3 & 2 \\ 0 & 1 & 2 & 3 \end{pmatrix}.$$

5. Obtain the triple-diagonal form (§4.24) of this matrix.

6. The largest proper value of a certain matrix A of order n is to be evaluated by iteration. Though the sequence $(A^2)^i x$ will give more rapid convergence than the sequence $A^i x$, the formation of A^2 requires n^3 preliminary multiplications. Assuming p iterations of A would be required, for what minimal value of p would it be more efficient first to obtain A^2?

CHAPTER 5

1. If $g(x)$ is of degree $n - 1$ or less, show that

$$\Sigma g(x_i)/\omega'(x_i) = 0.$$

2. Show that $$\Sigma x_i^n/\omega'(x_i) = 1.$$

3. Show that $$\Sigma x_i\omega''(x_i)/\omega'(x_i) = n(n + 1).$$

4. For any integer $\nu > 0$ show that

$$(\nu + 1)L_i^\nu(x_i) = \omega^{(\nu+1)}(x_i)/\omega'(x_i).$$

5. For $x = \cos \phi$ the functions

$$U_{n-1}(x) = 2^{1-n} \sin n\phi \csc \phi$$

are called Chebyshev polynomials of the second kind. Show that they are polynomials, obtain a recursion, and determine their zeros.

6. The values of $f(x_i)$ and their differences are to be tabulated, with the x_i equally spaced, but an erroneous value $f(x_0) + \epsilon$ is entered in place of $f(x_0)$. Show the effect of this error on the values of the successive differences.

7. The trigonometric functions are known exactly for certain values of the argument: 0°, ±30°, ±45°, ... Other values of the sine are to be obtained from these by interpolation. Use an error formula to ascertain how many figures are reliable if the interpolating polynomial is quadratic; if it is cubic.

8. Using the Chebyshev points, form the cubic interpolation polynomial for interpolating values of the sine over one quadrant.

CHAPTER 6

1. Taking $x_i = i$, use the method of §6.111 to construct the polynomials $\tau_r(x)$, $r = 0, 1, \ldots, 6$, orthogonal on the points $-3, -2, \ldots, +3$, with $W = I$.

2. Experimentally measured values y_i of $f(x_i)$ are given at points $x_i = i$. Values $f'(x_i)$ of the derivative are desired. A standard method is based upon finding the polynomial of some degree giving the best least-squares fit and differentiating. The result depends upon the number of points used and upon the degree of the polynomial. As an example, obtain formulas for $f'(0)$ in terms of $y_{-3}, y_{-2}, \ldots, y_3$.

3. An approximation of the form (6.0.1) to $f(x)$ is required giving the best least-squares fit to the data, subject to the restriction that the vector c of the coefficients γ_i is constrained to satisfy

$$Bc = z$$

exactly (neglecting rounding errors). Show that, with the auxiliary vector w, c is determined by the system

$$F^\mathsf{T}Fc + B^\mathsf{T}w = F^\mathsf{T}y,$$
$$Bc = z.$$

4. Obtain the expansion (6.2.2) for

$$f(x) = \cos (2 + x)\pi/8.$$

5. Find the best linear polynomial and the best quadratic polynomial, in the sense of least squares, for fitting the data

x	0	2	6	10	15	18	21
y	14.0	13.0	10.7	8.0	5.0	2.9	1.0

and find the residuals for each.

6. If $f(x)$ is odd with the period 2π, find the approximating trigonometric function given

x	0	$\pi/6$	$\pi/3$	$\pi/2$	$2\pi/3$	$5\pi/6$	π
y	0	2.5	4	4.5	4	2.5	0.

7. If $xu = \log(1+x)$, then $e^{xu} = 1 + x$. Consider this as an equation in u for fixed x. Then the equation

$$f(x) \equiv (1 + x - e^{xu})/x = 0$$

can be solved for u by the method of §3.2, yielding a sequence of rational fractions in x approximating $\log(1+x)$. Obtain the fifth term in this sequence. For what values of x does the sequence converge?

CHAPTER 7

1. Apply the Euler-Maclaurin formula (§7.36) to show that, if p is a positive integer,

$$1^p + 2^p + \cdots + n^p = n^{p+1}/(p+1) + n^p/2 + \sum_{\nu=2} B_\nu p(p-1)\cdots(p-\nu+2)n^{p-\nu+1}/\nu!.$$

2. Give a direct derivation of the recursion defined by Eqs. (7.1.18) and (7.1.19).

3. For $b = -a = 1$, $w(x) = 1$, calculate ω_0, ω_1, ω_2, ω_3, and the λ's associated with each.

4. Do likewise with $a = 0$, $b = \infty$, and $w(x) = e^{-x}$.

5. If $x_i = x_0 + ih$, $i = -3, -2, \ldots, +3$, obtain a formula of the form

$$\int_{x_{-3}}^{x_3} f(x)dx = \lambda_0 f_0 + \lambda_2(f_{-2} + f_2) + \lambda_3(f_{-3} + b_3) + R,$$

with R vanishing for polynomials up to a degree as high as possible, and find R in general.

6. By the method outlined in §7.3, obtain explicit expansions in terms of central differences for derivatives up to the fifth of $f(x)$ at $x = x_0$.

7. Let

$$I(h) = \int_a^b f(x)dx - R$$

represent the result of applying a numerical quadrature formula based on equally spaced abscissas (*e.g.*, Simpson's rule) to the evaluation of the integral of a particular function $f(x)$ between fixed limits. Show that I is an even function of h, expressible in the form

$$I(h) = I_0 + I_1h^2 + I_2h^4 + \cdots,$$

where I_0 is the exact value of the integral. Hence derive a formula for an improved approximation to I_0, given $I(h)$ and $I(h/2)$.

8. Obtain $\int_0^1 (1 + x)^{1/2} dx$ numerically using Gauss's method with a polynomial of third degree, and compare the result with the true value.

9. Use Simpson's rule with four subintervals to evaluate

$$\int_2^3 dx/\log x.$$

CHAPTER 8

1. Obtain a Monte Carlo estimate of the value of the integral in Prob. 8, Chap. 7. (Note that $y > (1 + x)^{1/2}$ is equivalent to $y^2 > 1 + x$.)

INDEX

A CATALOG OF SELECTED

DOVER BOOKS

IN SCIENCE AND MATHEMATICS

Astronomy

BURNHAM'S CELESTIAL HANDBOOK, Robert Burnham, Jr. Thorough guide to the stars beyond our solar system. Exhaustive treatment. Alphabetical by constellation: Andromeda to Cetus in Vol. 1; Chamaeleon to Orion in Vol. 2; and Pavo to Vulpecula in Vol. 3. Hundreds of illustrations. Index in Vol. 3. 2,000pp. 6⅛ x 9¼.
Vol. I: 0-486-23567-X
Vol. II: 0-486-23568-8
Vol. III: 0-486-23673-0

EXPLORING THE MOON THROUGH BINOCULARS AND SMALL TELESCOPES, Ernest H. Cherrington, Jr. Informative, profusely illustrated guide to locating and identifying craters, rills, seas, mountains, other lunar features. Newly revised and updated with special section of new photos. Over 100 photos and diagrams. 240pp. 8¼ x 11. 0-486-24491-1

THE EXTRATERRESTRIAL LIFE DEBATE, 1750–1900, Michael J. Crowe. First detailed, scholarly study in English of the many ideas that developed from 1750 to 1900 regarding the existence of intelligent extraterrestrial life. Examines ideas of Kant, Herschel, Voltaire, Percival Lowell, many other scientists and thinkers. 16 illustrations. 704pp. 5⅜ x 8½. 0-486-40675-X

THEORIES OF THE WORLD FROM ANTIQUITY TO THE COPERNICAN REVOLUTION, Michael J. Crowe. Newly revised edition of an accessible, enlightening book recreates the change from an earth-centered to a sun-centered conception of the solar system. 242pp. 5⅜ x 8½. 0-486-41444-2

A HISTORY OF ASTRONOMY, A. Pannekoek. Well-balanced, carefully reasoned study covers such topics as Ptolemaic theory, work of Copernicus, Kepler, Newton, Eddington's work on stars, much more. Illustrated. References. 521pp. 5⅜ x 8½.
0-486-65994-1

A COMPLETE MANUAL OF AMATEUR ASTRONOMY: TOOLS AND TECHNIQUES FOR ASTRONOMICAL OBSERVATIONS, P. Clay Sherrod with Thomas L. Koed. Concise, highly readable book discusses: selecting, setting up and maintaining a telescope; amateur studies of the sun; lunar topography and occultations; observations of Mars, Jupiter, Saturn, the minor planets and the stars; an introduction to photoelectric photometry; more. 1981 ed. 124 figures. 25 halftones. 37 tables. 335pp. 6½ x 9¼. 0-486-40675-X

AMATEUR ASTRONOMER'S HANDBOOK, J. B. Sidgwick. Timeless, comprehensive coverage of telescopes, mirrors, lenses, mountings, telescope drives, micrometers, spectroscopes, more. 189 illustrations. 576pp. 5⅝ x 8¼. (Available in U.S. only.)
0-486-24034-7

STARS AND RELATIVITY, Ya. B. Zel'dovich and I. D. Novikov. Vol. 1 of *Relativistic Astrophysics* by famed Russian scientists. General relativity, properties of matter under astrophysical conditions, stars, and stellar systems. Deep physical insights, clear presentation. 1971 edition. References. 544pp. 5⅝ x 8¼. 0-486-69424-0

Chemistry

THE SCEPTICAL CHYMIST: THE CLASSIC 1661 TEXT, Robert Boyle. Boyle defines the term "element," asserting that all natural phenomena can be explained by the motion and organization of primary particles. 1911 ed. viii+232pp. 5⅜ x 8½. 0-486-42825-7

RADIOACTIVE SUBSTANCES, Marie Curie. Here is the celebrated scientist's doctoral thesis, the prelude to her receipt of the 1903 Nobel Prize. Curie discusses establishing atomic character of radioactivity found in compounds of uranium and thorium; extraction from pitchblende of polonium and radium; isolation of pure radium chloride; determination of atomic weight of radium; plus electric, photographic, luminous, heat, color effects of radioactivity. ii+94pp. 5⅜ x 8½. 0-486-42550-9

CHEMICAL MAGIC, Leonard A. Ford. Second Edition, Revised by E. Winston Grundmeier. Over 100 unusual stunts demonstrating cold fire, dust explosions, much more. Text explains scientific principles and stresses safety precautions. 128pp. 5⅜ x 8½. 0-486-67628-5

THE DEVELOPMENT OF MODERN CHEMISTRY, Aaron J. Ihde. Authoritative history of chemistry from ancient Greek theory to 20th-century innovation. Covers major chemists and their discoveries. 209 illustrations. 14 tables. Bibliographies. Indices. Appendices. 851pp. 5⅜ x 8½. 0-486-64235-6

CATALYSIS IN CHEMISTRY AND ENZYMOLOGY, William P. Jencks. Exceptionally clear coverage of mechanisms for catalysis, forces in aqueous solution, carbonyl- and acyl-group reactions, practical kinetics, more. 864pp. 5⅜ x 8½. 0-486-65460-5

ELEMENTS OF CHEMISTRY, Antoine Lavoisier. Monumental classic by founder of modern chemistry in remarkable reprint of rare 1790 Kerr translation. A must for every student of chemistry or the history of science. 539pp. 5⅜ x 8½. 0-486-64624-6

THE HISTORICAL BACKGROUND OF CHEMISTRY, Henry M. Leicester. Evolution of ideas, not individual biography. Concentrates on formulation of a coherent set of chemical laws. 260pp. 5⅜ x 8½. 0-486-61053-5

A SHORT HISTORY OF CHEMISTRY, J. R. Partington. Classic exposition explores origins of chemistry, alchemy, early medical chemistry, nature of atmosphere, theory of valency, laws and structure of atomic theory, much more. 428pp. 5⅜ x 8½. (Available in U.S. only.) 0-486-65977-1

GENERAL CHEMISTRY, Linus Pauling. Revised 3rd edition of classic first-year text by Nobel laureate. Atomic and molecular structure, quantum mechanics, statistical mechanics, thermodynamics correlated with descriptive chemistry. Problems. 992pp. 5⅜ x 8½. 0-486-65622-5

FROM ALCHEMY TO CHEMISTRY, John Read. Broad, humanistic treatment focuses on great figures of chemistry and ideas that revolutionized the science. 50 illustrations. 240pp. 5⅜ x 8½. 0-486-28690-8

Engineering

DE RE METALLICA, Georgius Agricola. The famous Hoover translation of greatest treatise on technological chemistry, engineering, geology, mining of early modern times (1556). All 289 original woodcuts. 638pp. 6¾ x 11. 0-486-60006-8

FUNDAMENTALS OF ASTRODYNAMICS, Roger Bate et al. Modern approach developed by U.S. Air Force Academy. Designed as a first course. Problems, exercises. Numerous illustrations. 455pp. 5⅜ x 8½. 0-486-60061-0

DYNAMICS OF FLUIDS IN POROUS MEDIA, Jacob Bear. For advanced students of ground water hydrology, soil mechanics and physics, drainage and irrigation engineering and more. 335 illustrations. Exercises, with answers. 784pp. 6⅛ x 9¼. 0-486-65675-6

THEORY OF VISCOELASTICITY (Second Edition), Richard M. Christensen. Complete consistent description of the linear theory of the viscoelastic behavior of materials. Problem-solving techniques discussed. 1982 edition. 29 figures. xiv+364pp. 6⅛ x 9¼. 0-486-42880-X

MECHANICS, J. P. Den Hartog. A classic introductory text or refresher. Hundreds of applications and design problems illuminate fundamentals of trusses, loaded beams and cables, etc. 334 answered problems. 462pp. 5⅜ x 8½. 0-486-60754-2

MECHANICAL VIBRATIONS, J. P. Den Hartog. Classic textbook offers lucid explanations and illustrative models, applying theories of vibrations to a variety of practical industrial engineering problems. Numerous figures. 233 problems, solutions. Appendix. Index. Preface. 436pp. 5⅜ x 8½. 0-486-64785-4

STRENGTH OF MATERIALS, J. P. Den Hartog. Full, clear treatment of basic material (tension, torsion, bending, etc.) plus advanced material on engineering methods, applications. 350 answered problems. 323pp. 5⅜ x 8½. 0-486-60755-0

A HISTORY OF MECHANICS, René Dugas. Monumental study of mechanical principles from antiquity to quantum mechanics. Contributions of ancient Greeks, Galileo, Leonardo, Kepler, Lagrange, many others. 671pp. 5⅜ x 8½. 0-486-65632-2

STABILITY THEORY AND ITS APPLICATIONS TO STRUCTURAL MECHANICS, Clive L. Dym. Self-contained text focuses on Koiter postbuckling analyses, with mathematical notions of stability of motion. Basing minimum energy principles for static stability upon dynamic concepts of stability of motion, it develops asymptotic buckling and postbuckling analyses from potential energy considerations, with applications to columns, plates, and arches. 1974 ed. 208pp. 5⅜ x 8½. 0-486-42541-X

METAL FATIGUE, N. E. Frost, K. J. Marsh, and L. P. Pook. Definitive, clearly written, and well-illustrated volume addresses all aspects of the subject, from the historical development of understanding metal fatigue to vital concepts of the cyclic stress that causes a crack to grow. Includes 7 appendixes. 544pp. 5⅜ x 8½. 0-486-40927-9

ROCKETS, Robert Goddard. Two of the most significant publications in the history of rocketry and jet propulsion: "A Method of Reaching Extreme Altitudes" (1919) and "Liquid Propellant Rocket Development" (1936). 128pp. 5⅜ x 8½. 0-486-42537-1

STATISTICAL MECHANICS: PRINCIPLES AND APPLICATIONS, Terrell L. Hill. Standard text covers fundamentals of statistical mechanics, applications to fluctuation theory, imperfect gases, distribution functions, more. 448pp. 5⅜ x 8½. 0-486-65390-0

ENGINEERING AND TECHNOLOGY 1650–1750: ILLUSTRATIONS AND TEXTS FROM ORIGINAL SOURCES, Martin Jensen. Highly readable text with more than 200 contemporary drawings and detailed engravings of engineering projects dealing with surveying, leveling, materials, hand tools, lifting equipment, transport and erection, piling, bailing, water supply, hydraulic engineering, and more. Among the specific projects outlined-transporting a 50-ton stone to the Louvre, erecting an obelisk, building timber locks, and dredging canals. 207pp. 8⅜ x 11¼. 0-486-42232-1

THE VARIATIONAL PRINCIPLES OF MECHANICS, Cornelius Lanczos. Graduate level coverage of calculus of variations, equations of motion, relativistic mechanics, more. First inexpensive paperbound edition of classic treatise. Index. Bibliography. 418pp. 5⅜ x 8½. 0-486-65067-7

PROTECTION OF ELECTRONIC CIRCUITS FROM OVERVOLTAGES, Ronald B. Standler. Five-part treatment presents practical rules and strategies for circuits designed to protect electronic systems from damage by transient overvoltages. 1989 ed. xxiv+434pp. 6⅛ x 9¼. 0-486-42552-5

ROTARY WING AERODYNAMICS, W. Z. Stepniewski. Clear, concise text covers aerodynamic phenomena of the rotor and offers guidelines for helicopter performance evaluation. Originally prepared for NASA. 537 figures. 640pp. 6⅛ x 9¼. 0-486-64647-5

INTRODUCTION TO SPACE DYNAMICS, William Tyrrell Thomson. Comprehensive, classic introduction to space-flight engineering for advanced undergraduate and graduate students. Includes vector algebra, kinematics, transformation of coordinates. Bibliography. Index. 352pp. 5⅜ x 8½. 0-486-65113-4

HISTORY OF STRENGTH OF MATERIALS, Stephen P. Timoshenko. Excellent historical survey of the strength of materials with many references to the theories of elasticity and structure. 245 figures. 452pp. 5⅜ x 8½. 0-486-61187-6

ANALYTICAL FRACTURE MECHANICS, David J. Unger. Self-contained text supplements standard fracture mechanics texts by focusing on analytical methods for determining crack-tip stress and strain fields. 336pp. 6⅛ x 9¼. 0-486-41737-9

STATISTICAL MECHANICS OF ELASTICITY, J. H. Weiner. Advanced, self-contained treatment illustrates general principles and elastic behavior of solids. Part 1, based on classical mechanics, studies thermoelastic behavior of crystalline and polymeric solids. Part 2, based on quantum mechanics, focuses on interatomic force laws, behavior of solids, and thermally activated processes. For students of physics and chemistry and for polymer physicists. 1983 ed. 96 figures. 496pp. 5⅜ x 8½. 0-486-42260-7

Mathematics

FUNCTIONAL ANALYSIS (Second Corrected Edition), George Bachman and Lawrence Narici. Excellent treatment of subject geared toward students with background in linear algebra, advanced calculus, physics and engineering. Text covers introduction to inner-product spaces, normed, metric spaces, and topological spaces; complete orthonormal sets, the Hahn-Banach Theorem and its consequences, and many other related subjects. 1966 ed. 544pp. 6⅛ x 9¼. 0-486-40251-7

ASYMPTOTIC EXPANSIONS OF INTEGRALS, Norman Bleistein & Richard A. Handelsman. Best introduction to important field with applications in a variety of scientific disciplines. New preface. Problems. Diagrams. Tables. Bibliography. Index. 448pp. 5⅜ x 8½. 0-486-65082-0

VECTOR AND TENSOR ANALYSIS WITH APPLICATIONS, A. I. Borisenko and I. E. Tarapov. Concise introduction. Worked-out problems, solutions, exercises. 257pp. 5⅝ x 8¼. 0-486-63833-2

AN INTRODUCTION TO ORDINARY DIFFERENTIAL EQUATIONS, Earl A. Coddington. A thorough and systematic first course in elementary differential equations for undergraduates in mathematics and science, with many exercises and problems (with answers). Index. 304pp. 5⅜ x 8½. 0-486-65942-9

FOURIER SERIES AND ORTHOGONAL FUNCTIONS, Harry F. Davis. An incisive text combining theory and practical example to introduce Fourier series, orthogonal functions and applications of the Fourier method to boundary-value problems. 570 exercises. Answers and notes. 416pp. 5⅜ x 8½. 0-486-65973-9

COMPUTABILITY AND UNSOLVABILITY, Martin Davis. Classic graduate-level introduction to theory of computability, usually referred to as theory of recurrent functions. New preface and appendix. 288pp. 5⅜ x 8½. 0-486-61471-9

ASYMPTOTIC METHODS IN ANALYSIS, N. G. de Bruijn. An inexpensive, comprehensive guide to asymptotic methods–the pioneering work that teaches by explaining worked examples in detail. Index. 224pp. 5⅜ x 8½ 0-486-64221-6

APPLIED COMPLEX VARIABLES, John W. Dettman. Step-by-step coverage of fundamentals of analytic function theory–plus lucid exposition of five important applications: Potential Theory; Ordinary Differential Equations; Fourier Transforms; Laplace Transforms; Asymptotic Expansions. 66 figures. Exercises at chapter ends. 512pp. 5⅜ x 8½. 0-486-64670-X

INTRODUCTION TO LINEAR ALGEBRA AND DIFFERENTIAL EQUATIONS, John W. Dettman. Excellent text covers complex numbers, determinants, orthonormal bases, Laplace transforms, much more. Exercises with solutions. Undergraduate level. 416pp. 5⅜ x 8½. 0-486-65191-6

RIEMANN'S ZETA FUNCTION, H. M. Edwards. Superb, high-level study of landmark 1859 publication entitled "On the Number of Primes Less Than a Given Magnitude" traces developments in mathematical theory that it inspired. xiv+315pp. 5⅜ x 8½. 0-486-41740-9

CALCULUS OF VARIATIONS WITH APPLICATIONS, George M. Ewing. Applications-oriented introduction to variational theory develops insight and promotes understanding of specialized books, research papers. Suitable for advanced undergraduate/graduate students as primary, supplementary text. 352pp. 5⅜ x 8½. 0-486-64856-7

COMPLEX VARIABLES, Francis J. Flanigan. Unusual approach, delaying complex algebra till harmonic functions have been analyzed from real variable viewpoint. Includes problems with answers. 364pp. 5⅜ x 8½. 0-486-61388-7

AN INTRODUCTION TO THE CALCULUS OF VARIATIONS, Charles Fox. Graduate-level text covers variations of an integral, isoperimetrical problems, least action, special relativity, approximations, more. References. 279pp. 5⅜ x 8½. 0-486-65499-0

COUNTEREXAMPLES IN ANALYSIS, Bernard R. Gelbaum and John M. H. Olmsted. These counterexamples deal mostly with the part of analysis known as "real variables." The first half covers the real number system, and the second half encompasses higher dimensions. 1962 edition. xxiv+198pp. 5⅜ x 8½. 0-486-42875-3

CATASTROPHE THEORY FOR SCIENTISTS AND ENGINEERS, Robert Gilmore. Advanced-level treatment describes mathematics of theory grounded in the work of Poincaré, R. Thom, other mathematicians. Also important applications to problems in mathematics, physics, chemistry and engineering. 1981 edition. References. 28 tables. 397 black-and-white illustrations. xvii + 666pp. 6⅛ x 9¼. 0-486-67539-4

INTRODUCTION TO DIFFERENCE EQUATIONS, Samuel Goldberg. Exceptionally clear exposition of important discipline with applications to sociology, psychology, economics. Many illustrative examples; over 250 problems. 260pp. 5⅜ x 8½. 0-486-65084-7

NUMERICAL METHODS FOR SCIENTISTS AND ENGINEERS, Richard Hamming. Classic text stresses frequency approach in coverage of algorithms, polynomial approximation, Fourier approximation, exponential approximation, other topics. Revised and enlarged 2nd edition. 721pp. 5⅜ x 8½. 0-486-65241-6

INTRODUCTION TO NUMERICAL ANALYSIS (2nd Edition), F. B. Hildebrand. Classic, fundamental treatment covers computation, approximation, interpolation, numerical differentiation and integration, other topics. 150 new problems. 669pp. 5⅜ x 8½. 0-486-65363-3

THREE PEARLS OF NUMBER THEORY, A. Y. Khinchin. Three compelling puzzles require proof of a basic law governing the world of numbers. Challenges concern van der Waerden's theorem, the Landau-Schnirelmann hypothesis and Mann's theorem, and a solution to Waring's problem. Solutions included. 64pp. 5⅜ x 8½. 0-486-40026-3

THE PHILOSOPHY OF MATHEMATICS: AN INTRODUCTORY ESSAY, Stephan Körner. Surveys the views of Plato, Aristotle, Leibniz & Kant concerning propositions and theories of applied and pure mathematics. Introduction. Two appendices. Index. 198pp. 5⅜ x 8½. 0-486-25048-2

INTRODUCTORY REAL ANALYSIS, A.N. Kolmogorov, S. V. Fomin. Translated by Richard A. Silverman. Self-contained, evenly paced introduction to real and functional analysis. Some 350 problems. 403pp. 5⅜ x 8½. 0-486-61226-0

APPLIED ANALYSIS, Cornelius Lanczos. Classic work on analysis and design of finite processes for approximating solution of analytical problems. Algebraic equations, matrices, harmonic analysis, quadrature methods, much more. 559pp. 5⅜ x 8½. 0-486-65656-X

AN INTRODUCTION TO ALGEBRAIC STRUCTURES, Joseph Landin. Superb self-contained text covers "abstract algebra": sets and numbers, theory of groups, theory of rings, much more. Numerous well-chosen examples, exercises. 247pp. 5⅜ x 8½. 0-486-65940-2

QUALITATIVE THEORY OF DIFFERENTIAL EQUATIONS, V. V. Nemytskii and V.V. Stepanov. Classic graduate-level text by two prominent Soviet mathematicians covers classical differential equations as well as topological dynamics and ergodic theory. Bibliographies. 523pp. 5⅜ x 8½. 0-486-65954-2

THEORY OF MATRICES, Sam Perlis. Outstanding text covering rank, nonsingularity and inverses in connection with the development of canonical matrices under the relation of equivalence, and without the intervention of determinants. Includes exercises. 237pp. 5⅜ x 8½. 0-486-66810-X

INTRODUCTION TO ANALYSIS, Maxwell Rosenlicht. Unusually clear, accessible coverage of set theory, real number system, metric spaces, continuous functions, Riemann integration, multiple integrals, more. Wide range of problems. Undergraduate level. Bibliography. 254pp. 5⅜ x 8½. 0-486-65038-3

MODERN NONLINEAR EQUATIONS, Thomas L. Saaty. Emphasizes practical solution of problems; covers seven types of equations. ". . . a welcome contribution to the existing literature...."–*Math Reviews*. 490pp. 5⅜ x 8½. 0-486-64232-1

MATRICES AND LINEAR ALGEBRA, Hans Schneider and George Phillip Barker. Basic textbook covers theory of matrices and its applications to systems of linear equations and related topics such as determinants, eigenvalues and differential equations. Numerous exercises. 432pp. 5⅜ x 8½. 0-486-66014-1

LINEAR ALGEBRA, Georgi E. Shilov. Determinants, linear spaces, matrix algebras, similar topics. For advanced undergraduates, graduates. Silverman translation. 387pp. 5⅜ x 8½. 0-486-63518-X

ELEMENTS OF REAL ANALYSIS, David A. Sprecher. Classic text covers fundamental concepts, real number system, point sets, functions of a real variable, Fourier series, much more. Over 500 exercises. 352pp. 5⅜ x 8½. 0-486-65385-4

SET THEORY AND LOGIC, Robert R. Stoll. Lucid introduction to unified theory of mathematical concepts. Set theory and logic seen as tools for conceptual understanding of real number system. 496pp. 5⅜ x 8¼. 0-486-63829-4

TENSOR CALCULUS, J.L. Synge and A. Schild. Widely used introductory text covers spaces and tensors, basic operations in Riemannian space, non-Riemannian spaces, etc. 324pp. 5⅜ x 8¼. 0-486-63612-7

ORDINARY DIFFERENTIAL EQUATIONS, Morris Tenenbaum and Harry Pollard. Exhaustive survey of ordinary differential equations for undergraduates in mathematics, engineering, science. Thorough analysis of theorems. Diagrams. Bibliography. Index. 818pp. 5⅜ x 8½. 0-486-64940-7

INTEGRAL EQUATIONS, F. G. Tricomi. Authoritative, well-written treatment of extremely useful mathematical tool with wide applications. Volterra Equations, Fredholm Equations, much more. Advanced undergraduate to graduate level. Exercises. Bibliography. 238pp. 5⅜ x 8½. 0-486-64828-1

FOURIER SERIES, Georgi P. Tolstov. Translated by Richard A. Silverman. A valuable addition to the literature on the subject, moving clearly from subject to subject and theorem to theorem. 107 problems, answers. 336pp. 5⅜ x 8½. 0-486-63317-9

INTRODUCTION TO MATHEMATICAL THINKING, Friedrich Waismann. Examinations of arithmetic, geometry, and theory of integers; rational and natural numbers; complete induction; limit and point of accumulation; remarkable curves; complex and hypercomplex numbers, more. 1959 ed. 27 figures. xii+260pp. 5⅜ x 8½. 0-486-63317-9

POPULAR LECTURES ON MATHEMATICAL LOGIC, Hao Wang. Noted logician's lucid treatment of historical developments, set theory, model theory, recursion theory and constructivism, proof theory, more. 3 appendixes. Bibliography. 1981 edition. ix + 283pp. 5⅜ x 8½. 0-486-67632-3

CALCULUS OF VARIATIONS, Robert Weinstock. Basic introduction covering isoperimetric problems, theory of elasticity, quantum mechanics, electrostatics, etc. Exercises throughout. 326pp. 5⅜ x 8½. 0-486-63069-2

THE CONTINUUM: A CRITICAL EXAMINATION OF THE FOUNDATION OF ANALYSIS, Hermann Weyl. Classic of 20th-century foundational research deals with the conceptual problem posed by the continuum. 156pp. 5⅜ x 8½. 0-486-67982-9

CHALLENGING MATHEMATICAL PROBLEMS WITH ELEMENTARY SOLUTIONS, A. M. Yaglom and I. M. Yaglom. Over 170 challenging problems on probability theory, combinatorial analysis, points and lines, topology, convex polygons, many other topics. Solutions. Total of 445pp. 5⅜ x 8½. Two-vol. set. Vol. I: 0-486-65536-9 Vol. II: 0-486-65537-7

INTRODUCTION TO PARTIAL DIFFERENTIAL EQUATIONS WITH APPLICATIONS, E. C. Zachmanoglou and Dale W. Thoe. Essentials of partial differential equations applied to common problems in engineering and the physical sciences. Problems and answers. 416pp. 5⅜ x 8½. 0-486-65251-3

THE THEORY OF GROUPS, Hans J. Zassenhaus. Well-written graduate-level text acquaints reader with group-theoretic methods and demonstrates their usefulness in mathematics. Axioms, the calculus of complexes, homomorphic mapping, p-group theory, more. 276pp. 5⅜ x 8½. 0-486-40922-8

Math–Decision Theory, Statistics, Probability

ELEMENTARY DECISION THEORY, Herman Chernoff and Lincoln E. Moses. Clear introduction to statistics and statistical theory covers data processing, probability and random variables, testing hypotheses, much more. Exercises. 364pp. 5⅜ x 8½. 0-486-65218-1

STATISTICS MANUAL, Edwin L. Crow et al. Comprehensive, practical collection of classical and modern methods prepared by U.S. Naval Ordnance Test Station. Stress on use. Basics of statistics assumed. 288pp. 5⅜ x 8½. 0-486-60599-X

SOME THEORY OF SAMPLING, William Edwards Deming. Analysis of the problems, theory and design of sampling techniques for social scientists, industrial managers and others who find statistics important at work. 61 tables. 90 figures. xvii +602pp. 5⅜ x 8½. 0-486-64684-X

LINEAR PROGRAMMING AND ECONOMIC ANALYSIS, Robert Dorfman, Paul A. Samuelson and Robert M. Solow. First comprehensive treatment of linear programming in standard economic analysis. Game theory, modern welfare economics, Leontief input-output, more. 525pp. 5⅜ x 8½. 0-486-65491-5

PROBABILITY: AN INTRODUCTION, Samuel Goldberg. Excellent basic text covers set theory, probability theory for finite sample spaces, binomial theorem, much more. 360 problems. Bibliographies. 322pp. 5⅜ x 8½. 0-486-65252-1

GAMES AND DECISIONS: INTRODUCTION AND CRITICAL SURVEY, R. Duncan Luce and Howard Raiffa. Superb nontechnical introduction to game theory, primarily applied to social sciences. Utility theory, zero-sum games, n-person games, decision-making, much more. Bibliography. 509pp. 5⅜ x 8½. 0-486-65943-7

INTRODUCTION TO THE THEORY OF GAMES, J. C. C. McKinsey. This comprehensive overview of the mathematical theory of games illustrates applications to situations involving conflicts of interest, including economic, social, political, and military contexts. Appropriate for advanced undergraduate and graduate courses; advanced calculus a prerequisite. 1952 ed. x+372pp. 5⅜ x 8½. 0-486-42811-7

FIFTY CHALLENGING PROBLEMS IN PROBABILITY WITH SOLUTIONS, Frederick Mosteller. Remarkable puzzlers, graded in difficulty, illustrate elementary and advanced aspects of probability. Detailed solutions. 88pp. 5⅜ x 8½. 65355-2

PROBABILITY THEORY: A CONCISE COURSE, Y. A. Rozanov. Highly readable, self-contained introduction covers combination of events, dependent events, Bernoulli trials, etc. 148pp. 5⅜ x 8¼. 0-486-63544-9

STATISTICAL METHOD FROM THE VIEWPOINT OF QUALITY CONTROL, Walter A. Shewhart. Important text explains regulation of variables, uses of statistical control to achieve quality control in industry, agriculture, other areas. 192pp. 5⅜ x 8½. 0-486-65232-7

Math–Geometry and Topology

ELEMENTARY CONCEPTS OF TOPOLOGY, Paul Alexandroff. Elegant, intuitive approach to topology from set-theoretic topology to Betti groups; how concepts of topology are useful in math and physics. 25 figures. 57pp. 5⅜ x 8½. 0-486-60747-X

COMBINATORIAL TOPOLOGY, P. S. Alexandrov. Clearly written, well-organized, three-part text begins by dealing with certain classic problems without using the formal techniques of homology theory and advances to the central concept, the Betti groups. Numerous detailed examples. 654pp. 5⅜ x 8½. 0-486-40179-0

EXPERIMENTS IN TOPOLOGY, Stephen Barr. Classic, lively explanation of one of the byways of mathematics. Klein bottles, Moebius strips, projective planes, map coloring, problem of the Koenigsberg bridges, much more, described with clarity and wit. 43 figures. 210pp. 5⅜ x 8½. 0-486-25933-1

THE GEOMETRY OF RENÉ DESCARTES, René Descartes. The great work founded analytical geometry. Original French text, Descartes's own diagrams, together with definitive Smith-Latham translation. 244pp. 5⅜ x 8½. 0-486-60068-8

EUCLIDEAN GEOMETRY AND TRANSFORMATIONS, Clayton W. Dodge. This introduction to Euclidean geometry emphasizes transformations, particularly isometries and similarities. Suitable for undergraduate courses, it includes numerous examples, many with detailed answers. 1972 ed. viii+296pp. 6⅛ x 9¼. 0-486-43476-1

PRACTICAL CONIC SECTIONS: THE GEOMETRIC PROPERTIES OF ELLIPSES, PARABOLAS AND HYPERBOLAS, J. W. Downs. This text shows how to create ellipses, parabolas, and hyperbolas. It also presents historical background on their ancient origins and describes the reflective properties and roles of curves in design applications. 1993 ed. 98 figures. xii+100pp. 6½ x 9¼. 0-486-42876-1

THE THIRTEEN BOOKS OF EUCLID'S ELEMENTS, translated with introduction and commentary by Sir Thomas L. Heath. Definitive edition. Textual and linguistic notes, mathematical analysis. 2,500 years of critical commentary. Unabridged. 1,414pp. 5⅜ x 8½. Three-vol. set.
Vol. I: 0-486-60088-2 Vol. II: 0-486-60089-0 Vol. III: 0-486-60090-4

SPACE AND GEOMETRY: IN THE LIGHT OF PHYSIOLOGICAL, PSYCHOLOGICAL AND PHYSICAL INQUIRY, Ernst Mach. Three essays by an eminent philosopher and scientist explore the nature, origin, and development of our concepts of space, with a distinctness and precision suitable for undergraduate students and other readers. 1906 ed. vi+148pp. 5⅜ x 8½. 0-486-43909-7

GEOMETRY OF COMPLEX NUMBERS, Hans Schwerdtfeger. Illuminating, widely praised book on analytic geometry of circles, the Moebius transformation, and two-dimensional non-Euclidean geometries. 200pp. 5⅝ x 8¼. 0-486-63830-8

DIFFERENTIAL GEOMETRY, Heinrich W. Guggenheimer. Local differential geometry as an application of advanced calculus and linear algebra. Curvature, transformation groups, surfaces, more. Exercises. 62 figures. 378pp. 5⅜ x 8½. 0-486-63433-7

History of Math

THE WORKS OF ARCHIMEDES, Archimedes (T. L. Heath, ed.). Topics include the famous problems of the ratio of the areas of a cylinder and an inscribed sphere; the measurement of a circle; the properties of conoids, spheroids, and spirals; and the quadrature of the parabola. Informative introduction. clxxxvi+326pp. 5⅜ x 8½. 0-486-42084-1

A SHORT ACCOUNT OF THE HISTORY OF MATHEMATICS, W. W. Rouse Ball. One of clearest, most authoritative surveys from the Egyptians and Phoenicians through 19th-century figures such as Grassman, Galois, Riemann. Fourth edition. 522pp. 5⅜ x 8½. 0-486-20630-0

THE HISTORY OF THE CALCULUS AND ITS CONCEPTUAL DEVELOPMENT, Carl B. Boyer. Origins in antiquity, medieval contributions, work of Newton, Leibniz, rigorous formulation. Treatment is verbal. 346pp. 5⅜ x 8½. 0-486-60509-4

THE HISTORICAL ROOTS OF ELEMENTARY MATHEMATICS, Lucas N. H. Bunt, Phillip S. Jones, and Jack D. Bedient. Fundamental underpinnings of modern arithmetic, algebra, geometry and number systems derived from ancient civilizations. 320pp. 5⅜ x 8½. 0-486-25563-8

A HISTORY OF MATHEMATICAL NOTATIONS, Florian Cajori. This classic study notes the first appearance of a mathematical symbol and its origin, the competition it encountered, its spread among writers in different countries, its rise to popularity, its eventual decline or ultimate survival. Original 1929 two-volume edition presented here in one volume. xxviii+820pp. 5⅜ x 8½. 0-486-67766-4

GAMES, GODS & GAMBLING: A HISTORY OF PROBABILITY AND STATISTICAL IDEAS, F. N. David. Episodes from the lives of Galileo, Fermat, Pascal, and others illustrate this fascinating account of the roots of mathematics. Features thought-provoking references to classics, archaeology, biography, poetry. 1962 edition. 304pp. 5⅜ x 8½. (Available in U.S. only.) 0-486-40023-9

OF MEN AND NUMBERS: THE STORY OF THE GREAT MATHEMATICIANS, Jane Muir. Fascinating accounts of the lives and accomplishments of history's greatest mathematical minds–Pythagoras, Descartes, Euler, Pascal, Cantor, many more. Anecdotal, illuminating. 30 diagrams. Bibliography. 256pp. 5⅜ x 8½. 0-486-28973-7

HISTORY OF MATHEMATICS, David E. Smith. Nontechnical survey from ancient Greece and Orient to late 19th century; evolution of arithmetic, geometry, trigonometry, calculating devices, algebra, the calculus. 362 illustrations. 1,355pp. 5⅜ x 8½. Two-vol. set. Vol. I: 0-486-20429-4 Vol. II: 0-486-20430-8

A CONCISE HISTORY OF MATHEMATICS, Dirk J. Struik. The best brief history of mathematics. Stresses origins and covers every major figure from ancient Near East to 19th century. 41 illustrations. 195pp. 5⅜ x 8½. 0-486-60255-9

Physics

OPTICAL RESONANCE AND TWO-LEVEL ATOMS, L. Allen and J. H. Eberly. Clear, comprehensive introduction to basic principles behind all quantum optical resonance phenomena. 53 illustrations. Preface. Index. 256pp. 5⅜ x 8½. 0-486-65533-4

QUANTUM THEORY, David Bohm. This advanced undergraduate-level text presents the quantum theory in terms of qualitative and imaginative concepts, followed by specific applications worked out in mathematical detail. Preface. Index. 655pp. 5⅜ x 8½. 0-486-65969-0

ATOMIC PHYSICS (8th EDITION), Max Born. Nobel laureate's lucid treatment of kinetic theory of gases, elementary particles, nuclear atom, wave-corpuscles, atomic structure and spectral lines, much more. Over 40 appendices, bibliography. 495pp. 5⅜ x 8½. 0-486-65984-4

A SOPHISTICATE'S PRIMER OF RELATIVITY, P. W. Bridgman. Geared toward readers already acquainted with special relativity, this book transcends the view of theory as a working tool to answer natural questions: What is a frame of reference? What is a "law of nature"? What is the role of the "observer"? Extensive treatment, written in terms accessible to those without a scientific background. 1983 ed. xlviii+172pp. 5⅜ x 8½. 0-486-42549-5

AN INTRODUCTION TO HAMILTONIAN OPTICS, H. A. Buchdahl. Detailed account of the Hamiltonian treatment of aberration theory in geometrical optics. Many classes of optical systems defined in terms of the symmetries they possess. Problems with detailed solutions. 1970 edition. xv + 360pp. 5⅜ x 8½. 0-486-67597-1

PRIMER OF QUANTUM MECHANICS, Marvin Chester. Introductory text examines the classical quantum bead on a track: its state and representations; operator eigenvalues; harmonic oscillator and bound bead in a symmetric force field; and bead in a spherical shell. Other topics include spin, matrices, and the structure of quantum mechanics; the simplest atom; indistinguishable particles; and stationary-state perturbation theory. 1992 ed. xiv+314pp. 6⅛ x 9¼. 0-486-42878-8

LECTURES ON QUANTUM MECHANICS, Paul A. M. Dirac. Four concise, brilliant lectures on mathematical methods in quantum mechanics from Nobel Prize-winning quantum pioneer build on idea of visualizing quantum theory through the use of classical mechanics. 96pp. 5⅜ x 8½. 0-486-41713-1

THIRTY YEARS THAT SHOOK PHYSICS: THE STORY OF QUANTUM THEORY, George Gamow. Lucid, accessible introduction to influential theory of energy and matter. Careful explanations of Dirac's anti-particles, Bohr's model of the atom, much more. 12 plates. Numerous drawings. 240pp. 5⅜ x 8½. 0-486-24895-X

ELECTRONIC STRUCTURE AND THE PROPERTIES OF SOLIDS: THE PHYSICS OF THE CHEMICAL BOND, Walter A. Harrison. Innovative text offers basic understanding of the electronic structure of covalent and ionic solids, simple metals, transition metals and their compounds. Problems. 1980 edition. 582pp. 6⅛ x 9¼. 0-486-66021-4

HYDRODYNAMIC AND HYDROMAGNETIC STABILITY, S. Chandrasekhar. Lucid examination of the Rayleigh-Benard problem; clear coverage of the theory of instabilities causing convection. 704pp. 5⅜ x 8¼. 0-486-64071-X

INVESTIGATIONS ON THE THEORY OF THE BROWNIAN MOVEMENT, Albert Einstein. Five papers (1905–8) investigating dynamics of Brownian motion and evolving elementary theory. Notes by R. Fürth. 122pp. 5⅜ x 8½. 0-486-60304-0

THE PHYSICS OF WAVES, William C. Elmore and Mark A. Heald. Unique overview of classical wave theory. Acoustics, optics, electromagnetic radiation, more. Ideal as classroom text or for self-study. Problems. 477pp. 5⅜ x 8½. 0-486-64926-1

GRAVITY, George Gamow. Distinguished physicist and teacher takes reader-friendly look at three scientists whose work unlocked many of the mysteries behind the laws of physics: Galileo, Newton, and Einstein. Most of the book focuses on Newton's ideas, with a concluding chapter on post-Einsteinian speculations concerning the relationship between gravity and other physical phenomena. 160pp. 5⅜ x 8½. 0-486-42563-0

PHYSICAL PRINCIPLES OF THE QUANTUM THEORY, Werner Heisenberg. Nobel Laureate discusses quantum theory, uncertainty, wave mechanics, work of Dirac, Schroedinger, Compton, Wilson, Einstein, etc. 184pp. 5⅜ x 8½. 0-486-60113-7

ATOMIC SPECTRA AND ATOMIC STRUCTURE, Gerhard Herzberg. One of best introductions; especially for specialist in other fields. Treatment is physical rather than mathematical. 80 illustrations. 257pp. 5⅜ x 8½. 0-486-60115-3

AN INTRODUCTION TO STATISTICAL THERMODYNAMICS, Terrell L. Hill. Excellent basic text offers wide-ranging coverage of quantum statistical mechanics, systems of interacting molecules, quantum statistics, more. 523pp. 5⅜ x 8½. 0-486-65242-4

THEORETICAL PHYSICS, Georg Joos, with Ira M. Freeman. Classic overview covers essential math, mechanics, electromagnetic theory, thermodynamics, quantum mechanics, nuclear physics, other topics. First paperback edition. xxiii + 885pp. 5⅜ x 8½. 0-486-65227-0

PROBLEMS AND SOLUTIONS IN QUANTUM CHEMISTRY AND PHYSICS, Charles S. Johnson, Jr. and Lee G. Pedersen. Unusually varied problems, detailed solutions in coverage of quantum mechanics, wave mechanics, angular momentum, molecular spectroscopy, more. 280 problems plus 139 supplementary exercises. 430pp. 6½ x 9¼. 0-486-65236-X

THEORETICAL SOLID STATE PHYSICS, Vol. 1: Perfect Lattices in Equilibrium; Vol. II: Non-Equilibrium and Disorder, William Jones and Norman H. March. Monumental reference work covers fundamental theory of equilibrium properties of perfect crystalline solids, non-equilibrium properties, defects and disordered systems. Appendices. Problems. Preface. Diagrams. Index. Bibliography. Total of 1,301pp. 5⅜ x 8½. Two volumes. Vol. I: 0-486-65015-4 Vol. II: 0-486-65016-2

WHAT IS RELATIVITY? L. D. Landau and G. B. Rumer. Written by a Nobel Prize physicist and his distinguished colleague, this compelling book explains the special theory of relativity to readers with no scientific background, using such familiar objects as trains, rulers, and clocks. 1960 ed. vi+72pp. 5⅜ x 8½. 0-486-42806-0

CATALOG OF DOVER BOOKS